Keeping Chickens
FOR
DUMMIES®

Keeping Chickens
FOR
DUMMIES®

by Pammy Riggs,
Kimberly Willis and
Rob Ludlow

WILEY

A John Wiley and Sons, Ltd, Publication

Keeping Chickens For Dummies®

Published by
John Wiley & Sons, Ltd
The Atrium
Southern Gate
Chichester
West Sussex
PO19 8SQ
England

E-mail (for orders and customer service enquires): cs-books@wiley.co.uk

Visit our Home Page on www.wiley.com

For general information on our other products and services, please contact our Customer Care Department within the U.S. at 877-762-2974, outside the U.S. at 317-572-3993, or fax 317-572-4002.

For technical support, please visit www.wiley.com/techsupport.

Wiley also publishes its books in a variety of electronic formats and by print-on-demand. Some content that appears in standard print versions of this book may not be available in other formats. For more information about Wiley products, visit us at www.wiley.com.

British Library Cataloguing in Publication Data: A catalogue record for this book is available from the British Library

ISBN 978-1-119-99417-6 (paperback), ISBN 978-1-119-97178-8 (ebook), ISBN 978-1-119-99418-3 (ebook), ISBN 978-1-119-99419-0 (ebook)

Printed and bound in Great Britain by TJ International, Padstow, Cornwall

10 9 8 7 6 5 4 3 2 1

About the Authors

Pammy Riggs and her family moved away from mainstream life more than twenty years ago to create a different kind of reality, transforming a bare, wet Devon field into Providence Farm (www.providencefarm.co.uk) – a thriving small organic farm and a cosy low-carbon home – planting woodland and reinstating wild meadows along the way.

In the warm light of this liberation of her life, Pammy now nurtures a burning desire to share her invaluable knowledge and vast experience. Pammy speaks and teaches at various educational institutions and has created a brand new home for her writing, courses and ideas for getting reconnected with a more natural and creative life at www.rootedin.co.uk. Her passion is to empower us all to get out from behind our computers and desks and get our hands dirty in the rich soil of life.

Kimberly Willis lives with her husband, Steve, on a small farm in the thumb area of Michigan. When not writing, she works at the MSU Extension office in Lapeer County, Michigan, as a horticulturalist and doubles as the resident chicken expert. Kim has raised a number of breeds of chickens and other types of poultry for over 30 years. She has shown poultry locally. She is an advocate for eating locally and sells her fresh brown eggs to friends and family. She is a proud member of www.backyardchickens.com.

Kim is also a garden writer and has numerous articles in print and online. Your can read her work at www.life123.com or www.squidoo.com/gardeninggranny or her blog at www.gardeninggranny.blogspot.com.

Rob Ludlow, his wife Emily, and their two beautiful daughters, Alana and April, are the perfect example of the suburban family with a small flock of backyard chickens. Like countless others, what started out as a fun hobby raising a few egg-laying hens has almost turned into an addiction.

Originally, Rob started posting his experiences with chickens on his hobby website – www.Nifty-Stuff.com – but after realising how much his obsession with chickens was growing, he decided to concentrate his efforts on a site devoted completely to the subject. Now, Rob owns and manages www.backyardchickens.com (BYC), the largest and fastest-growing community of chicken enthusiasts in the world.

Rob hopes to work with BYC's tens of thousands of members to promote a change of the old concept, 'a chicken in every pot', to a new version, the BYC vision – 'a chicken in every yard'!

Dedications

Pammy Riggs: To my darling Mum, Joy Seales, who gave me all that Scottish education and time mucking about in Scottish farmyards – valuable compost for a fruitful animal-filled life.

Kimberly Willis: I would like to dedicate this book to my husband, Steve, who took over the cooking and dishes so I could concentrate on my latest book.

Rob Ludlow: To the three most important girls in my life – Emily, Alana and April – who not only support but also contribute to my joy of raising backyard chickens.

Authors' Acknowledgements

Pammy Riggs: Thank you, Dummies, for choosing me and hand-holding me through this first book. Thanks, also, to *Country Smallholding* magazine for launching my writing by giving me a regular 'Chicken Whisperer' column, and last but not least thanks to my husband, Ritchie Riggs, whose irreverent humour and grounded common sense is the sounding board I value.

Kimberly Willis: I would like to acknowledge all the help that Jennifer Connolly has given me with my first Dummies book as well as acknowledge all the hard work that Christy Pingleton has done on this book. I would also like to acknowledge the fun chicken website, www.backyardchickens.com, which gave me much insight on the concerns new chicken owners have.

Rob Ludlow: Thanks to my brother Michael for getting me started with chickens, and to Mike Baker and Kristin DeMint for their help with the project. Especially huge thanks to the incredibly smart, patient and helpful staff at www.backyardchickens.com, and the thousands of friendly BYC community members.

Publisher's Acknowledgements

We're proud of this book; please send us your comments through our Dummies online registration form located at www.dummies.com/register/.

Some of the people who helped bring this book to market include the following:

Commissioning, Editorial, and Media Development

Development Editor: Steve Edwards

Commissioning Editor: Claire Ruston

Assistant Editor: Ben Kemble

Copy Editor: Anne O'Rorke

Proofreader: Andy Finch

Production Manager: Daniel Mersey

Cover Photos: © Vincent O'Byrne/Alamy

Cartoons: Ed McLachlan

Composition Services

Project Coordinator: Kristie Rees

Layout and Graphics: Carl Byers, Lavonne Roberts, Corrie Socolovitch

Proofreader: Lindsay Amones

Indexer: Claudia Bourbeau

Publishing and Editorial for Consumer Dummies

    Kathleen Nebenhaus, Vice President and Executive Publisher

    Kristin Ferguson-Wagstaffe, Product Development Director

    Ensley Eikenburg, Associate Publisher, Travel

    Kelly Regan, Editorial Director, Travel

Publishing for Technology Dummies

    Andy Cummings, Vice President and Publisher

Composition Services

    Debbie Stailey, Director of Composition Services

Contents at a Glance

Table of Contents

Introduction

· ·

*A*ll over the country, more and more people are discovering the joy of chickens. Some people are practical, wanting to explore producing their own food; some are nostalgic, longing for a taste of 'the good life'; and other people are drawn in by the sight of some pretty chicks.

Chickens are a special part of the authors' lives. Everyday we listen to the questions and concerns that people have about chickens, and we take great enjoyment in the chickens that we own, too. We're thrilled that more and more people in towns, cities and villages across the country want to keep chickens, but that means the body of people who need information about chickens just keeps on growing. Because we can't be there to answer questions in person, we decided it was time for a modern, comprehensive chicken book that provides quick answers to all your questions – and here it is!

About This Book

This chicken book is different from some of the others out there – you can find the answers you're looking for more easily here because of the way we've organised the book. Go ahead, flip through the book and see for yourself. Bold headings direct your eyes to the very sections you need, and you don't have to read the whole book for a quick answer.

We've written this book to give you a broad overview of all aspects of keeping chickens, from laying hens to meat chickens. Today you may be interested in discovering how to care for some cute, fluffy chicks you fell in love with at a country fair, and you can find that information here. In five months or so, when they begin laying eggs, you need information on what to do and how to manage hens. That information is here too. And if you get tired of those chicks because they all turn out to be big, fat, noisy cockerels, well, we give you thorough instructions on how to turn them into a roast chicken dinner. So put this book on your bookshelf in a prominent place so that you can refer to it again and again.

We're careful to use modern, scientifically correct information on chicken care and direct your attention to sources of additional information when necessary. But we also throw in lots of good, everyday, down-to-earth advice that can only come from owning and enjoying our own chickens.

Conventions Used in This Book

To help you find what you want quickly, we use a few conventions through-out the book. For example:

- All web addresses appear in `monofont` so that you can spot them easily.

  Because of the way the book is printed, some web addresses may break to the next line. We don't add any characters such as a hyphen if this happens, and so copy the address just as you see it.

- New terms appear in *italic* and are closely followed by easy-to-understand definitions.

- We use **bold** to highlight the keywords in lists.

What You're Not to Read

Of course, we think each and every word that we've painstakingly written is effective and important, but because you don't have to read this book from cover to cover, rest assured that we tell you what type of info you can skip without guilt. Maybe we should say 'what you don't *need* to read' instead. Sometimes we have a little story we want to share, but it isn't really something you need to know. So we put it in a sidebar – those grey boxes filled with text. Feel free to skip any sidebar you want.

Sometimes we also have some more technical material for those of you who want a little more detail. These parts aren't crucial to your chicken-keeping experience, and are marked with a 'Technical Stuff' icon.

Foolish Assumptions

To get this book flowing, we factored in some assumptions about you, the reader. Here's how we sized you up:

- Although you've come across chickens before, you aren't an expert on chickens yet and need some very basic information.

- You want to find out more about keeping chickens or eating the chickens you do have.

- You like animals and want to treat them with kindness and have a good knowledge of their needs.

- You don't want to rear chickens on a huge scale like 500 laying hens or 2,500 broilers. We assume you want information on small home flocks.

✔ You have some very basic carpentry or craft skills. Although we do provide some basic guidance for building chicken housing in this book, we
don't have enough room to instruct you on building skills. So if you don't
have the skills, we give you permission to call on a friend who does.

How This Book Is Organised

To access information easily and quickly, you need organisation that's logical
and precise. So we break this book into parts and then chapters, which group
together information on a particular aspect of chicken-keeping, say, looking
after baby chicks. We then break chapters into sections, with bold headings
that help you to find what you're looking for.

Here's just a tiny taste of the book's organisation.

Part 1: Choosing Chickens

These chapters explore your reasons for wanting to keep chickens, talk about
basic chicken biology (Chapter 2) and give you some information on breeds
of chickens (Chapter 3). This part also takes you through planning and purchasing your flock (Chapter 4).

Part II: Housing Your Flock

This part covers basic chicken-keeping skills, whether you want eggs or meat.
In Chapter 5 we talk about chicken housing that's used for various types of
chickens, and in Chapter 6 we go a bit further, giving you suggestions for
designing chicken housing yourself. In Chapter 7 we cover the basics of getting your chicken housing set up and keeping it in good working order.

Part III: Caring For Your Flock: General Management

Obviously, caring for your chickens takes some time and effort, and this part
covers the basics on care. Chapter 8 discusses something very important to
the chickens – feeding. Yes, chickens eat anything, but you need to know the
best way to feed your chickens.

Of course, food isn't the only thing that keeps your chickens healthy. You have to protect them from predators and pests (Chapter 9), plus you have to take an active role in preventive health (Chapter 10) and recognise and treat illness (Chapter 11).

Part IV: Breeding: The Chicken and the Egg

This part focuses on one thing: making more chickens. Chicken reproduction is a fascinating journey. In Chapter 12 we discuss mating, and in Chapter 13 we discuss incubating eggs, both artificially and nature's way. Of course, the chicken reproductive process doesn't stop just at the eggs . . . they do have to hatch. So, in Chapter 14 we cover how to care for those chicks when they do.

Part V: Considering Special Management Issues

The chapters in this part focus on the specialised things you need to know about keeping chickens for a particular purpose, such as to give you eggs or meat. In Chapter 15, we discuss managing laying hens so that they can produce good eggs for you consistently. Chapter 16 covers the finer points of growing chickens for meat and Chapter 17 looks at dispatching chickens and storing their meat.

Part VI: The Part of Tens

This last part, known in a *For Dummies* book as The Part of Tens, gives you some quick-reference reviews or information we didn't include elsewhere. In Chapter 18 we offer tips for raising healthy chickens, and in Chapter 19 we talk about some common chicken misconceptions – talk about foolish assumptions!

Icons Used in This Book

Icons are special symbols set in the margins near paragraphs of text in the book. They're meant to draw your attention. Some people use them as a way to access certain pieces of important information.

This book uses the following icons.

Tips are special time- or money-saving pieces of advice. They come from our years of experience with chickens.

This icon urges you to remember this piece of information because it's important. Sometimes a referral to another chapter for more precise information may be nearby.

A warning icon means that we're mentioning something that may pose a danger to you or your chickens. Pay attention to warning icons because they contain important information.

This icon provides some technical information that may or may not interest you. You can skip this paragraph if you want, without missing any essential information.

Where to Go from Here

So, the time has come to get reading. Of course, eventually you want to read every scrap of this book, but you may need to know some things – now!

Here are some ideas of where you may want to begin, depending on your situation:

- If you're one of those rare people who likes to be well prepared before you start a project such as keeping chickens, you may want to start with Chapter 1.
- If you're sitting here with the book in one hand and a box of chicks at your feet, you may want to start with Chapter 14 to get more info on caring for those chicks.
- If you have some chickens and they aren't laying the eggs you expect, flip to Chapter 15.
- If you have chickens that look a bit peaky, check out Chapter 11 to work out and treat whatever your chickens may be suffering from.
- If someone gave you some cute fluffy chicks for Easter that turned into ten fighting and crowing cockerels by autumn, try Chapter 16, which discusses how to turn them into something more valuable and a lot less noisy – meat for the freezer.

Part I
Choosing Chickens

'So I said to the poultry egg dealer,
I want chicks who can stand up to a <u>really</u>
cold winter this year.'

In this part . . .

The chapters in this part focus on some basic chicken information, such as chicken biology and different breeds of chickens. We try to infuse you with our love of chickens but give you enough information to make sure that chicken-keeping really is for you. If you're new to chicken-raising, you may be anxious about buying chickens, and so in Chapter 4 we discuss what you need to know about acquiring chickens.

Chapter 1

Enjoying Chicken-Keeping

. . . .

. . . .

*W*e'll come straight out with it – we love chickens and we hope that you're reading this book because you love chickens, too; because, as a chicken-keeper, you'll have their welfare at the forefront of your mind. We discuss a very basic issue in this chapter – one that you need to consider before you do anything else. Should you even keep chickens? Chickens make colourful, moving lawn ornaments and they can even provide you with your breakfast. But they do take some attention and expense, and you need specific knowledge to care for them properly.

Consider this chapter as chicken family planning. If you read the information here and still believe that you're ready to start your chicken family, you have the whole rest of the book to get all the information you need to begin your adventure.

Dealing with the Legal Issues

Various rules have been put in place over the years to combat the problems associated with chicken-keeping. In this section, we look at those rules – and how to overcome them responsibly.

Plenty of small home-owned flocks are happily clucking and scratching around gardens all over the country. All the legal stuff you have to think about before kick-starting a chicken-keeping hobby can sound a bit daunting, but don't let it put you off if you're keen. 'Forewarned is forearmed', and knowing about the potential problems of a crowing cockerel, for example, may just save you from experiencing the hassle first hand.

Knowing what info you need

To know whether you can legally keep chickens, first you need to know what can stop you from doing so. Therefore, before you go to the expense of setting up your chicken-keeping operation or get the kids too excited about the new hobby, check the following:

✔ **Covenants written into your house deeds.** About 100 years' ago people kept chickens in urban backyards as commonly as in farmyards. In fact, chickens became so common that they began to pose problems in densely populated areas when people didn't look after them properly, which caused bad smells and attracted vermin.

To combat these problems, rules against chicken-keeping were written into some house deeds as new dwellings were being built around the turn of the last century. Although this situation is thankfully quite rare nowadays, you need to check that your house deeds don't contain such rules. If you live in a terraced row or a street of similar houses, the chances are that all the houses are bound by the same rules.

If you bought your house before you got interested in chicken-keeping, rules concerning chicken-keeping may not be something that you checked. Anything that says keeping poultry isn't allowed on your property should be written into the deeds quite clearly, and so now is the time to read them thoroughly.

Restrictions in house deeds may often be outdated, for the current era in which people keep chickens for pleasure rather than for bulking out a meagre diet, and rules may be relaxed in your area or people may be unaware they exist. If your chicken-keeping creates any problems for other people, however, those rules can be used against you.

✔ **Local council by-laws.** Consult with your local council to check that no by-laws exist that prevent anyone in your area from keeping livestock at their property.

✔ **Tenancy agreements.** If you rent, your landlord may have written a 'no pets' clause into your tenancy terms. If so, check whether this covers chickens too – it may be that 'outdoor' pets aren't viewed as a problem.

✔ **The law.** Laws exist that govern what you're allowed to build (relevant if you're thinking of going for a solidly constructed chicken palace). Other relevant laws concern pollution of ground water from poor storage of manure (the Environment Agency police this area — see its website at `www.environment-agency.gov.uk` for more information) and obligate you to treat animals well. They also restrict who you can and can't sell chicken-related produce to. Rules also exist that concern noise levels – the Environmental Health Department of your local council has a duty to investigate any noise pollution complaints. Remember that cockerels can be very loud!

If your flock consists of 50 or more chickens, you're obliged to register with the Department for the Environment, Food and Rural Affairs (DEFRA), and to bring to its attention any unusual deaths or diseases in wild birds near your chicken-keeping venture. If your chickens have a disease problem, your veterinary surgeon informs DEFRA if he thinks that the problem is significant. (Chapter 11 deals with 'notifiable diseases'.)

Finding the info

House deeds are usually held for surety against a mortgage by the solicitor or bank that arranged your mortgage for you. The information is available, but you may have to give the office where the deeds are held some notice if you want to study them or copy some of the details. For a fee, and if your property is registered, you may be able to find the information you need at www.landregisteronline.gov.uk.

If you're unlucky enough to find restrictions in your house deeds, you can look at getting the rules changed – if you've come to an agreement with neighbours, and particularly if someone else living in your street already has chickens. Check to see if you can get a free half-hour consultation with a solicitor to find out whether making the change is worthwhile and possible, being an outdated situation. Someone else living nearby may already have done so, which can help your case along.

For checking the rules regarding ground water, rivers and streams, the Environment Agency is the place to go. Environmental Health Officers have a few roles, too – food safety, pest and vermin control as well as noise pollution all come under their jurisdiction. You can contact these people through your local county council, where you can also check for any applicable by-laws that may affect you and contact the Planning Officers if you need to talk through any building regulations. Take advantage of their knowledge – they can be very helpful.

If your flock gets to a size where you need to register it with DEFRA, you can do so via its telephone helpline (0845 33 55 77) or website (www.defra. gov.uk). We don't advise rushing into keeping 50 birds straight away, but it doesn't take much to reach that number after you start to breed chicks.

Looking ahead to restrictions that may affect you later on

Most people start their flock with egg-layers and then progress to growing some chicks of their own to replace elderly laying hens or because they want to see a few cute fluffy chicks running around the place. Baby chicks soon turn into adults and then start laying themselves or become noisy cockerels, which poses a new challenge to the home flock owner. (Chapters 12 and 14 look at breeding chickens and rearing chicks.)

If you have as many as ten hens laying well, you may be getting five dozen eggs a week – far more than most households can eat! You may decide to sell some eggs to help towards the cost of keeping your chickens, but you need to be aware of the laws that come into play as soon as you start to sell any number of eggs to a third party. These laws affect to whom you may sell and what you can legally call your eggs. We explain these laws in Chapter 15.

Dealing with your extra cockerels before they start to crow in chorus and keep the neighbourhood awake means that you must find out the correct and legal methods of dispatching them and dealing with any waste, and be familiar with the rules about eating the meat from these birds. Chapter 17 looks at this area.

Don't let rules and regulations put you off keeping chickens for pleasure, but bear in mind the way this hobby can escalate, which brings these rules into play. That way, when you're ready to increase your flock size, you're also ready to undertake the legal responsibilities that go with it.

Assessing Your Capabilities: Basic Chicken Care and Requirements

Chickens can take as much time and money as you care to spend, but you need to consider the *minimum* time, space and money commitments you need to put into keeping chickens. In the next sections, we give you an idea of what those minimums are.

Time

When we speak about time here, we're referring to the daily caretaking chores. Naturally, getting housing set up for your birds takes some time, particularly if you're building a chicken house. Give yourself plenty of time to finish before you acquire the birds. You have to judge how much time that's going to be depending on the scope of the project, your building skills and how much time each day you can devote to it. (See Chapter 6 for more on constructing your own chicken housing.)

Count on a minimum of 15 minutes every morning and evening to care for chickens in a small flock, if you don't spend a lot of time just observing their antics – they can be incredible time-wasters. Even if you install automatic feeders and drinkers (see Chapter 8), as a good chicken-keeper you need to check on your flock twice a day. If you have laying hens, collect eggs once a day, which doesn't take long.

Try to attend to your chickens' needs before they go to bed for the night and after they're up in the morning. Ideally chickens need 14 hours of light and 10 hours of darkness. In the winter you can adjust artificial lighting so that it accommodates your schedule. Chickens find it very stressful if you turn on lights to do chores after they're sleeping.

In addition to your daily tasks, you need to allocate additional time once a week for basic cleaning chores. If you have just a few chickens, this can be less than an hour. Cleaning includes such things as removing manure, adding clean litter, scrubbing water containers and refilling feed bins. Depending on your chicken-keeping methods, you may need to put in additional time every few months for more intensive cleaning chores.

More chickens doesn't necessarily mean spending more daily time on them until you get to very large numbers – a pen full of 25 meat birds may only increase your caretaking time a few minutes in comparison to a pen of four laying hens – but the way in which you keep chickens can increase the time needed to care for them. For example, if you keep chickens for showing and you house them in individual cages, feeding and watering them takes at least five to ten minutes per cage.

Space

Each adult full-sized chicken needs an area of floor space of at least 0.3 square metres (3.2 square feet) for shelter – more if it's available – and another 0.3 square metres (3.2 square feet) at the very minimum as outside run space, if the chicken isn't going to be running loose much. So a chicken shelter for four hens, for example, needs to be about 0.6 metres wide by 1.3 metres long (2 by 4.25 feet) and the outside pen another 0.6 metres wide by 2 metres long (2 by 6.5 feet), so that your total space is 0.6 metres by 3.3 metres (2 by 11 feet) (which covers an area of about 2 metres/22 feet square – these dimensions don't have to be exact). For more chickens you need to provide more space, and you need a little additional space to store feed and maybe a place to store or compost the used litter and manure. Of course, the more space you can provide for your chickens, the better.

As far as height goes, the chicken coop doesn't have to be more than 1 metre (3.3 feet) high, but you may want something bigger than a coop to be tall enough for you to walk upright inside it.

Besides the actual size of the space, you need to think about location, location, location. You probably don't want your space in the front garden unless it's secluded or well fenced off from the street, and you probably want the chicken house to be as far from your neighbours as possible to lessen the chance that they may complain.

Money

Unless you plan on purchasing rare breeds that are in high demand, the cost of purchasing chickens doesn't break most budgets. Adult hens that are good layers cost around £10, and chicks of most breeds cost a couple of pounds each. The cost of adult fancy breeds kept as pets ranges from a few pounds to much, much more, depending on the breed. Sometimes you can even get free chickens if you find a commercial laying flock and accept second-year layers that someone's getting rid of. The Battery Hen Welfare Trust (www. bhwt.org.uk) organises 'ex-bats' for collection for a nominal fee and tries to fit you up with something in your area.

Housing costs are extremely variable, but are one-time costs. If you have a corner of a barn or an old shed to convert to housing and your chickens will be free-ranging most of the time, your housing start-up costs are going to be very low – maybe just £25 to get the basic equipment. If, however, you want to build a fancy chicken shed with a large outside run, your cost can run into hundreds of pounds. If you want to buy a pre-built structure for a handful of chickens, count on £100 plus.

The best way to plan your housing costs is to first decide what your budget can afford. Next, look through Chapters 5 and 6 of this book to find out about types of housing. Then do a 'comparison shop' to see what building supplies would cost for your chosen housing (or pre-built structures) and see how it fits your budget. Don't forget to factor in shipping costs for pre-built units.

You may incur a few other one-time costs for essential equipment such as feeders, drinkers and nest boxes. For four hens, clever shopping should get you these items for less than £40.

An ongoing cost of keeping chickens is buying in their feed. Commercial chicken feed is reasonably priced and generally comparable to common brands of dry dog and cat food, and the choice of feeds available is big. Because the chickens' health needs to be your first consideration, try to avoid getting poor grade feed. The lower the price, the more likely a feed is to include Genetically Modified (GM) soya and grain. You have to pay more to buy a higher grade food that avoids GM material. The quantity you buy in one go also affects the price.

The decisions you make about feed affect the quality of the food you get from your chickens. If they can't get out to range and forage for themselves with your system and they rely on the food you give them for their nutrients, going for the cheapest feeds can be a false economy – chickens can only make their meat and eggs with what you provide. If you have loads of space where the chickens can find free food they won't be eating as much of the feed you provide anyway.

How many chickens you have and how you keep them determine how much feed you use: count on about ¼ kilo (½ pound) of feed per adult, full-sized bird per day. We estimate the cost of feed for three to four layers to be around $10–12 per month, but feed costs do rise from time to time. If you can find a local source of grain or even grow part of your chickens' food yourself, you can keep the costs to a minimum.

Focusing Your Intentions: Specific Considerations

Dozens of reasons exist why you may want to keep chickens. Some people decide to take up chicken-keeping because they're nostalgic for the chickens they remember from childhood. Other people do it because they've heard that chickens control flies and ticks and turn the compost pile. Some children want to keep chickens for a school project or a Scout or Guide badge, or may reach an age where they demand pets and it seems logical to opt for a useful animal that doesn't live indoors. Some people want to produce their own quality eggs or organic meat, whereas others just want to provoke the neighbours!

If you're not exactly sure why chicken-keeping appeals to you, think about what you want to get out of it in advance and decide whether keeping chickens really suits your temperament and lifestyle. Impulse chicken-buying isn't a good idea, and so a little forethought is the way to tackle the decision. If chicken-keeping turns out not to be your bag, you may waste a lot of time and money and have some unhappy chickens on your hands. For that reason, in this section we show you all the options before you take that final step into what we think is a great, fun hobby for all the family.

Egg layers, meat birds and pet/show chickens have slightly different housing and care requirements. Having a purpose in mind as you select breeds (see Chapter 3) and develop your housing keeps you from making expensive mistakes and ensures that your chicken-keeping experience is more enjoyable.

Keeping chickens for several different purposes is fine – some for eggs and others as show birds for example – but thinking about your intentions in advance makes good sense.

Producing eggs (and, therefore, keeping layers)

The word 'egg' can refer to the female reproductive cell, a tiny bit of genetic material barely visible to the naked eye, but in this chapter *egg* refers to the large, stored food supply around a bit of female genetic material that's wrapped inside a hard shell. Because the mother deposits and detaches eggs as an embryo develops, embryos aren't able to obtain food from her body through veins in the uterus. Their food supply must be enclosed with them as they leave the mother's body. (Chapter 13 looks at incubating eggs and hatching chicks in detail.)

The egg that people enjoy with their breakfasts was really meant to be food for a developing chick. Luckily for us, though, hens continue to deposit eggs regardless of whether or not they've been fertilised to begin an embryo, and so you don't need to have a cockerel in your flock. In fact, if you keep chickens in your garden in a built-up area, you're better off not keeping a cockerel. We explain why in the earlier section 'Dealing with the Legal Issues'.

If you want *layers* (in other words, hens that you keep for laying eggs), you need housing that includes nest boxes for them to lay their eggs in and a way for you to easily collect those eggs. Layers are adult birds (they have to be mature to lay eggs) that appreciate some outdoor space, and if you have room for them to do a little roaming around the garden, your eggs have darker yolks from all the goodies they find on their scavenging adventures, and you need to provide less feed for them. (You can read more about caring for layers and collecting eggs in Chapter 15.)

Thinking about home-grown meat

Don't expect to save lots of money raising your own chickens for meat unless you regularly pay a premium price for organic, free-range chickens at a butcher's shop or farmers' market. Most homeowners raising chickens for home use end up paying as much per kilo/pound as they would buying chicken on sale at the local supermarket, perhaps more if you buy the cheapest chickens in store. But that's not why you may want to rear them.

People want to raise their own chickens for meat because they can control what food the birds eat, how they're treated during their relatively short lives and how they're ultimately dispatched. They want to take responsibility for the way some of their food is produced and pride in knowing how to do it. Many people are also concerned about the inhumane conditions commercial meat chickens are reared in and the way people's food is handled before it reaches them. Specifically, some people want to slaughter chickens in ways that conform to kosher or halal (religious) laws. For these reasons, many people are now rearing their own or buying locally grown, humanely raised chickens and getting used to a 'new' taste in chicken.

Knowing the practicalities of home-grown meat

Raising chickens for meat isn't easy, especially at first, but it isn't so hard that you can't master it. For most people the hardest part is the killing, and so the good news is that small-scale poultry abattoirs can do that job for you, for a fee. These abattoirs aren't always easy to find, however, and so you need to understand what skills are needed and the various regulations surrounding killing poultry for meat in case you don't manage to locate one near you and you have to do it yourself.

The good news is that average keepers who have a little space and enough time can successfully raise all the chicken they want to eat in a year. And with modern meat-type chickens, you can be eating home-grown chicken 14 weeks after you get the chicks, or even sooner. So, unlike raising a beef animal or pigs, you can grow your own meat in less than four months.

The major differences between how you rear your own meat birds and how they're 'factory farmed' lie in the numbers of birds in one place, the amount of space the birds have while growing, their access to the outdoors and what they're fed. You can ensure that your birds have a diet based on plant protein if you like, or organic grains or pasture. Most home-grown chickens are also slaughtered under more humane and cleaner conditions than commercial chickens.

If you want to rear meat birds, here's what you need to think about:

✔ **Emotional challenges.** If you're the type of person who gets emotionally attached to animals you care for, or you have children who are very emotional about animals, think carefully before you purchase meat birds. Although traditional meat breeds can end up all right as pets, you really shouldn't leave the broiler-strain birds beyond the ideal dispatching time because their health suffers as a result.

We like our birds and we don't like to kill things, but we love eating our own organically and humanely raised meat. To get around the emotional issues we use a small-scale poultry slaughterhouse (called an *abattoir*). Abattoirs come in a few different categories based mostly on their size and *throughput* – that is, the number of birds being killed there each year. To stay within the ethics of raising your own chickens, and if you want help with the job of dispatching and preparing your birds, you have to seek out a small-scale abattoir. The bigger poultry abattoirs don't want your birds in their systems, which are so vast and mechanised that you can't be sure of getting your own bird back at the end anyway. If you can find a good, small, poultry abattoir near you, treasure it, and do your best to be punctual and pay straight away. It adds to the cost of the final product but it isn't much, and the price is well worth it if you prefer not to do the job yourself.

That being said, we do know how to kill and *eviscerate* (the posh word for gutting, dressing or preparing) a bird, and we advise everyone who raises meat birds to find out how to do it. A day may come when you need the skill, and knowing about the process makes you aware of all the factors that go into producing meat, including the fact that a life was sacrificed so that you can eat meat. You appreciate the final product and all the skills it takes to produce it even more. In Chapter 17 we discuss dispatching and dressing. Read the chapter, and then think about whether you can do what's necessary if you have to. Whatever type of bird you choose to keep – a meat bird, a layer, a show bird or a pet – a time may come when you need to dispatch one humanely. This skill goes with the chicken-keeping territory.

✔ **Space.** You need enough space to raise at least 10 to 25 birds to make meat production worthwhile. If you live in an urban area with room for only a few chickens, producing meat probably isn't for you. Even in slightly roomier suburban areas you need to consider your situation carefully before rearing meat birds. In these areas you may find rearing meat birds in confinement easier than letting chickens free-range or pasture, but you're unlikely to want to do that.

If you live in a rural area, however, feel you have plenty of room and think that you can do your own slaughtering, you can go ahead and raise your own meat chickens, or at least give it a go to find out whether it's for you. Start with a small batch and see how you do. You can find excellent poultry 'dispatch and dressing courses' run by smallholder associations around the country that equip you with the right skills for the job. Search around online for one.

Don't think that growing your own meat chickens saves you any money. It almost never does. In fact, the fewer birds you rear, the more costly each one becomes. Economy of scale – for example, being able to buy and use 1,000 kilos (2,200 pounds) of feed instead of two 25-kilo (55-pound) sacks – helps costs, but most people can't do that. You grow your own meat for the satisfaction, flavour and to take responsibility for the chicken meat you want to eat.

Getting used to a different taste

Be prepared for a big difference in taste between eating your own reared meat birds and buying 'factory farmed' chickens from the supermarket. You can rear chickens that taste just like the chickens you buy in the supermarket by keeping the meat birds confined, but most home flock owners want to keep free-range or pastured meat chickens.

If you're not used to eating chicken meat from free-range birds, expect to get used to a new flavour – one that your great grandparents would have recognised. Raising chickens on a diet that includes grass and other foraged food produces a firmer meat that has more muscle or dark meat and a different, more 'chickeny' flavour. For most people, the flavour's better. A difference also exists in the texture of the meat from birds that have freedom to walk, run and flap. Supermarket chicken meat, unless it's labelled 'free-range' or 'organic', comes from birds that have had no chance to exercise and run around; they're killed at a very young age and the processing sometimes includes adding water to the meat, which then takes on a bland taste and spongy feel. Home-reared chicken is very different, and you may take some time to get used to it.

Some people object to the limited genetics that form the basis of commercial chicken production and the way the broiler hybrids grow meat so fast at the expense of their own health. Their meat is fatter and softer, with more on the breast than with carcasses of other types of chickens. The good news for keepers of home flocks is that new strains of meat birds can be obtained that grow more slowly and are suitable for free-ranging. Research thoroughly when you're ordering to be sure that you buy the right breed of meat chicks for you. (Chapters 3 and 4 look at breeds and buying chickens.)

Showing for adults and children

Showing chickens is a rewarding hobby for adults and an easy way for youngsters to begin raising livestock (and possibly earn a reward!). Chickens, being easy to handle and care for, can also be a good hobby for mentally handicapped adults. A few chickens can provide hours of entertainment, and collecting eggs is a pleasing reward. If you want pet birds, you can buy chicken breeds that tame easily and come in unusual feather styles and colours.

If you're considering keeping chickens as show birds or as pets, consider the following requirements:

- ✔ **Purchasing cost.** Excellent specimens of some show breeds can be quite expensive.

- ✔ **Space.** For showing, you often need to raise several birds to maturity to pick the best specimen to show, and you may want to keep your own cockerel, who certainly crows! These requirements demand extra room. If you live in an urban area, you need to confine your chickens so they don't bother neighbours or get killed in traffic.

- ✔ **Time.** People often keep show birds in individual cages, particularly in the lead up to a show, which increases the amount of time needed to care for chickens.

If you live in a rural area, you can indulge your chicken fantasy to the fullest, maybe getting one of everything! Just use common sense and don't get more birds than you can care for.

Taking Neighbours into Consideration

Your neighbours are people who are in sight, sound and smelling distance of your chickens. Even if keeping chickens is legal in your urban or suburban area, getting your neighbours' approval and continued tolerance of your hobby makes good sense. If your chicken keeping hobby is so quiet and unobtrusive that your neighbours don't think about them, they won't complain. If they know about them but get free eggs, they probably don't complain and if they like your hobby they may be useful to you when you want to go on holiday. A constant battle with neighbours who don't like your chickens, however, may lead to the council banning your chickens or even banning everyone's chickens.

Regardless of your situation, the following list gives you some ideas to keep you in your neighbours' good books:

- ✔ **Try to hide housing or blend it into the landscape.** If you can disguise your chicken quarters in the garden or hide them behind the garage, so much the better. Don't locate your chickens close to the property boundary or a neighbour's patio area if at all possible. (Chapters 5 and 6 have more about designing and choosing chicken housing.)

- ✔ **Keep your chicken housing neat and clean.** Your chicken shelter should be neat and immaculately clean. (Chapter 7 tells you more about housekeeping.)

- ✔ **Store or dispose of manure and other wastes properly.** Consider where you're going to store or dispose of manure and other waste. You can't use poultry manure in the garden without giving it some time to age, because it burns plants. It makes good compost, but a pile of chicken manure composting may offend some neighbours. You may need to bury waste or make an arrangement to give it away to gardening enthusiasts.

- ✔ **Consider doing without cockerels.** Hens don't make much noise, but a crowing cockerel can cause a lot of disruption. Although you may love the sound of a cockerel greeting the day, the noise can be annoying to some people, and neighbours who call the local council with a complaint get taken seriously. Cockerels can exceed accepted noise pollution limits and whether you live in a crowded town or a remote village in the countryside, noise pollution is noise pollution and can legally be stopped.

Contrary to popular belief, you can't stop cockerels from crowing by locking them up until well after dawn. Cockerels can and do crow at all times of the day and even at night.

- ✔ **Get a bantam cockerel, if you must have a cockerel – even if you have full-sized hens.** He'll crow, but not as loudly. Don't keep more than one cockerel; they tend to encourage each other to crow more.

- ✔ **Keep your chicken population low.** Keeping a smaller flock keeps the impact your birds will have on your neighbours to a minimum.

- ✔ **Confine chickens to your property.** Even if you have a huge garden, you may want to keep your chickens confined to lessen neighbours' complaints, because foraging chickens can roam a good distance. Chickens can easily destroy a newly planted vegetable garden, uproot young perennials and pick the blossoms off the annuals. They can make walking across the lawn or patio barefoot a sticky situation. Aggressive cockerels can scare or even harm small children and pets. And if your neighbours come out one morning and find your chickens roosting on the top of their cars, they're not going to be happy.

Cats rarely bother adult chickens, but even small dogs may chase and kill them. In urban and suburban areas, dogs running loose can be a big problem for chicken-owners who allow their chickens to roam. Free-ranging chickens can also be the target of malicious mischief by kids. And of course, chickens rarely survive an encounter with a car.

You can fence your property if you want to, but remember that light-weight hens and bantams can easily fly up onto and go over a 1.3-metre (4.3-foot) high fence. Some heavier birds may also discover how to hop over the fence. Chickens are brilliant at wriggling through small holes if the grass looks greener on the other side, too.

✔ **Control pests aggressively.** In urban and suburban areas you must have an aggressive plan to control pest animals such as rats and mice. If neighbours see your chickens as the source of these pests, they may complain. (Read Chapter 9 for tips on controlling pests.)

✔ **Share the chicken benefits.** Take some eggs to your neighbours or allow their kids to feed the chickens. A gardening neighbour may like to have your manure and soiled bedding for compost. Do what you can to make chickens seem like a mutually beneficial endeavour.

✔ **Never dispatch a chicken in view of the neighbours.** Neighbours may go along with you having chickens as pets or for eggs but have strong feelings about raising them for meat. For these reasons, never dispatch any chickens where neighbours can see it or draw attention to the fact that you do. You need a private, clean area with running water, to kill humanely. If you're dispatching birds at home you also need a way to dispose of blood, feathers and other waste. This waste smells and attracts flies and other pests. We strongly advise those of you who raise meat birds and have close neighbours to send your birds out to be slaughtered if you can, or at least be very discreet.

Don't assume that because you and your neighbours are good friends, they won't care or complain about you keeping chickens if they annoy them in any way.

Chapter 2

Understanding Basic Chicken Biology and Behaviour

In This Chapter

▶ Examining basic chicken anatomy

▶ Determining whether a chicken is healthy

▶ Taking a look at how chickens behave

*S*electing and raising healthy birds involves understanding breed varia-tions, identifying and treating illnesses and talking about your chickens (particularly when looking at health problems or discussing breeding). To do all this, you need to be able to identify a chicken's various parts. With that in mind, this chapter looks at basic chicken biology.

You also need to know how to tell whether a chicken is healthy or not, and so we describe what a healthy bird looks like. And in order to care adequately for and interact with birds, you need to know a bit about their daily routines – how they eat, sleep and socialise – as well as their moulting and reproduction cycles and behaviour. This chapter is your guide.

Familiarising Yourself with a Chicken's Physique

Domestic breeds of chickens derive from wild chickens that still crow in the jungles of Southeast Asia. The Red Jungle fowl is thought to be the primary ancestor of domestic breeds, but the Gray Jungle fowl has also contributed some genes. Wild chickens such as these are still numerous in many parts of southern Asia, and chickens have escaped captivity and gone feral or 'wild' in subtropical regions in other parts of the world. So we've a pretty good idea of the original appearance of chickens and their habits.

Wild hens weigh about 1.5 kilos (3.3 pounds), and wild cockerels weigh up to 2 kilos (4.4 pounds). Wild chickens are slender birds with an upright carriage. Some of that slender body shape remains but many body variations occur in the 200-plus breeds of chickens that exist today, which range in size from 0.5-kilo (1.1-pound) bantams to 7-kilo (15.4-pound) giants. When you bite into a juicy, plump chicken breast from one of the modern meat breeds, you're experiencing one of these body variations first-hand.

The many dog breeds provide examples of what humans can do by selectively breeding for certain traits. Dog breeds from the Chihuahua to the St Bernard derive from the wolf. During domestication, not only did the size change, but also the colour, hair type and body shape altered in numerous ways. Chickens may not have as many body size variations as dogs, but they do have a few, and humans have been able to manipulate the genetics of wild breeds of chickens to come up with the sizes, colours, shapes and 'hairstyles' of chickens that exist today.

The Red Jungle fowl cockerel is quite gaudy, with brilliant red plumage on the back, golden neck feathers and black tail and body feathers shot through with iridescent greens and blues. The female is duller, in shades of brown that camouflage her as she sits on her eggs. Although domestic chickens do come in a wide range of colours and patterns (in Chapter 3 we discuss some of the chicken breeds that have been developed), modern chickens generally keep these distinct colour differences between male and female. Males remain flashy and the females are more soberly garbed. In some breeds, such as the White Leghorns and many other white and solid-colour breeds, both sexes may be the same colour, but even when the chickens are a solid colour, differences in the comb and shape of feathers help distinguish cockerels from hens. Just to complicate matters, in the Seabright and Campine breeds the hens look like the cockerels, with only a slight difference in the shape of the tail. (To see the differences between cockerels and hens, refer to Figure 2-1.)

Figure 2-1:
Comparing
a cockerel
and hen.

Identifying a Chicken's Many Parts

As a chicken-keeper, being able to describe the various parts of a chicken's body properly is important. Calling a chicken leg 'the drumstick' or referring to the 'wishbone' probably isn't the best way to communicate with other chicken owners or with a veterinarian. You need to be able to identify parts you don't see in the supermarket. And when you're looking for a new breed of chick, or buying some fertile eggs or reading advertisements for chickens, you need to know what sellers are talking about when they discuss things such as wattles and spurs. We provide that info in this section.

Checking out similarities and differences

Although some chicken body parts vary in looks, almost all chickens have all the body parts we discuss. For example, although comb shape and size can vary, both by breed and sex, all chickens have a comb.

One exception to chickens having the same body parts is the lack of a tail in Araucana chickens and some chickens that are mixed with them. No one quite knows why this is; it's another mutation that humans chose to selectively breed. Another interesting difference is that some breeds have four toes, and some have five. When a breed has an extra toe, it always points to the back. All chickens have two legs. In the future we may have chickens with four wings, but for the time being, all chickens have two wings. All chickens have feathers too, although the look of the feathers can vary in quality and quantity.

Male and female chickens look the same under the tail – you can't sex a chicken from its external organs – but males (cockerels) usually have exaggerated body parts such as combs and wattles, differently shaped feathers on the tail and neck, and an iridescent colouration to the feathers of the tail, neck and wings that females (hens) generally lack. Cockerels are also slightly heavier and taller than hens of the same breed.

Honing in on the head and neck

The most significant parts of a chicken's head are the comb, the eyes and ears, the beak and nostrils, the wattles and the neck. The following sections provide a closer look at each of these parts, from the top of the head down.

The comb

At the very top of the chicken's head is a fleshy red area called the *comb*. (The combs of Silky chickens, a small breed, are a very dark – almost black – maroon red.) The comb acts like the radiator of a car – circulating blood through the comb's large surface area to release heat, which helps to cool the chicken. The comb also has some sex appeal for chickens.

Both male and female chickens have combs, but they're larger in males. Different breeds have different types of combs. Depending on the breed, the comb may be:

- Big and floppy
- Medium-sized and upright
- Doubled
- Shaped like tiny horns
- Crumpled-looking and close to the head (called a *rosecomb*)

These differences in combs have come about from breeders selecting for them. Large combs are prone to frostbite in cold weather, and parts of them may turn black and fall off. Because of this, chicken breeds with small combs close to the head were often developed in cold countries. Conversely, large floppy combs may help chickens cool down in hot, humid weather.

When baby chicks hatch they have tiny combs that get larger as they mature. The shape of the comb may not be totally apparent in a young chicken, but you should be able to tell whether the comb is upright, rose-combed or double.

Figure 2-2 illustrates some example types of combs.

The eyes and ears

Chickens have small eyes – yellow with black, grey or reddish-brown irises – set on either side of the head. Like many birds, they can see colours. They have eyelids that close from the bottom upwards, unlike humans, and they sleep with their eyes closed.

Chicken ears are small openings on the sides of the head. A tuft of feathers may cover the openings and a bare patch of skin that's usually red or white surrounds them. A fleshy red lobe hangs down at the bottom of the patch. In some breeds, the skin patch and lobe may be blue or black. The size and shape of the lobes vary by breed and sex.

A chicken may occasionally have blue or black skin elsewhere, but the skin around the ear is still red or white. This colouring can help you decide whether a mixed-breed hen will lay white or brown eggs, if that's important to you. If a chicken has red ear skin, it generally lays brown eggs. If the skin patch around the ear is white, it usually lays white eggs. The three breeds that lay blue or greenish-coloured eggs (the Araucana, the Ameraucana and the Easter Eggers) have red ear-skin patches.

Strawberry comb   Rose comb

Buttercup comb   Upright comb

Figure 2-2:
Some
different
types of
chicken
combs.

The beak and nostrils

Chickens have beaks with which to pick up food and to groom themselves;
they run their feathers through their beaks to smooth them into shape. Beaks
consist of thin, horn-like material, and in most breeds of chicken the beak is
yellow; a few breeds have dark blue or grey beaks. The lower half of a chick-
en's beak fits inside the upper half of the beak. You should see no gap where
daylight shows between the beak halves when the bird is breathing normally,
and neither beak half should be twisted to one side or the other.

At the top of the beak are the chicken's two *nostrils,* or nose openings, that
are surrounded by a raised tan patch called the *cere.* The nostrils should be
clean and open but in some birds the nostrils may be partially hidden by the
bottom of the comb. Birds with topknots (feathers on the head) have much
larger nostril caverns than those without. A chicken's sense of smell is probably as good as a human's, according to the latest research.

Inside the beak is a triangular-shaped tongue. The tongue has tiny barbs on
it, which catch and move food to the back of the mouth. Chickens don't have
teeth, have only a few taste buds and their sense of taste is limited.

Beaks are present on baby chicks, and a thickened area on the end of the
beak, called the *egg tooth,* helps them chip their way out of the eggshell.

The wattles and the neck

Under the beak are two more fleshy lobes of skin, one on each side, called the *wattles.* The wattles of males are larger than in females, and their size and shape differ according to breed. The wattles are usually red, although in some breeds they can be blue, maroon, black or other colours.

Chicken's necks are long and slender. Made for peeking over tall foliage to look for predators, the neck is covered with small, narrow feathers, called *hackle feathers,* that all point downwards. (See the section 'Finding out about feathers', later in this chapter, for more info.)

Weighing up the bulk of the body

A chicken's body is rather U-shaped, with the head and tail areas higher than the centre. The fleshy area from beneath the neck down to the belly is called the breast. Some breeds – generally those that are raised for meat – have plumper breasts. Birds, of course, don't have mammary glands.

The area of the back between the neck and the tail is called the *saddle,* which can be a colourful area in male birds. Wings are attached to the body on both sides here, just below the neck. Chicken wings are jointed in two places, with bones shaped a lot like human arm bones.

Finally, you come to the tail. Wild male chickens carry their tails in a slight arch, and the tails tend to be narrow and long, to better slide through the undergrowth where they live. Some domesticated breeds have been bred with no tail feathers at all, but many have been bred with exaggerated tail arches; wide, fan-like tails and extremely long tail feathers. Male tail feathers are often very colourful, but the cockerel's fancy sickle tail feathers have no effect on his function other than to make him look grand, and tail feather differences were selected by humans just for show.

Hens, or female chickens, tend to have small tails, in a fan-type pattern or a narrow arch pattern. Males have these feathers too but they're usually hidden from sight under the more flamboyant arched ones. The feathers may be the same colour as the body or a contrasting colour. They're not as colourful as the male's but help with balance and are used to indicate different moods. A tail held high, for example, may indicate indignation and a drooped tail sickness.

Looking at the legs and feet

Most chickens have four toes, each of which has a nail. Three toes point towards the front and one points to the back. Some breeds have a fifth toe.

This toe is at the back, just above the backward pointing toe, and doesn't touch the ground. It usually curves upwards.

All chickens also have a spur. A *spur* is like a toe that doesn't touch the ground; instead, it sticks straight out of the inner part of the chicken's leg. Hard and bony, the spur starts out as a bud and continues to grow through-out a cockerel's life to the point where adult cockerels may have large, sharp spurs that they can use for defence and fighting. The spur is only a small bud in most hens, and is hardly visible in small hens. If an old hen gets spurs, they stay small.

Checking out chicken skin

The skin of most chickens is yellow or white, with the exception of the Silky, which has black skin, no matter what the feather colour is (the colour of a chicken's feathers doesn't determine the colour of the skin). Skin colour can be influenced by what the chicken eats. In chickens that are fed a lot of corn and chickens that free-range and eat a lot of greens, for example, yellow skin turns a darker, more golden colour and white skin turns creamy.

Chicken's flesh is thin, loose, tears easily and is covered with follicles from which the feathers grow. However, in one breed – the Transylvanian Naked Neck – no feathers grow on the neck. This breed also gets called 'Turkens' because although they aren't crossed with turkeys they do look like them.

Thick, overlapping plates of skin called *scales* cover the lower part of the leg. As chickens age, the skin on the lower leg looks thicker and rough. Many breeds of chickens have yellow legs, regardless of skin or feather colour, but some breeds have white, grey or black legs.

Finding out about feathers

Feathers cover most of a chicken's body. The legs are bare on most breeds, but some have feathers growing down the legs and even on their toes as well. Other variations of feathering include *muffs* – puffs of feathers around the ear lobes; *beards* – long, hanging feathers beneath the beak; and *crests* or *topknots* – puffs of feathers on the head that may fall down and cover the eyes.

The look and feel of feathers varies by breed. Some breeds of chickens, called *soft-feathered,* have loose, fluffy feathers whereas other breeds, called *hard-feathered,* appear smooth and sleek. You may also hear about birds called *Frizzles.* In Frizzles, a feather mutation causes the shaft of the feather to curl or twist, making the feathers on the bird stick out all over in a random fash-ion. Talk about a bad hair day! The Frizzle mutation can occur in a number of chicken breeds, and even in geese.

Once a year – usually in the autumn – birds shed their feathers, starting at the head. This shedding period is called the *moult,* and it takes about seven weeks to complete. This moult period is stressful to chickens, and so we discuss moult in more detail in Chapter 10. Chickens can lose a feather at any time and grow a new one, but new feathers are more plentiful during the moulting period.

Types of feathers

All chickens have two basic types of feather:

- **Contour feathers:** The outer feathers that form the bird's distinctive shape. They include wing and tail feathers and most of the body feathers.

- **Down feathers:** Which lack the barbs and strong central shaft that the outer feathers have, remaining fluffy. They form the layer closest to the body, providing insulation. Silky chickens have down feathers that are as long as those of normal chickens but their outer feathers also lack barbs, and so Silkies look fluffy all over.

Feathers also vary according to what part of the chicken they cover. The following list associates these various types of feathers with the chicken's anatomy:

- **On the neck.** The row of narrow feathers around the neck constitutes the *hackles.* Hackle feathers can stand up when the chicken gets angry, which is where the phrase 'Getting one's hackles up' comes from. These feathers often vary in colour from the body feathers. In most male chickens the hackle feathers are pointed, colourful and iridescent. Female hackle feathers have rounded tips and look duller.

- **On the belly and midsection.** Small, fluffy feathers cover the belly and remaining body areas of the chicken. In many cases, the underside of the bird is lighter in colour. In showing terms, these and any similar quality feathers around the thighs are called *fluff.*

- **On the wings.** Wings have three types of feathers. The top section, closest to the body, consists of small, rounded feathers called *coverts.* The middle feathers, called *secondaries,* are longer. The longest and largest feathers, called *primaries,* are on the end of the wing. Each section overlaps the other just slightly.

- **On the legs.** Chicken thighs are covered with soft, small feathers. In most breeds, the feathers end halfway down the leg, at the hock joint. In some breeds, however, the legs have fluffy feathers right down to and covering the toes. These breeds are referred to as *booted.* This term sometimes describes the name of the breed, as in 'Booted Bantam'.

- **On the tail.** Cockerels have long, shiny, attractive tail feathers. In many breeds, the top three or four tail feathers, called *sickle feathers,* are narrower and may arch above the rest of the tail. Hens have tail feathers too, but they're short and plainly coloured and they don't arch.

Anatomy of feathers

Feathers grow out of follicles in the chicken's skin. Groups of tiny muscles around each feather follicle in the skin allow the bird to fluff itself up by raising and lowering the feather.

The feathers themselves are made of *keratin,* the same stuff that makes up your fingernails and hair. Each feather has a hard, central, stem-like area called a *shaft.* The bottom of the mature shaft is hollow where it attaches to the skin and is called a *quill.* Immature feathers have a vein in the shaft, which bleeds profusely if the feather is cut or torn. The age of a chicken has nothing to do with whether a feather is mature or not.

Immature feathers are sometimes called *pin-feathers* because they're tightly rolled when they start growing and look like pins sticking out of the chicken's skin. The thin, white, papery coating that covers them gradually wears off or the chicken grooms it off by running the pin-feathers through its beak. When the cover comes off, the feather expands, and when the feather expands to its full length, the vein in the shaft dries up.

On both sides of the shaft are rows of *barbs,* and on each barb are rows of *barbules.* The barbules have tiny hooks along the edge that lock or zip them to their neighbours to make a smooth feather. When chickens preen themselves, they're smoothing and locking the feather barbs together.

The colour of feathers comes both from pigments in the feather and from the way the keratin that forms the feathers is arranged in layers. Blacks, browns, reds and yellows generally come from pigments. Iridescent greens and blues usually come from the way light reflects off the layers of keratin, which is similar to the way light is reflected off an opal or pearl. Male chickens generally have more iridescent colours than females.

The feathers of some breeds of chickens are a soft blue-grey colour, which results from pigmentation. This colour is tricky to breed deliberately because breeding blue birds to blue birds doesn't necessarily result in all blue-coloured birds.

Looking a Picture of Health

Although we cover chicken health more fully in Chapter 10, we briefly discuss what a healthy chicken looks like in this section. Having this information may stop you from mistaking illness or deformity for the normal appearance of a chicken. For example, you may be quite shocked to feel a huge swelling in your chicken's throat when it's off to roost for the night, especially because that swelling was definitely not there when you let the chicken out in the morning. This swelling is, however, quite normal – healthy, in fact – and shows that your chicken has spent the day eating well. The swelling is the *crop* – an elastic pouch designed to store food that's digested overnight. The swelling is gone in the morning.

The following quick pointers can help you determine whether a chicken is healthy and normal:

- ✓ **Activity Level.** Here again, differences exist between breeds, but a healthy chicken is rarely still during the daylight hours. Some breeds are more nervous and flighty; others are calm but busy. In very warm weather, all chickens become less active.

- ✓ **Eyes.** Chicken eyes should be clear and shiny, with no discharge or swelling around them. When a chicken is alert and active, its eyelids shouldn't be showing.

- ✓ **Feathers.** In general, a chicken shouldn't be missing large patches of feathers. One exception to this is hens kept with a cockerel. These hens often have bare patches on the back and behind the head that are normal and caused by mating. However, you should never see open sores or swelling where the skin is bare.

  If you take on ex-battery hens, they're normally quite bare when you first re-home them. Commercial egg farms get rid of the hens when they do because their laying slows right down, perhaps stopping altogether for a while, because they're in the middle of their first moult, and are never quite so prolific afterwards. With some TLC they soon feather up again.

  A healthy bird has its feathers smoothed down when it's active, though some breed differences do exist – for example, a Frizzle with its twisted feathers never looks smooth. A bird with its feathers fluffed out that isn't sleeping or taking a dustbath is probably ill.

- ✓ **Feet and Toes.** A chicken's three front toes should point straight ahead, and the feet shouldn't turn outwards. The *hock joints* (like knee joints that bend backwards) shouldn't touch, and the toes shouldn't point in towards each other. Chicken feet shouldn't be webbed (*webbing* is skin connecting the toes), although occasionally webbed feet show up as a genetic defect, and you shouldn't see any swellings on the legs or toes. Check the bottom of the foot also for swelling and raw, open areas.

- ✓ **Mental State.** Chickens should appear alert and avoid strangers if in a lighted area. Unless they've been tamed, inactive birds that allow easy handling are probably ill. Chickens in the dark, however, are very passive, which is normal.

- ✓ **Mouth.** Chickens breathe with their mouths closed, except in very hot conditions. If cooling the bird doesn't result in it breathing with its mouth closed, it's ill.

- ✓ **Nose.** Both nostrils should be clear and open, with no discharge.

✔ **Vent.** The feathers under the tail of the chicken around the *vent* or *cloaca,* the common opening for faeces, mating and passing eggs, shouldn't be matted with faeces, or the area have any surrounding sores or wounds.

✔ **Wings.** Chickens of most breeds carry their wings close to the body, but a few breeds have wings that point downwards. (Study the breed characteristics to see what's normal for your breed.) The wings shouldn't droop down or look twisted. Sometimes droopy wings signify illness in the bird.

If a bird damages a wing and the wing heals wrongly, it doesn't affect the laying or breeding ability of the bird. However, some birds hatch out with bad wings, which is usually the result of a genetic problem. Avoid using these birds for breeding.

Keeping an Eye on Chicken Behaviour

Watching a flock of chickens can be as entertaining as watching children in the playground. Chickens have very complex social interactions and a host of interesting behaviours. And like most domesticated animals, chickens prefer to be kept in groups. A group of chickens is called a *flock.*

Knowing a little about chicken behaviour is crucial to keeping chickens. In this section, we briefly discuss some typical chicken behaviour so you can decide whether the chickens you keep are normal or totally mad.

We hope that discovering a little bit about chicken behaviour may sway those of you who are sitting on the fence about whether to keep some chickens. Chicken-keeping is a fun hobby, even if you're keeping them for serious meat or egg production. When the power goes off, you can go back to ancestral times, sit out in the garden and watch the chickens instead of TV!

Processing information

Being called a 'bird brain' is supposed to indicate that you're a bit dim. However, although birds' brains may be organised more like reptile brains than mammal brains, plenty of evidence indicates that birds, including chickens, are actually pretty clever. (And after you've kept chickens for a while, you may be inclined to agree!) Scientists have recently discovered that although the 'thinking' area of a bird's brain may look different to that of a mammal, birds are capable of thought processes that some species of mammals can't achieve. And chickens' brains are able to repair a considerable amount of damage, too – something mammal brains can't do.

An example of their intelligence is that birds, including chickens, understand the concept of counting. You can train birds to count items to achieve a reward, whereas most mammals who are said to count are in fact responding to signals from the trainer. Birds can also trick or deceive other birds, and even other animals, which means that they must be able to understand the outcome of a future or planned action.

Some birds mimic the sounds of other birds and animals; few other animals mimic sounds. You can't, however, teach chickens to talk as you can with some bird species, and they don't mimic other animals, and so chickens probably fall about mid-range on the intelligence scale of birds.

Because vision is very important to chickens' survival, their brains have a large optic area. A chicken can, for example, spot a hawk or hawk-like object from a good distance away, and the brain immediately tells the chicken to run for cover or freeze, whichever is most effective. They also discover how to spot and avoid other predators quite quickly. We've known chickens that can differentiate between different dogs, spotting the distinctive differences and knowing whether they're friends or foes.

Chicken eyes are also adept at spotting the tiniest seed or the slightest movement of a bug. We've seen them pick up ant eggs and pick the seed off a bit of dandelion fluff. Whereas human eyes can miss an expertly camouflaged caterpillar – or even a big fat one – beady chicken eyes quickly spot it.

Communicating with each other

Chickens are very vocal creatures and they communicate with each other frequently; they're rarely quiet for long unless sleeping, and even then they snore! Chickens make a wide range of sounds that are somewhat open to human interpretation, but we attempt to define some of the sounds as follows:

- ✔ **Cackling.** Hens make a loud calling noise after they lay an egg, and many times other hens join in too. It can go on for a few minutes. Some people call it a signal of pride; others say it's a yell of relief!

- ✔ **Chucking or clucking.** Both cockerels and hens make a chuck-chuck or cluck-cluck sound as a conversational noise. It occurs at any time and can be likened to people talking amongst themselves in a group. Who knows what they discuss.

- ✔ **Crowing.** The loud 'cock-a-doodle-do' a cockerel makes is the chicken noise people know best. Cockerels crow when they become sexually mature, and they don't just do it in the morning. They crow all day long and during the night, too.

The crow is a territorial signal and is how the cockerel announces his presence to the world as ruler of his kingdom. Different cockerels have different crows – some are loud, some softer, some hoarse sounding, some shrill and so on.

- **Growling.** All chickens can make a growling noise. Hens commonly make this noise when they're sitting on eggs and someone disturbs them. It's a warning sound and may be followed by an attack or peck.

- **Perp-perping.** Cockerels make a soft perp-perp noise to call hens over to a good supply of food. Hens make a similar noise to alert their chicks to a food source.

- **Rebel yelling.** These noises are hard to describe, but chickens give out a loud squeal of alarm when they spot a hawk or other predator. In response, other chickens scatter for cover.

- **Squawking.** Grab a chicken of either sex or scare one and you probably hear this loud sound. Sometimes other chickens run when they hear the noise, and other times they're attracted, depending on the circumstances.

- **Other noises.** The preceding sounds are some of the more common chicken noises. Baby chicks peep and trill, hens make a sort of crooning sound when they're nesting and some hens seem to be humming when they're happy and contented. Cockerels make aggressive fighting noises. Sit around a chicken run long enough, and you hear the whole range of sounds.

Investigating chicken table manners

Chickens are notorious for eating almost anything – bugs and worms, seeds and vegetation, and meat. (Although they can't break bones into pieces, they pick the meat off them.) We've even seen them eat frogs and small mice. They pick through the faeces of other animals for edible bits and scratch up the compost pile looking for choice nuggets.

Because chickens' taste buds aren't well developed, tastes that we consider bad don't faze them. This tendency can be their downfall if they eat polystyrene, paint chips, fertiliser or other things that look like food to them. Therefore, good chicken-keepers need to protect their charges from eating things such as pesticide-coated vegetation, treated seeds, plastic, polystyrene beads and other harmful items.

Food takes only about 2½ hours to pass completely through a chicken's digestive system. The food a chicken picks up in its beak is first sent to the *crop*, a pouch-like area in the neck for storage. The crop is stretchy and allows the scavenging chicken to snatch up sudden food finds and store them for a slower ride through the rest of the digestive system. From the crop, food passes to the stomach, where digestive enzymes are added. (Skip back to 'Looking a Picture of Health' earlier in this chapter to find out about swollen crops.)

You may notice chickens picking up small rocks or pieces of gravel, sometimes called *grit*. These pieces go into the gizzard, just beyond the stomach, and help the chicken break down food in the way human teeth do. When chickens roam freely, they get plenty of grit for digestion. If confined, you may need to provide it – we discuss that in Chapter 8.

Both male and female chickens actively hunt for food a good part of the day. Hens sitting on eggs are an exception: they leave the nest for only brief periods of time to feed. Chickens that are confined still go through the motions of hunting for food, scratching and picking through their bedding and chasing the occasional fly. If food is plentiful, chickens may rest in the heat of the day or stop to take a dustbath. Chickens don't eat at night or in the dark.

Most breeds of chickens are equally good at finding food when given the chance, with a few exceptions. The large, heavy, broiler-type meat birds are like sumo wrestlers – they prefer to park their huge bodies in front of a trough and just sit and eat. They don't do well if they have to range to find a large quantity of food or if their food doesn't consist of high-energy, high-protein items.

Taking a kip

When chickens sleep, they really sleep. Total darkness makes chickens go into a kind of stupor, which makes them an easy mark for predators at this point; they don't defend themselves or try to escape. Chickens also sit still through rain or snow if they go to sleep in an unprotected place. If you need to catch a chicken, however, go out with a torch a couple hours after darkness has fallen and you should have no problem, providing you know where they roost.

Because chickens are vulnerable when they sleep, they prefer to *roost* (perch) as high off the ground as they can when sleeping. The more 'street savvy' birds pick a spot with overhead protection from the weather and hunting owls. Chickens like to roost in the same spot every night, and so when they're used to roosting in your chicken house they try to go back home at nightfall, even if they've managed to escape that day or you allow them to roam. For this reason, keeping new chickens confined to the hen house for a good 24 hours before you allow them to roam free is a sensible move. That way, they recognise it as home and return to roost every evening at dusk.

Socialising together

With chickens, it's all about family. Chickens have very special and firm rules for all family or flock members. In the wild they form small flocks of no more than 12 to 15 birds, and each wild flock has one cockerel.

Ranking begins from the moment chicks hatch or whenever chickens are put together. Hens have their own ranking system separate from the cockerels. Every member of the flock soon knows its place, although some squabbling and battling may ensue whilst sorting out the ranking process. Small domestic flocks make life easier for chickens. In large flocks of 25 or more chickens and more than one cockerel, fighting may periodically break out as both hens and cockerels try to maintain the 'pecking order'.

In the hen-ranking system, the dominant hen eats first, gets to pick where she wants to perch or lay eggs, and is allowed to take choice morsels from the lesser-ranked hens. The second-ranked hen bows to none but the first, and so on. In small, well-managed flocks with enough space, the hens are generally calm and orderly as they go about their daily business.

Cockerels establish a ranking system too if a flock has more than one. A group of young cockerels without hens fight, but generally an uneasy truce based on rank becomes established. Cockerels in the presence of hens fight much more intensely and the fight may end in death for one of the cockerels. If more than one cockerel survives in a mixed-sex flock, the defeated one becomes a hanger-on – always staying at the edge of the flock and keeping a low profile.

If a lot of hens have a lot of space, such as in a free-range situation, each cockerel may establish his own separate flock or harem and pretty much ignore the other cockerel, except for occasional spats. How aggressive a cockerel is depends on the breed, as well as on individuals within a breed. We've had some very aggressive cockerels, to the point where they've been a serious nuisance or even a danger to the human caretaker. When a cockerel becomes aggressive towards humans, it's really time to think about removing him before he does someone serious harm.

A cockerel always dominates the hens in his care. (Sorry, no women's lib movement exists in the chicken world!) He gets what he wants when he wants it. And what he doesn't want is a lot of squabbling amongst his flock. When he's eating, all the hens can eat with him, and no one's allowed to pull rank. If squabbling among hens gets intense at other times, he may step in and resolve the problem.

A cockerel can be much smaller and younger than the hens in the flock. If he's too immature he may become 'a hen-pecked husband' and is a sorry looking bird, but as long as he's mature he 'rules the roost'. It's not all about terrorising the ladies, though. The cockerel is also the stern but loving protector of his family and a guide, as well as the hens' lover. He stands guard over them as they feed, shows them choice things to eat (usually letting them have the first bites) and even guides them to good nesting spots.

Cockerels tend to have a favourite hen – usually, but not always, the dominant hen in the flock – but they treat all their ladies pretty well. They may mate more frequently with the favourite, but all hens get some attention.

Romancing the hens

Cockerels have a rather limited courtship ritual compared to some birds, and the amount of 'romancing' varies among individuals too.

When a cockerel wants to mate with a hen, he usually approaches her with a kind of tiptoe-like walk and may strut around her a few times. Usually a hen approached this way crouches down and moves her tail to one side as a sign of submission. The cockerel jumps on the hen's back, holds on to the back of her neck with his beak, and rapidly thrusts his cloaca against hers a few times. He then dismounts, fluffs his feathers and walks away. Boastful crowing may also take place soon after mating, although crowing isn't reserved just for mating. The hen stands up, fluffs her feathers and walks away as well. Both may preen their feathers for a few minutes after mating.

A young cockerel may mate several hens within a few minutes of each other, but usually mating is spread out throughout the day. A cockerel may mate a hen even if he's infertile: fertility reduces as cockerels age, and cold weather also causes a drop in fertility.

Living without a cockerel – the celibate hen

Hens don't need a cockerel to complete their lives, or even to lay eggs for that matter. A hen is hatched with all the eggs she's ever going to have, and she lays those eggs for as long as she lives, or until she's out of eggs, whether a cockerel is around or not. The number of eggs a hen lays over her lifetime varies by breed and the individual. After the third year of life though, a hen lays very few eggs.

Of course, without a cockerel, no babies hatch from those eggs, but the eggs that you eat for breakfast don't need to be fertilised in order to be laid. Hormones control the egg cycle whether a cockerel is present or not. And fertilised eggs don't taste differently – nor are they more nutritious – to unfertilised eggs.

If you're wondering whether your hens would be happier with a cockerel around, our guess is that they probably would be, because it fits the more natural family lifestyle of chickens. But hens are pretty self-sufficient, and if they've never known life with a cockerel, they really don't know what they're missing.

Going broody

At about the time an egg is to be laid, usually early in the morning, the hen's hormone levels rise, and she seeks out a nest and performs some nest-making behaviours: moving nest material around with her beak, turning around in the nest to make a hollow and sometimes gently crooning a lullaby. Several hens may crowd into a nest box at one time, and they seem to be stimulated by the laying of hens around them.

After the egg is laid, hormone levels generally drop, and the hen hops off the nest and announces her accomplishment with a loud cackling sound. Other hens may join in the celebration. Morning is a noisy time in the hen house! After laying an egg, the hen then generally goes about her daily routine.

A hen that *does* have the instinct to sit on (or *incubate*) eggs and hatch them, and is in the mood to do so, is called a *broody* hen. Many modern breeds of chickens no longer incubate eggs because the maternal instinct has been bred out of them. Thus, you have to artificially incubate (put in an incubator) the eggs to perpetuate those breeds. The instinct was bred out of them because when a hen is sitting on a clutch of eggs or raising her young, she doesn't lay eggs.

When a hen does go broody, she tries to sneak off and hide her eggs in a secret nest. If she can't do this, she commandeers one of the nest boxes in the coop. The hen then lays about 10 to 12 eggs before she starts sitting in earnest. Fertilised eggs are fine in their suspended animation stage until she decides to sit. The reason the hen doesn't start sitting on the eggs until about ten have built up is that nature intends for the chicks to hatch all at once, so that mother hen has an easier time caring for them.

We discuss the egg formation process in more detail in Chapter 15, and we describe natural and artificial incubation processes in Chapter 13.

Eggs take 21 days to hatch. When the chicks have hatched, mum leads them out into the world, showing them how to eat and drink and defending them to the best of her ability. A hen can be very aggressive when defending her young, and many tales are told about hens giving their lives to defend their babies. At night or when the weather's cold, the chicks snuggle under her for warmth. Chicks stay with their mothers at least until they're feathered and often for four or five months.

Taking a bath – different for a chicken

An interesting behaviour of chickens is their bathing habits. They hate getting wet, but absolutely love a dustbath. Wherever loose soil – or even loose litter – is present on the floor of the chicken house, you find chickens bathing.

Chickens scratch out a body-sized depression in the soil and lie in it, throwing the soil from the hole into their fluffed-out feathers and then shaking to remove it. They seem very happy when doing this, and so it must feel good. In nature, this habit helps to control parasites.

In the garden or lawn, these dustbath holes can be quite damaging, but you can do little about it except put up a fence. If you confine your chickens all the time, they really appreciate a box of sand or wood ash from the fire to bathe in.

Chapter 3

Getting Familiar with Chicken Breeds

A chicken *breed* is a group of chickens that look similar, have similar genetics and when bred together produce more chickens similar to them. Breeds don't just emphasise visual differences, however. Just like most domesticated animals, chickens have been bred over time for their ability to fit certain needs that humans have for them, including desires to produce birds that:

✔ Lay more eggs

✔ Lay different-coloured eggs

✔ Grow faster

✔ Provide more breast meat

✔ Behave more calmly

✔ Are resistant to disease

✔ Are particularly beautiful

In this chapter, we introduce you to breed terminology and discuss how chicken-keepers categorise breeds. Then we explain the most common breeds within each category and give you a mini-biography about each one – we keep them short and sweet, promise!

If you're interested in breeding purebred chickens or showing chickens, digging a little deeper into breed characteristics and finding out about colour varieties are essential. Every few years the Poultry Club of Great Britain publishes a book called *British Poultry Standards,* which contains colour photos and complete breed descriptions. You may be able to find the book at a library or your nearest poultry club. If you have no luck, you can purchase it from The Poultry Club, Keeper's Cottage, 40 Benvarden Road, Dervock, Ballymoney, Co. Antrim, BT53 6NN, or order it through the website – www.poultryclub.org.

Knowing What You Need to Know: A Brief Synopsis

When researching chicken breeds, you're likely to come across unfamiliar terms, which we decipher in this section. The breeds themselves are categorised into groups; this section introduces those groups. If, after reading it, you know what particular purpose you want your chickens for, flip forward through this chapter to find out more about breeds in that group. For example, if you want to keep chickens for eggs, go straight to the breeds under 'Filling Your Basket: Laying Breeds', and so on.

Speaking the lingo: Common breed terminology

As you explore the world of chicken breeds, you're likely to come across a handful of commonly used terms: purebreds, hybrids, strains, mutations and mixed breeds. In the following sections, we tell you what they mean and what you need to know to get up to speed in no time.

Purebreds

Purebred chickens are those that have been bred to similar chickens for a number of generations and that share a genetic similarity. Breeding two chickens from the same breed usually produces offspring that look much like the parents. If you own a little flock and want to produce new chicks that are the same as your original flock every year, or if you're interested in showing chickens, purebreeds are the ones for you.

Although more than 200 breeds of chickens are known to exist, with many colour varieties within those breeds, fewer than half of these breeds are common. In fact, the vast majority of chickens alive today belong to one of just a few breeds – White (or Pearl) Leghorns, Rhode Island Reds, Cornish and Plymouth Rocks – and crosses of those breeds that comprise the commercial chicken industry for eggs and meat. However, thanks to small-scale chicken-keepers or 'fanciers' like you, who raise them and see value in maintaining genetic diversity, other breeds are kept from disappearing.

Hybrids

Many common and well-known chicken 'breeds' aren't breeds at all, but hybrids. *Hybrids* are the result of crossing two purebreds (see the preceding section), which animal breeders do with the aim of producing a chicken with good traits from both parents, along with increased health and productivity (known as 'hybrid vigour').

Hybrids are for *end use* – that is, they're good for only one purpose, which makes hybrids a good choice if you want the most tender meat in the shortest time or lots of big eggs. They make a less sensible choice if you want to breed birds to produce a new flock because the results are unpredictable. Some babies look like one parent or the other, and others look like neither. Thus, to maintain a supply of hybrid birds, you need to keep two separate purebreds as parent stock.

Strains

Both purebreds and hybrids can be further defined as particular strains. A *strain* is usually one breeder's selection and is based on how that breeder feels the stock should look. Purebred strains represent the basic breed characteristics but may be slightly bigger, more colourful, hardier and so on. In the case of hybrids, the birds that result from mating two particular purebred birds may also be called strains if only a single company or breeder produces them.

Strains are often given names or numbers by which to identify them, especially if breeders produce them in large numbers for commercial use. The Cornish and White Rock hybrid, for example, has a number of strains. Some strains grow faster, some survive heat better, some have white skin and so forth, and can make good choices for you depending on your needs. The same genetics firm may offer several strains.

Mutations

Occasionally a *mutation* pops up, causing a chicken to look or act differently to the birds it was bred from. A mutation is nature rearranging genetic material accidentally and can be good, bad or unimportant. Sometimes, careful breeders can turn good mutations into breeds.

Mixed breeds

In the same way that many mongrels exist in the world of dogs, a number of mongrels exist in the world of chickens. *Mixed-breed chickens* are those whose ancestry isn't known, and they're a combination of many breeds.

Mixed-breed chickens can be a great way to start a home flock. In fact, mixed-breed chickens are often the result of a chicken owner starting with a variety of purebred chickens and letting them breed, and are probably what most old-fashioned farmyards chickens were. If you just want average layers or chickens for that countryside feel, go with mixed breeds. The only problem is that if you don't know the parents of a bird that turns out to be beautiful or very productive, you have a hard time breeding more birds like it.

Over time, flocks of mixed-breed birds that are allowed to reproduce indiscriminately tend to produce smaller-sized, less productive and perhaps less healthy birds. The chickens tend to revert back to the size, colour, behaviour and laying habits of their wild ancestors.

Categorising breeds

From the earliest times, humans and animals have been categorised in one way or another – hunters, gatherers, herders and so on – and chickens are no different. To make choosing the breeds you want to raise easier, chickens are grouped into the following categories:

- ✔ **Dual-purpose breeds.** Dual-purpose breeds lay reasonably well, are calm and friendly and have sufficient meat on their bones for excess birds to be used for eating. Dual-purpose breeds are useful for smallholders.

- ✔ **Egg layers.** In the first year of production, laying breeds may lay 290 to 300 eggs and only slightly less for the next year, and they don't sit on their own eggs. They don't make good meat birds, although they can be eaten.

- ✔ **Meat breeds.** Meat birds were developed to have deeper, larger breasts, a larger frame and fast growth. Most of the chickens known as meat breeds are actually hybrids, although some purebreds such as the Cornish Game are considered to be great meat birds. Meat-type birds generally don't lay very well and are difficult to breed.

- ✔ **Show breeds.** Some people keep chickens for purely ornamental reasons – for their beautiful colours or unusual feathers. These show breeds generally don't lay a lot of eggs, although the eggs they do produce are certainly edible. Most ornamental-type birds don't make very good meals, although you can eat any chicken.

- ✔ **Bantam breeds.** Bantams are miniature versions of bigger chicken breeds or small-sized chickens that never had a larger version. If no large version exists, they're called *true bantams*. Bantams range from 500 grams (1.1 pounds) to about 2 kilograms (4.4 pounds).

Within almost all breeds several colour variations exist. If you're unsure about what colour they should be, check out a good reference book such as the *British Poultry Standards,* mentioned at the beginning of this chapter.

For many breeds, colour variance between male and female chickens is normal. Even if the chickens are a solid colour, males generally have different tail feather structure, larger combs and wattles and some iridescence to their feathers. Go to Chapter 2 for an explanation of the sexual differences.

Wanting It All: Dual-Purpose Breeds

Home flock owners often want chickens that not only give them a decent amount of eggs but are also meaty enough so that they can use excess birds for meat. The eggs from dual-purpose breeds taste the same as those from laying breeds; you just don't get as many. The meat tastes the same if the dual-purpose birds are grown in the same way as you keep meat birds, but their breasts are smaller, and the birds grow much more slowly.

The following list identifies several breeds classified as dual-purpose, some of which used to be considered as meat breeds until the modern meat hybrids were developed (see the later section 'Filling Your Plate: Best Breeds for the Table'):

- **Barnevelders.** The Barnevelder is an old breed that's making a come-back because of its dark brown eggs. Barnevelders are fair layers and are heavy enough to make a good meat bird, although they grow slowly. These calm, docile, fluffy-looking and soft-feathered birds come in black, white and blue-laced, as well as other colours.

- **Brahmas.** The Brahma is a large, fluffy looking bird that lays brown eggs. Brahmas are very good sitters and mothers and as such are often used to hatch other breeds' eggs. They're good meat birds, but they mature slowly. Their feet are feathered, and they come in several colours and colour combos. They withstand cold weather well, and they're calm and easy to handle.

- **Orpingtons.** Another well-feathered bird, Orpingtons deserve their popularity as a farm breed. These large, meaty birds are also pretty good layers of brown eggs. They sit on the eggs and make good mothers when the eggs hatch. Buff or golden-coloured Orpingtons are the most popular, but they also come in blue, black and white. They're calm and gentle, and can forage pretty well but don't mind confinement.

✔ **Plymouth Rocks.** These birds are an excellent old American breed, good for both eggs and meat. White Plymouth Rocks are used for hybrid meat crosses, but several other colour varieties of the breed, including buff, blue and the popular striped black-and-white birds called Barred Rocks, make good dual-purpose birds. They're pretty good layers of medium-sized brown eggs that sit on their eggs and are excellent home meat birds. These usually calm, gentle birds enjoy a free-range lifestyle.

✔ **Sussex.** With their characteristic Columbian markings of black-tipped hackles, wings and tail feathers, these birds are excellent as home-owned flocks. They're pretty good layers with a good reputation for sitting on eggs, and the males make excellent meat birds. They come in several colours – buff, speckled and red Sussex, with Light Sussex the most common. These docile, friendly birds are easily tamed.

Figure 3-1 illustrates the Barred Plymouth Rock, a common dual-purpose breed.

Figure 3-1:
A good common dual-purpose breed – the Barred Plymouth Rock.

Filling Your Basket: Laying Breeds

Different breeds of chickens lay eggs of different colours and sizes, and choosing different chicken breeds for this reason has become fashionable lately. If you want hens that lay a certain colour of eggs, you can read breed descriptions or take a look at the colour of the skin patch around the ear for a giveaway clue. Hens that have white skin around the ear generally lay white eggs and hens that have red skin around the ears generally lay brown eggs, in any number of shades. (You can't tell what shade of brown from looking at the ear patch.) The breeds that lay greenish-blue eggs usually have red ear-skin patches.

All egg colours have exactly the same nutritional qualities and taste; the colour is just a pigment on the outside of the shell.

Breeding out motherly instincts

In the wild, chickens lay eggs for only a short time. They normally lay about ten eggs, and then they stop laying and incubate the eggs. During the time that they're sitting on the nest and for a couple of months afterwards while they raise the chicks, they don't lay more eggs.

To get domestic chickens to lay more eggs than their counterparts in the wild, breeders chose chickens that were less inclined to sit on their eggs or take care of their chicks. After a time, they developed breeds that seldom sat on eggs and laid a lot more eggs than their ancestors. When people found that they were able to incubate eggs artificially, instead of keeping some motherly hens around to raise the offspring of the more productive layers, selection for laying really took off.

White-egg layers

Although individual exceptions do exist, white-egg layers tend to be more nervous and harder to tame than brown-egg layers, but having a range of different coloured eggs to collect every day is all the rage and so white egg laying breeds are becoming popular.

If you want white eggs as part of your chicken experience (see Figure 3-2), the following birds are the best breeds to choose from:

- ✔ **Ancona.** Anconas lay large white eggs. These birds are black-feathered, with some feathers having a white tip that gives the bird a 'dotted' appearance. Similar in shape to the Leghorn, Anconas are flighty and wild-acting. Originally from Italy, they're becoming rare and harder to find.

- ✔ **Hamburg.** One of the oldest egg-laying breeds, Hamburgs are very prolific layers of white eggs. They come in spangled and pencilled gold or silver, or solid white or black. Hamburgs of all colours have slate-blue leg shanks and rose combs. They're active birds and good foragers, but not especially tame.

- ✔ **Minorca.** Minorcas are large birds that lay lots of large to extra-large white eggs. They come in black, white or buff (golden) colours. Minorcas can have single or rose combs. They're very active and good foragers, but they're not easy to tame. The Minorca is another bird that's becoming hard to find.

- ✔ **White or Pearl Leghorn.** This bird accounts for at least 90 per cent of the world's white-egg production. Lightweight, and with a large, red single comb, Leghorns also come in other colours that don't lay as many eggs, but are fine for home flocks. Leghorns tend to be nervous and don't do as well in free-range or pastured situations as other breeds. California Whites are a hybrid of Leghorns and a Barred breed and are quieter than Pearl Leghorns. Other hybrids are also available.

Brown-egg layers

For home flocks, brown-egg layers are the most popular chickens and are easy to come by (see Figure 3-3). The brown eggs laid by these birds can vary from light tan to deep chocolate brown, sometimes even within the same breed, and as hens get older their eggs tend to be lighter coloured. Some of the best brown-egg layers are:

- ✔ **Australorp.** A true breed rather than a hybrid, Australorps lay lots of medium-sized, light-brown eggs. Before Isa Browns came along, they were the brown-egg-laying champions. Both the hens and the cockerels are solid black birds with single combs. They're calm, mature early and some sit on eggs. They were developed in Australia from meat birds, and the cockerels make moderately good eating.

- ✔ **Black Rock.** Crossed from a Rhode Island Red and a Barred Plymouth Rock, the Black Rock is good for smallholders. This long-lived bird produces a lot of brown eggs over a few years, can withstand tough weather and free-ranges well. Black Rocks are common in home owned flocks as well as more commercial free-range flocks.

- ✔ **Hubbard Isa Brown.** This hybrid comprises the world's largest population of brown-egg layers. The Hubbard Isa Brown is a combination of Rhode Island Red and Rhode Island White chickens – which are considered to be separate breeds, not colours – and possibly other breeds. They were developed by a large genetics firm and are available all over the world.

  Isa Brown hens are red-brown in colour with some white under-feathering and occasional white tail feathers, whereas the cockerels are white. They lay large to extra-large brown eggs that range in colour from light to chocolate brown. These calm and gentle birds are easy to work with, and are also good foragers. Although they may be production birds, they have great personalities and are very people-friendly. The disadvantages are that you can't keep them for breeding (they don't breed true), and the cockerels don't make good meat birds.

- **Maran.** Sometimes referred to as Cuckoo Marans, the *cuckoo* bit refers to a colour type (irregular bands of darker colour on a lighter background), although they actually come in several colours, including silver, golden, black, white, wheat, copper and others. Marans were once rare but are now popular for their very dark brown eggs, even though they don't taste any different to any other egg! Not every Maran lays equally dark eggs, though, and the eggs vary in size from medium to large. Most Marans are good layers, but some of the other breeds in this list are better. The various strains exhibit a lot of variation in the breed in terms of temperament and whether they brood. Not all poultry organisations recognise these birds as a pure breed.

- **New Hampshire Red.** New Hampshire Reds are very similar to Rhode Island Reds, and people often confuse them with one another. True New Hampshire Reds are lighter red, and the males have black tail feathers. They're more likely to brood eggs than Rhode Island Reds. New Hampshire Reds are usually calm and friendly but very active. Two strains of the bird exist: some are very good brown-egg layers, whereas others don't lay as well but are better meat birds. Although uncommon in Britain, New Hampshire Reds have global appeal.

- **Rhode Island Red.** Rhode Island Reds were developed in the United States from primarily meat birds, with an eye towards making them productive egg layers as well. As a result, they lay lots of large brown eggs. Both sexes are a deep red-brown colour and can have a single or rose comb. These hardy, active birds generally aren't too wild, but the cockerels tend to be aggressive. Breeders have crossed Rhode Island Reds with many other breeds because of their useful attributes, such as disease resistance.

- **Warren, Goldline, Hy-line and Lohmann Brown.** All these chickens are variations of the same breeding that produced Hubbard Isa Browns. Some are developed from New Hampshires or other heavy breeds rather than Rhode Island Reds. These prolific layers of brown eggs don't sit on eggs, and they don't make good meat birds because of their light frames. They're usually calm and friendly. If you're not going to breed birds and just want good egg production, any of these chickens fill the need.

- **Welsummer.** This breed is popular for its very dark brown eggs, which are medium to large in size. The hens are partridge coloured (dark feathers with a gold edge), whereas the cockerels are black with a red neck and red wing feathers. An old, established breed, Welsummers are friendly, calm birds that are good at foraging. Some sit on eggs.

If you take on ex-battery hens, they all come from one of the common brown hybrid layers.

Figure 3-3:
A good
brown-egg-
layer
breed –
the Rhode
Island Red.

Coloured-egg layers

Like the brown-egg layers that we describe in the preceding section, coloured-egg layers are also popular with home flock owners (see Figure 3-4) and are a bit of a novelty. They originate from breeds of chickens kept by the indigenous tribes of South America and have since travelled widely. Despite pictures that show eggs in a rainbow of colours (no chickens lay pink or red eggs!), their eggs are shades of blue and blue-green. Brown-egg layers whose eggs are a very creamy light brown are sometimes said to lay yellow eggs.

If the thought of getting a good supply of coloured eggs appeals to you, the following birds make good choices:

✔ **Araucana.** Originally from Chile, like its rumpless cousin, the tailed Araucana lays blue-green eggs in generous amounts. These birds sport abundant muffs around the head and a small pea comb. Their soft feathers come in many colour varieties, from white, through lavender to cuckoo, with many other variations, including spangled. They have clean legs and a strong stance.

✔ **Easter Egger.** Easter Eggers are mongrels in the world of chickens because no one is quite sure of their background. They're usually a combination of Araucanas and Rumpless Araucanas and perhaps some other South American blue-egg layers or other layers. Easter Eggers can be a bit more prolific in egg-laying, but the egg colour and temperament of the birds, as well as their adult body colour, range all over the place. They lay shades of blue and blue-green eggs and are a bit of a fashion item since coloured eggs have become popular.

✔ **Rumpless Araucana.** You seldom see this breed of chicken in its pure-bred form. Many chickens sold as Araucana are actually mixes, and so beware. Rumpless Araucanas have no tail feathers: they have *tuffs,* small puffs of feathers at the ears, instead of large 'muff' clumps, or they're clean-faced. They don't have beards, but do have pea combs and most have willow (grey-green) legs. They lay blue to blue-green eggs in similar numbers to their tailed cousins. They're calm and make good broodies.

Figure 3-4:
A coloured-egg-layer breed – Rumpless Araucana.

Filling Your Plate: Best Breeds for the Table

In the not-so-distant past, most meat chickens were young males that were the excess offspring from heavy, generally brown-egg-laying or show birds. People kept them just long enough to make a good meal, which usually meant about five to six months of feeding and caring for the young cockerels. And in your great-grandparents' day, any hen that stopped laying was also used for meat.

About 50 years' ago, breeders began to create hybridised strains of chickens specifically for meat (see Figure 3-5). These strains grew fast, gained weight quickly on less feed than other chickens and had more breast meat – something consumers seem to want. Over the years, these hybrids – particularly one hybrid that was a cross between White Rocks and White Cornish breed chickens – have come to dominate the commercial meat market, often bearing the name of the company that breeds them, – Ross Cobb, Hubbard and Sasso, for example. In fact, almost every chicken you can buy in a supermarket is a strain of the previously mentioned cross.

The majority of commercial meat birds are concentrated on one hybrid and its different strains, but other good meat breeds work well for home flock owners. Many home breeders are trying to develop meat breeds that grow quickly and have good meat yield, and yet are hardy and active enough to do well in pasture or free-range situations.

The meat of these other meat breeds may taste slightly different to what you're used to if they're free-range and take longer to grow. The meat is a little firmer, and they have less breast meat. Some people describe it as an 'old-fashioned' chicken flavour, which is hard to describe (but basically 'more chickeny'), unless you've eaten both types. However, if you give the chickens commercial feed and don't allow them to get too old before dispatching, the older-type meat birds taste very much like hybrid meat birds. The advantage of hybrids is that they grow about twice as quickly and on less feed than conventional meat breeds.

The following list contains some of the best birds for the table:

- **Cornish.** The Cornish is an old breed known for its wide stance and big breast. Several colours other than the white ones are used to make meat hybrids, including red, buff and laced patterns. The birds have a rose comb, and their feathers are tight and sleek. Cornish birds are poor layers of small white eggs and are difficult to breed from; some do brood. They can be aggressive with each other; however, they're fairly tame with keepers. As only fair foragers, these birds are better raised in pens.

- **Hubbards and Sassos.** These birds originate in the big commercial companies and from some of the same breeding stock. Designed for the newer 'free-range' meat market, Hubbards and Sassos can do very well as home-produced meat, ready in about 12 weeks, if you keep and manage them well. They free-range but must have access to a good compound feed as well. They don't lay well or breed true, and so are to be kept only for meat production.

- **Jersey Giants.** This American breed comprises the world's largest chickens. They come in black or white and are meaty birds, but they grow slowly. Jersey Giants are fairly good layers of medium-sized brown eggs and are calm, good at foraging and make good broodies.

- **Ross Cobbs.** Almost all chickens sold for meat today are a hybrid of White Cornish and White Rock chickens, known as Ross Cobbs, or very similar types. Different strains exist, even some where feather colour has been introduced. Expert breeders tinkered with this hybrid until they now have an extremely fast-growing bird with a huge breast. In this way, Ross Cobbs do a remarkable job of converting feed into meat and are often dispatched as young as five weeks' old for the commercial market. They have soft, tender-textured, bland-flavoured meat and lots of white breast meat. Both sexes make good eating and grow at nearly the same rate. If you've only eaten shop-bought chicken, you certainly know how they taste.

These Cornish–Rock crosses lost some things to achieve these 'meat goals'. They can't reproduce normally, and so you can't breed from them if you let them reach sexual maturity. They're extremely closely related genetically, with three or so large firms controlling the production of parent stock worldwide. They have to be managed carefully and fed a high-protein diet to avoid problems with their legs and hearts. Their special nutritional needs mean that they aren't good for free-range or pastured poultry. Basically, they're footballs with feathers – they're inactive, and they spend most of their time eating. Therefore, Ross Cobbs are an excellent choice if you want quick, plump fairly bland meat, but not if you want chickens for anything else.

Figure 3-5:
Examples of meat breeds – Jersey Giant (left) and a white hybrid meat bird (right).

Choosing Breeds for Showing

Certain breeds of chickens exist today mainly for their keepers' pleasure (see Figure 3-6 for a peek at a few). People may have used them as layers or meat birds in the past, but better breeds came along and replaced them. The practice of showing chickens keeps many of these breeds alive. Some of the most beautiful chickens may not be good layers or meat birds, and so owners justify keeping them around by raising them to show, which keeps their gene pool available. They make excellent lawn ornaments and pets too.

The following list introduces you to some excellent breeds to keep for showing:

✔ **Cochin.** Cochins are big, fluffy balls of feathers, with feathers covering the feet. These popular show birds originally came from China and come in buff, black, white and partridge colours. As excellent brooding hens that love to raise families, keepers often use cochins to hatch other breeds' eggs. Their own eggs are small and creamy tan. Cochins are calm and friendly but may be picked on in a flock with active breeds.

✔ **Old English Game/Modern Game.** Both of these breeds used to be bred for fighting but are now used for show. People either like or hate the look of these birds. They stand very upright, with long necks and legs and tight, sleek feathers and come in numerous colours. Both breeds are very active and aggressive birds. Modern Games are larger and heavier. Both types lay small, white eggs. Old English Games are good brooders and mothers; Modern Games are less so.

✔ **Polish.** Polish chickens are small, silly-looking birds with a floppy crest of feathers that covers their eyes. Their crests may block their vision and cause them to seem a little shy or stupid, or to be bullied by other chickens, and so if you aren't showing them, trim their crests so they can see better. Some Polish also have beards. These chickens come in several colours, one of the most popular being a black-bodied bird with a white crest. They lay small, white eggs that they generally don't sit on.

Figure 3-6: Common show and pet breeds – Old English Game (top left), Cochin (top right) and Polish (bottom).

Keeping Perfect Pets: Bantam Breeds

People have always enjoyed seeing miniature versions of domestic animals, and in chickens small breeds called *bantams* are among the best types to keep when you don't have much room. Almost every standard chicken breed has its bantam variety, but some bantam breeds exist only in the small size. When the bantam has a full-size counterpart, the breed description is basically the same, except that the bantams are much smaller. When a chicken has no full-sized representative, it's called a *true bantam* (see Figure 3-7).

Bantams are seldom heavy layers, but the small eggs are still good to eat, taste exactly like other chicken eggs and are the same colour as those laid by a full-sized example of the breed. Bantams don't, however, make good meat birds.

Because most bantam breeds look just like the full-sized version of the breed, we only describe two bantams here:

- ✔ **Japanese.** The Japanese is a true bantam without any large counterpart, and has tiny, short legs and a high arched tail, often in a colour that contrasts with the body. It comes in many colours and colour combinations and weighs about 500 grams (1.1 pounds). Most Japanese bantams are friendly, although the cockerels sometimes become aggressive. Japanese bantams lay white eggs.

- ✔ **Silky.** The Silky is probably one of the most popular pet chickens. Silkies may be quite small or a bit larger, but whereas their size varies, they all have distinctive, fur-like feathers. They're so cute you want to cuddle them, and they don't mind being handled. Some have beards, and they come in black, blue, white, buff and other colours. The skin of Silkies is black, and their combs and earlobes dark maroon red, with a turquoise area around the ear. Silkies are the ultimate mothers. They love to sit on eggs and happily raise anyone's babies. They're sometimes picked on by more active breeds. Silkies lay white eggs.

Figure 3-7:
Bantams –
the Silky
(top) and the
Japanese
true bantam
(bottom).

Supporting Heritage and Rare Breeds

As with many areas of life, chicken-keeping has its fads and fashions. For example, the current craze is for chickens that lay very dark brown eggs. Welsummers and Marans have all increased in numbers lately, due to this dark-brown-egg craze. The downside of fads is that a number of breeds lose popularity and go into decline. Determining which breeds are rare anymore is difficult because so many poultry breeds are getting scarce and the genetic pool of chickens is becoming very limited.

If helping to preserve a particular heritage or rare breed appeals to you, perhaps because of their looks, because you see the need to preserve the genetics of a breed or simply because you like the idea of having chickens that not many other people keep, consider lending these breeds your helping hand:

✔ **Faverolle.** As well as being very pretty, Faverolles are also efficient egg layers. Their eggs tend to be medium-sized and creamy coloured rather than pearl white. Faverolles come in two colours: white and salmon. (The salmon colour is actually black and white with fawn (salmon-coloured) wings and back areas.) The Faverolle has a 'muff' of feathers around the ears, a beard and lightly feathered feet with a fifth toe. These birds are calm and tame, good for home flocks and sometimes sit on eggs.

✔ **Ixworth.** This all-white, large, goose-stepping breed was created in 1932 and is named after a village in Suffolk. The Ixworth is an excellent meat bird and the females are good quality layers. This breed has a bantam counterpart.

✔ **La Fleche.** The La Fleche is known as the Devil Bird because its comb is shaped like a horn. La Fleches are good layers of large, creamy white eggs. They come in black, blue, white and cuckoo colours. Fairly good as meat birds, they're quick to mature but are now quite rare.

✔ **Langshan.** Langshans, which originated in Germany, are large birds with a tall erect carriage due to their long legs, and lay large cream-coloured eggs. They enjoy a free-range life and readily become tame with a bit of attention. Breeders have to train their birds to stand correctly in the show pen and so Langshans are only for the most devoted of chicken-keepers.

✔ **Penedesenca.** Still fairly rare in the United Kingdom, Penedesencas were developed in Spain and are good layers of small-sized, very dark brown eggs. The birds come in several colours and have an unusual comb, shaped like a crown. These birds are rather wild; they withstand heat well and are good free-range birds.

✔ **Transylvanian Naked Necks.** Originating and still quite popular in Austria and Hungary, in some countries this chicken is called a 'Turken' because it looks like a chicken crossed with a turkey. When you get past the shock of the naked neck these chickens are astoundingly good birds of a fair size. Their qualities have been exploited by the commercial poultry meat world. Tougher than they look, these birds are able to free-range and withstand cold weather.

Most rare breeds aren't prolific layers and are rare for a reason. This scarcity is reflected in the price you pay for them.

Chapter 4

Buying Chickens

· ·

In This Chapter

▶ Determining the age, gender and number of birds to buy

▶ Working out your start-up costs

▶ Starting with chicks and adults

▶ Finding out where to buy birds and what to look for

· ·

*M*aking up your mind to become a chicken owner is one thing, but before you do so you need to make some important decisions. A little planning goes a long way towards ensuring a good experience with your first attempt at keeping chickens.

In this chapter, we help you decide whether you want to start with chicks or adult birds, and how many and what type of chickens to buy. We talk about how to go about buying chicks and adult birds, including hints on where to buy chickens and how to choose healthy birds.

Planning Your Flock

Chickens are social flock creatures, and so for them to be totally happy you need to plan to keep at least two. How many chickens it takes for *you* to be happy is another matter.

Check the rules and regulations of the area you live in to see whether you're allowed to keep chickens. Very rarely, in built up areas, a clause in the house deeds prohibits hens from being kept at the property (see Chapter 1).

In the sections that follow, we help you decide how many chickens you need to buy and whether to start with eggs, chicks or adults, as well as help you decide where to get them.

Deciding what you'll start with: Eggs, chicks or adults

So you've made the decision to keep chickens. We hope that you have an idea of why you want them, too (if not, head to Chapter 1, because you need to know why you want them before you can start making purchase decisions). Next you need to decide how to begin your chicken-raising experience.

You can choose between starting with eggs, chicks or adults. In this section we cover the pros and cons of all three options – one of which is the right choice for you. However, the best way forward usually is to start off with baby chicks (a few days old), off-heat chicks (a few weeks old) or fully matured adults.

Starting with fertile eggs

Buying fertile eggs from a breeder or an Internet sales site is one way of starting your chicken-raising adventure, but if you're planning to follow this route you may want to reconsider. Fertile eggs are difficult to store correctly and even more difficult to transport – they're likely to break or become chilled or overheated. They must be packed so that they don't experience too much jostling and shaking, because too much shaking kills the embryo in suspended growth inside the egg. In fact, sellers rarely guarantee that any will hatch, but they generally still charge nearly the same for eggs as for chicks. Finally, incubation with even the best models of incubators can be a tricky process best left to more experienced hands.

That said, if you want to keep certain rare or heritage breeds, fertile eggs that you have to hatch yourself may be your only option. If this is the case, or you insist on starting with fertile eggs, you can improve your chances of hatching the eggs by finding a local source. Set up your incubator before the eggs arrive and place them in it promptly on arrival, but don't expect more than about 50 per cent to hatch.

Before buying expensive hatching eggs of rare breeds, you may want to practise hatching less-expensive eggs of common breeds to get some experience.

Starting with chicks

Many people start with chicks because they fall in love with some at an auction or someone gives their kids some chicks for Easter. Beginning your chicken-raising experience with baby chicks is probably the most economical and practical way to start, for the following reasons:

✔ You can choose from a wide assortment of breeds.

✔ Chicks are less likely to carry disease and parasites than older birds, especially if you buy them from a well-run hatchery.

✔ You don't have to guess how old the birds are.

✔ Some breeders sex the chicks for you, so you don't have to guess.

✔ Chicks usually cost less than older birds.

Although you can have day-old chicks delivered to your door, this service is usually only available to commercial growers. A better option is to source your chicks from a reputable breeder within driving distance of your home.

If you live near a hatchery and can pick up the chicks, you may be able to purchase just a few. You may also have local chicken breeders in your area who can sell you just a few chicks. Your local feed merchant is likely to have an advertisement board with information about nearby breeders.

For many people, though, chicks may not be the best way to begin keeping chickens, however cute they are. Baby chicks have their drawbacks:

✔ **They need TLC.** All baby animals take extra time and effort to care for.

✔ **They're fragile.** Small children and pets can easily damage or kill them.

✔ **They require special brooding equipment, which costs money.** You need to keep the chicks warm and protected.

✔ **Their quality is hard to judge.** Judging the quality of a chick is very difficult, and so if you want to show chickens, you need to keep a lot of chicks until they become adults in order to pick the best of the bunch.

✔ **They need time to mature.** Pullet (female) chicks take at least five months to mature and begin to lay eggs. Pullets of some breeds may take several months longer.

✔ **Their sex is hard to determine.** Although hatcheries do a pretty good job of sexing chicks, you may not end up with as many hens as you expect, which means you have extra cockerels to deal with. If you buy your chicks from private breeders, their skill at sexing chicks may be poor, and you may not get any hens at all.

✔ **They need extra protection.** If you just want chickens for living lawn ornaments, you can't allow them to roam safely for several months. You may want to consider adult birds if you're the impatient type.

To find out more about what rearing baby chicks requires, turn to Chapter 14.

Instead of newly hatched chicks, a few hatcheries and some local breeders offer what they call *off-heat chicks* – chicks of a few weeks old that no longer need the heat from a *brooder* (a heated enclosure for rearing young chicks) to survive. Older chicks around 6–8 weeks are easier to sex, depending on their breed, and may enable you to get a better idea of their show quality. They take less time to reach productive age – a plus if you're buying layers. If you're after meat chickens, be aware that because they grow so fast they're seldom offered as off-heat birds, unless you can come to an arrangement with the seller.

Adult birds cost considerably more to buy than chicks because they've cost the breeder in food and heat, and so fewer breeds are available to buy as off-heat birds than as baby chicks. If you want off-heat chicks, order them early in the season because the supply may be limited.

Make sure that you know the age of the birds on offer, particularly if you want to avoid the use of a brooder.

Starting with adults

If you worry that baby chicks may be too difficult to rear, or that you want egg production right away, or want to be able to assess the show qualities of the birds you buy, your best bet is to start with adult birds.

Buying adult birds has pros and cons. Here are some of the cons:

- Many people struggle to tell the age of an adult chicken.
- Older hens lay fewer or even no eggs. Remember this if you're choosing ex-battery hens.
- Old cockerels may not be fertile.
- Some people have a hard time sexing adult birds, although with most breeds, a little experience soon helps that problem.
- Adult birds may have been exposed to many diseases and are more likely than chicks to have parasites. You need to examine the birds carefully before you buy.

On the plus side:

- You can quickly assess the quality and colour of an adult bird.
- Adult birds require less fussing to get them established in their new home.
- Young adult layers generally sold as *point of lay,* or just *POL* (the stage when they are nearly ready to lay their first eggs, at around 18–22 weeks old), quickly start providing you with breakfast.

Choosing the gender

The gender you choose depends on your purpose for raising chickens. Use the following list of reasons to decide whether you need cockerels, hens or both:

- **Breeders.** If you want to breed more than two purebred breeds of chickens or different colour varieties, you need more than just one male bird. Plan your housing so that you have two or more separate flocks.

- **Egg producers.** If you want layers, order sexed pullets at POL or buy adult hens. They cost more, but the extra outlay is worth it.

- **Meat birds.** Many people order only *cockerels* (male birds) because they grow faster and larger than *pullets* (young females). Cockerels can also be cheaper than pullets in some breeds, but in the broiler strains they often cost more because they grow faster. You don't have to worry about broiler-type cockerels fighting unless you fail to dispatch them before their fighting instinct kicks in.

- **Show birds or pets.** People rarely keep only cockerels, unless they just want a pet or they're raising show birds. (In most breeds of chickens, the male is the most colourful and makes the best show bird.) Although cock birds can become aggressive as they age, a single pet cockerel kept without hens rarely becomes aggressive.

Hens don't need a cockerel around to produce eggs or to live a fairly normal life, although they seem to appreciate having a male around. You don't have to keep a cockerel with your hens, particularly if you live close to neighbours and think that they may be bothered by a cockerel crowing. But if you like the sound of crowing, think that cockerels are handsome or feel that it's more natural for the chicken family to have one, you only need one cockerel with your flock unless you're breeding chickens (which we describe in Chapter 12).

Selecting the right number of chickens

No matter how many chickens you intend to have eventually, start off small if you're new to chicken-keeping. Get some experience caring for the birds and see whether you really want to have more. Even if you have some experience, you may want to go to larger numbers of birds in stages, making sure that you have proper housing and enough time to care for the birds at each step.

Chickens are social and don't do well alone, and so you need to start with at least two birds: two hens or a cockerel and a hen. (Two males fight!) Beyond two birds, the number of birds you choose to look after depends on your needs and situation:

✓ **Layers.** You can expect one young hen of an egg-laying strain to lay about six eggs a week, two to lay a dozen eggs and so on. If the birds aren't from an egg-laying strain but you still want eggs, count on three or four hens for a dozen eggs a week. Work out how many hens you need based on how many eggs your family uses in a week – just remember to add in more hens if you don't get an egg-laying strain.

✓ **Meat birds.** Raising just a few chickens for meat doesn't really pay, but if your goal is to produce meat and space is limited, you can raise meat birds in batches of 10 to 25 birds, with each batch of broiler strains taking about 10 to 12 weeks to grow to dispatching size. If space and time to care for the birds aren't problems, determine how many chickens your family eats in a week and base your number of meat birds on that figure.

If 12 weeks is necessary to raise meat chickens for the table and your family wants two chickens a week, you probably want to buy your meat chickens in batches of 25 and start another group as soon as you dispatch the first. Or if you want a rest between batches, raise 50 to 60 meat chicks at a time and start the second batch about three months after the first. Remember that in a decent freezer with a minimum temperature of –18 degrees Celsius (0 degrees Fahrenheit), frozen chicken retains good quality for about six months.

✓ **Pet and show birds.** When you keep chickens for pet and show purposes, you're only limited by your housing size and the time and resources you have to care for them. Full-size birds need about 0.3 square metres (3.2 square feet) of shelter space per bird; bantam breeds somewhat less. Just don't overcrowd your housing.

If you're going to breed chickens to preserve a breed or produce show stock, plan on at least two hens for each cockerel but not more than ten. In some large breeds with low fertility, you may need a ratio of five or six hens per cockerel.

Counting the Costs

Anytime you start a hobby or begin producing food for the household, you run into start-up and maintenance costs, and keeping chickens is no exception. The good news, however, is that chickens are more economical to purchase as pets or food-producing livestock than most other animals. Unless you're looking for expensive rare breeds, most people can start a small flock of 4 to 25 chickens for less than £80.

Regardless of whether you're starting with 4 or 25 chickens, use the following list of tips to keep costs down when purchasing your flock:

✔ If you're mail-ordering chicks and need fewer than the minimum number you're required to order, try to find someone to share an order with you. Some small producers allow people to order chicks in small numbers, and they combine those orders to meet the minimums.

✔ Some people who want just a few laying hens order a few pullet chicks alongside a more established egg producer's order. The commercial breeders who advertise in the back of country pursuits magazines may be able to point you in the direction of one of their regular customers where you can pick up your chicks and get the number you require.

✔ For meat birds, many people order only cockerels because they grow faster and larger than pullets. Cockerels can also be cheaper than pullets in some of the rarer breeds but in the broiler strains they often cost more. So ordering commercial meat chicks *as hatched* – meaning chicks whose sex hasn't been determined – generally saves you money; with *as hatched* chicks, both sexes grow equally well.

✔ For meat birds, although some people still dispatch a few chickens at a time as the need arises, raising in batches makes better economic sense. You use the same amount of electricity for the brooder, have to buy bedding and feed and so on, so raising 10 to 25 chicks at a time isn't much more expensive than raising 2 or 3 meat birds.

✔ If you have to order chicks by mail, try to order from a hatchery close to you. The closer the hatchery, the lower the transport costs.

✔ Day-old chicks are the most economical way to buy chickens. Fertile eggs may cost almost the same as chicks, but after the expense of purchasing an incubator, running it and generally only having half of the eggs actually hatch, chicks come out ahead in cost savings.

✔ Pay to have chicks vaccinated at the hatchery; for the hatchery to do it costs less than for you to buy vaccines or pay a vet. (For more about vaccinating chicks, see Chapter 10.)

✔ Buy adult birds in the autumn because young birds have just finished growing and people are selling their excess young birds. People are also thinking about winter feed costs, and so the birds are less expensive in the autumn than in spring when supply is low and demand is high for older birds.

When purchasing adult hens for egg-laying, do some cost-comparison shopping and be wary of people selling hens at low prices. Old, non-laying hens – that eat as much as hens who lay well – are hard to distinguish from young hens. (For tips on how to tell whether a hen is laying, see the section, 'Where to buy adult chickens', later in the chapter.)

Starting with Chicks

Chicks come in several colours and sizes, and when you're looking at a tantalising website or gazing down at a pen full of cute babies at the local auction, you may struggle to pick just a few. But remember: although the chicks are small now, they quickly need more room, and so don't buy more than you can take good care of. Of course, starting out with good-quality chicks helps. In the sections that follow, we give you the information you need to find and select healthy chicks from a variety of sources.

Where to buy chicks

A number of places sell baby chicks and we discuss these sources in the following sections. Some baby chicks are available in all but the coldest months of the year, but you have more opportunities in the spring and early summer to find the chickens you want.

Breeders and hatcheries

A well-laid-out website is a dangerous thing: all those interesting breeds and cute pictures! You can order chicks from such websites, or by phone, from various hatcheries across the United Kingdom.

In most hatcheries, eggs are hatched according to demand, so when you order 25 Rhode Island Red chicks, the hatchery adds 25 or so Rhode Island Red eggs to the incubator, and they hatch in about 21 days. This description is a simplification of the process, but it serves as a reminder that you need to allow some time between when you order chicks and when you want to receive them.

Chicks come in different colours. If you order chicks of multiple breeds but similar colours that may be difficult to tell apart, ask the hatchery when you place your order to separate the breeds with a cardboard divider.

The hatchery drivers deliver baby chicks on the day after they hatch and they normally arrive within 24 hours (being in transit for longer than that stresses the chicks). They can survive well without food or water for two or three days after hatching. Remnants of the egg yolk are attached to them, which slowly get absorbed in the first few days after hatching and make eating and drinking unnecessary.

Baby chicks can be safely transported by car or truck at most times of the year. They're packed closely into boxes so that the combined body heat helps warm them – this is the reason for the usual 25-chick minimum order. However, even in groups of 25, baby chicks can become chilled or over-heated during shipping, and so order your chicks when the weather is mild, just to be on the safe side.

In the past, most courier firms would transport live cargo. Nowadays, chick transport is generally arranged and run by the hatchery using its own staff, who are briefed in the transport needs and welfare of the chicks. Most hatcheries have regular drop-off points and routes, so choose the date for your chicks to be shipped that's closest to the time you decide you want them.

Your chicks should arrive in a contented state, ready for their first food and drink. If they're peeping loudly when they arrive, they're stressed and don't get off to the best start. For this reason, source your chicks from as close to home as possible to avoid this happening.

Don't think that you need to order a lot of extra chicks in case many of them die in transit or after you receive them. Hatcheries commonly add a free chick or two to shipments to cover shipping losses, and if you take care of your chicks correctly you shouldn't lose many of them.

We've had chicks delivered many, many times and have only had one bad experience, when about half of the chicks arrived dead. The box had been delayed an extra day because of a big snowstorm late in the season.

When you collect your chicks or get them delivered, immediately open the box and inspect them. Let your hatchery know if the box contains a lot of dead chicks. Some hatcheries don't guarantee safe arrival, but many do. In the event that something has gone wrong, take a photograph as evidence, and make sure that you count the chicks before you complain. Some hatcheries add extra chicks to the order to account for losses.

To find out which hatcheries are reputable in your area of the country, ask other chicken owners where they got their birds or visit an online site such as www.backyardchickens.com to see what the forum members recommend.

Some UK hatcheries

The following list of hatcheries isn't a recommendation of their quality or service and isn't inclusive. It's just a starting point to help you find the chicks you're looking for.

- **Black Rock Poultry:** phone 01505 613075
- **Brotherton Farm West Lothian:** phone 01506 873982
- **Cyril Bason:** phone 01588 673204; website www.cyril-bason.co.uk
- **Farthingwood Poultry:** phone 07599 974174; website www.farthingwood poultry.co.uk
- **Harepathstead Poultry:** phone 01404 823415; website www.harepathstead poultry.co.uk
- **Newland Poultry:** phone 07830 176254; website www.newlandgrange.com
- **Storrs Poultry:** phone 01226 764822; website www.storrspoultry.co.uk

Hatcheries also sell supplies for chickens, and some sell fertile eggs.

Local breeders

Local breeders have small hatcheries or use hens to hatch eggs, and you visit their establishments in person to pick out chicks. You can usually find a local breeder by looking on notice boards at your local chicken food shop or in the breeders' directory and advertisement sections of smallholding magazines.

When you visit a local breeder, you can see how the chickens are kept and whether they look healthy and happy. You may well be able to purchase just a few chicks – maybe chicks that are already off to a good start. You have the help and advice of someone experienced with raising chickens.

The disadvantages of purchasing locally are that the breeds you want may not be available and the times of the year when you can get particular chicks may be limited. Some breeders have chicks to sell most of the time; others want to incubate eggs only when you request chicks, and so you may need to plan ahead.

Many local breeders aren't able to vaccinate chicks for several diseases the way large hatcheries can, and some may not know how to sex chicks, because chick sexing is a specialised skill.

Oh, that guy with the box of chicks at the car boot sale? You don't want to buy chicks from him because you don't know how they were hatched. You have to take his word for what breed and sex they are and you're likely to end up with diseased or sickly chicks.

Specialist poultry farms

Many specialist poultry farms have chicks for sale in the spring – more and more of them place advertising boards at the farm gate. Some take orders for chicks and others bring in batches of popular breeds to sell. These farms don't make a lot of money from the bought-in chicks, but they're hoping that you also purchase starter feed, brooder lamps, water-holders and so on while you're there.

At the specialist poultry farms that take pre-orders, first look at a list or catalogue and decide what chicks you want, and then order them. Often you don't have to order the 25-chick minimum, which can be very helpful (the farm combines several orders to make up the shipping minimum). You generally pre-pay, and a few weeks pass between the day you place your order and the day you pick up the chicks. The farm can care for your babies if you can't pick them up during office hours, but don't expect it to care for them longer than a day.

With large, combined orders, a farm generally separates customers' individual orders by cardboard barriers in the shipping box. Sometimes chicks manage to breach these barriers, and if the different breeds shipped look alike as chicks you may find, as they get older, that you got the wrong breed or sex.

Some farms carry extra chicks for sale to impulse buyers. Check before you buy to see whether anyone knows the breed and sex of the chicks. It's unlikely, but if the farm just bought a batch of cute, cheap, mixed chicks, they're probably mostly cockerels, and not the type that make good meat birds. You want to avoid them.

If the farm can tell you that they're broiler chicks or pullets from a laying strain of bird, you can then judge whether they suit your needs. Even if they're a mix of laying breed pullets, for example, they may be a good buy. Look for lively chicks and make sure that they're being kept in a warm, clean environment, protected from too much handling. Remember how many you've decided you need and don't overbuy because of the cuteness factor or because they're on sale!

When to buy chicks

If your chicks are going to be delivered by car or truck, try to order them when weather conditions aren't likely to be too extreme. Bad weather may affect them during shipping, and the cost of running a brooder for chicks is higher when the weather is still very cold.

If you're looking for a rare breed or must have a certain breed, shop early in the year because some rare breeds and very popular breeds sell out early in the spring. In many cases, if you pay for the chicks, you can request delivery at a later date, but at least you've reserved the birds you want. Remember that many rare breeds don't lay as many eggs as more common breeds, and so the hens that lay them are fewer also. Rare is Rare for a reason!

If you're buying laying birds started as chicks, remember that they don't begin laying for at least five months. You may want to order them as early as you can so that they mature while days are still reasonably long. If they're going to be five months old in the middle of winter, egg-laying may be delayed until the days start getting longer.

Purchase meat birds so that when they come out of the brooder (when they're about one month old) you have a good grass supply if they're to free-range. Three to four months is needed to raise broiler-type birds to a good size for the table and you want to make best use of the summer weather.

What to look for

Most people, when they buy a car, a shrub for the garden or anything that counts as an investment of some kind, check over their potential purchase to make sure that it's in good condition before parting with their hard-earned cash. Buying chicks is no different – whether you're buying them as pets, layers or meat birds, you need to be sure of what you're buying.

The tips in the following sections apply mainly if you're going to buy chicks from a local breeder or farm. If you buy chicks from a distant source, you're pretty much stuck with what you have, but the tips we provide here may tell you whether something's wrong and you need to call the hatchery.

Checking the colour

Looking at the colour of chicks is important – it may give you clues as to the sex or breed of the bird.

Not all baby chicks are yellow. Some are brown, brown striped, grey, black or reddish. If you're buying mail-order chicks, check the breeder's website to see whether the description tells or shows you the colour of the chick. Even if you're going to pick chicks from a local breeder, you may want to check out some chick websites so that you've an idea of what they should look like before you go.

If you don't know what colour your chicks are supposed to be, you can generally assume that dark adult birds come from black, brown or grey chicks. The chicks of chickens with variegated feathers, such as a partridge colour, generally have faint dark stripes on a yellow or brown background. White, red and buff chickens usually have yellow chicks. But the colour of chicks can vary from breed to breed and even among chicks from the same breed.

When you get your chicks, the colour may tell you whether the right chicks were delivered. If the colour seems wrong, check to see whether the hatchery reserved the right to send substitute breeds if it hadn't got the breed you wanted. (Usually you're notified if a hatchery sends you a substitute.) If you have questions, call the hatchery. If you're buying locally, you can just ask the seller about the chicks' colour.

Determining sex

Checking the sex of chicks before you buy is important for ensuring that you get the hens or cockerels – or the mixture of both (as hatched) – that you want. Unless the chicks are from a sex-linked line, though, telling the sex of newly hatched chicks simply by looking at them is impossible, and so don't let people tell you that they can.

Sex-linked chicks are hybrids of two breeds. The male chicks are one colour and the females another, which makes sexing quite easy. Sex-linking only works for some breeds, though – for example, a Rhode Island male and a Light Sussex female have paler down for the male chicks and buff/brown down for the females.

Some people are pretty good at picking out the sex of chicks as they grow by observing their combs and some of the feathers. The combs of cockerels, for example, generally grow a little faster than those of pullets. Cockerel tail feathers may also look different than those of pullets. As they get their adult feathers, sexing young birds becomes much easier.

Telling the gender of chicks even at a few days of age is also possible by looking inside the vent area, but it isn't easy. You have to look inside the *cloaca* (the area where waste and eggs are passed) with a strong light, sharp eyes and some training. Most people, even breeders, have trouble doing it without harming the chick, and so hatcheries employ experienced people to sex chicks if they offer sexed chicks other than sex-linked breeds. That's why sexed chicks cost more.

If you're picking the chicks from a group, you pretty much have to take your chances on whether you get males or females. If you get to see the chicks within their first 24 hours after hatching, you can tell their sex fairly accurately by studying the feather formation on the wing tips (see Chapter 12). If the chicks are a few weeks old, the breeder may have a pretty good idea which are males and females, but do expect some surprises.

Looking at type of comb and other breed characteristics

The comb of chicks is small but visible. When you examine the head of a baby chick, you should be able to tell whether the comb will be single, rose or some other comb formation (see Chapter 2 for more info about combs); its shape is as its breed type. You can't tell whether the size and placement are as they should be until the bird is older.

Chicks of breeds with crests and topknots should have a puffy bump on the head or even a tiny topknot right from hatching. Sometimes this looks like a whorled area on the top of the head. The beginnings of muffs and beards may also be seen on chicks from breeds with those traits. Of course, you won't be able to judge their quality until the chicks are grown.

Feather-footed breeds are a bit more difficult to spot as chicks. Some have fluff growing down their legs, but in others, the difference from bare-legged chicks is hard to spot.

If the breed you're looking for is supposed to have five toes, look for the fifth toe on the back of the leg – and don't confuse it with the bump that becomes the spur. It should look much like the front toes. The colour of a chick's legs should match the breed's characteristic colour.

Assessing health

Whether you buy chicks, hatch eggs or adopt adult birds, having a healthy flock begins with choosing healthy birds. Serious breeders usually have their flocks tested and vaccinated for prominent diseases, but ask what vaccines have been given for other diseases. If you're offered the option, have the hatchery vaccinate your chicks for Marek's disease. Doing so costs a bit more, but is well worth it – vaccinating chicks is very difficult for home-flock owners. (For more information on vaccines and testing for disease, see Chapter 10.)

Healthy baby chicks:

- ✔ **Are active.** Of course, like all baby animals they do sleep more than adults, but if disturbed, they quickly get up and move away.

- ✔ **Are cheerfully – but not too – noisy.** Chicks that are very noisy are unhappy and stressed from being cold or hungry and thirsty. When they arrive in a delivery box, the stress is evident from the shrill cheeps. However, if you place them in the right temperature with food and water, they quickly calm down.

- ✔ **Have two bright, clear eyes.**

- ✔ **Have clean back ends, or vents, that aren't pasted up with faeces.**

- ✔ **Have straight beaks that aren't twisted to one side or the other.** Some hatcheries trim the end of the beak to prevent chicks from picking at each other, so don't be alarmed if the beak tip is missing. Smaller breeders probably don't do this, however, because it's not a necessary operation.

- ✔ **Have straight toes, not bent – or worse – missing ones.**

Newly hatched chicks may have a slight lump on the belly where the egg yolk was, and that's okay. But the belly area shouldn't look sore and red. The chicks shouldn't have any wounds or bloody areas.

If you look at a content group of chicks in a proper brooder, some stay under or near the heat lamp sleeping peacefully, whereas others eat, drink or walk around. They're quiet except for an occasional peep. If a chick is sitting by itself looking droopy, it may not be healthy. If a chick is touched and responds very little, it probably isn't healthy. If it's lying on its back with its legs in the air, it's definitely unhealthy!

Chicks that are panting, with their beaks open, are too warm, or sick. If they appear normal after being cooled down, they should be fine. If the chicks are as far from the heat source as possible, the temperature's probably too hot. If they're piled on each other near the heat source and peeping loudly, it's probably too cold. If chicks are very noisy but they aren't obviously hot or cold and they have food and water, something else is wrong. Although you can fix the temperature or hunger problem, avoid purchasing chicks if you can't tell what's wrong with them.

You certainly can't judge the health of a chick from looking at a website – you have to depend on the seller to send you good, healthy chicks. If you get your chicks through mail order, and many are dead or appear weak and drowsy when they arrive, contact the supplier right away – and take photographic evidence.

Handling chicks

As a chicken owner, you need to master the proper ways of catching and holding chickens of all ages and sizes (Chapter 10 has additional info on the safe handling of chicks). Children especially need to know how to correctly catch and hold chickens too, if you allow them to do so – many poor chicks have been strangled by the loving grasp of children. Don't allow children under the age of five to hold chicks without close supervision. Never let them catch the chicks; instead, have them sit down and hand them the chicks to hold briefly.

Never let children kiss chicks or chickens, or rub them on their faces. It makes a cute picture, but is a dangerous practice for your health. All chickens – even cute, fluffy ones – can carry salmonella and other nasty bacteria and viruses even though they appear perfectly healthy. For this reason, make sure that children don't touch their faces or mouths after handling chickens or eat anything until they've thoroughly washed their hands in hot, soapy water.

Also ensure that small children don't rest their faces on brooder or cage edges to get a better look. If they handle feed and water dishes, pick up eggs or help with other chores, make sure that they promptly wash their hands. You too should remember this advice, because children aren't the only ones who can come down with salmonosis or other diseases. Wash your hands with soap and water before eating, smoking or putting your hands near your mouth, nose or eyes.

Starting with Adults

If you decide that buying adult chickens is the best route for you, read on to find out where to get them and how to choose healthy birds.

Where to buy adult chickens

The best place to buy adult birds is from a reputable breeder. You may also find good birds for sale by attending a poultry show at a county fair, and by visiting one of the growing number of specialist poultry farms springing up on the outskirts of towns and cities. Some very good smallholder organisations exist where you can be sure to fine genuine help. The Devon Association of Smallholders (or DASH), for example (see www.devon smallholders.co.uk), shares information between interested chicken-keepers. Animal auctions are another resource for birds, but use extreme caution in these cases.

Local feed stores, online forums and newspapers can help you locate sources of adult birds near you. Specialist magazines provide a rich source of information too.

When to buy adult chickens

You can buy adult chickens at any time of year, but autumn isn't the best time because chickens moult then and may look a bit scruffy. When a chicken *moults,* it loses its feathers and replaces them – a process that can take as long as seven weeks. Although birds that are moulting aren't exactly sick, moulting does put them under stress, and if that stress is compounded by a change of environment or travelling, the chickens become more susceptible to illness. In addition, you can't get a good idea of the birds' feather colour and quality.

If re-homing ex-battery hens appeals to you, you can do this at any time of the year. The hens are removed from their commercial premises at the time of their first moult (which may not necessarily happen in autumn with these birds); that's why they often look particularly bald and scruffy when they arrive. With some tender loving care they learn normal chicken behaviour and grow their feathers back within six weeks.

What to look for

If you're buying adult birds without seeing them first (which we don't recommend), perhaps from someone selling point of lay birds, be sure to get a guarantee of health and age.

Sellers should, at the minimum, be able to tell you the breed, sex, age and correct colour of the bird or birds that they're selling you, but knowing what to look for yourself when seeing the birds in the flesh maximises your chances of buying a happy, healthy chicken.

Assessing health

If you're choosing your birds in person from a seller, look for active, alert birds in clean surroundings. Sick birds may look fluffed up and listless. Also remember to make the following checks:

- ✔ **Check the face.** Nasal discharge or runny eyes are signs of illness.

- ✔ **Check the breathing.** Unless the weather's very hot where the chickens are, the birds shouldn't be breathing with open beaks or making rattling noises.

- ✔ **Check that the bird doesn't have wounds, sores or large bare patches.** Hens that have been with a cockerel may have a small bare area at the back of the neck and on the back from mating. This area doesn't mean the birds aren't healthy.

    The exception to this rule is if you've chosen an ex-battery hen – these hens may look quite rough when you first take charge of them. Buying through a reputable source such as the Battery Hen Welfare Trust (check out the website at www.bhwt.org.uk before taking this route) means that although your hens may look bare in places, they've been given at least a basic health check before you collect them.

- ✔ **Handle the chicken you intend to buy to see how the flesh feels under the feathers.** Birds that are too thin or overly fat can have problems.

- ✔ **Check the vent (under the tail).** The feathers should be clean – not caked with diarrhoea.

- ✔ **Look to see whether all the toes are present, and check the comb for damage.**

✔ **Look through the feathers for parasites, paying particular attention to the area around the vent.** If you see a crusty white deposit or you can see lice running around the base of the feathers, avoid chickens from that source completely – you'd be importing problems that you may later regret.

✔ **Check the bird's temperament when buying a cockerel on his home turf.** Enter the pen with him and see whether he acts aggressively towards you. Aggressive cockerels are a pain to work with and can even harm small children, and so no matter how beautiful he is, avoid him. Be aware that cockerels come with no guarantee that that they're going to remain placid. Defending territory is in their nature, and moving house may trigger this instinct in him.

Here's how healthy adult chickens look:

✔ Bright, clear eyes.

✔ Clean nostrils, with no discharge.

✔ Alert and active.

✔ Plump and glossy comb and wattles. In cockerels, large blackened areas of the comb indicate frostbite, which may cause temporary infertility.

✔ No swellings or lumps on the body. Don't mistake a full crop on the neck for a lump.

✔ Smooth legs, with shiny skin. They have four or five toes, depending on the breed. They don't have any swellings or lumps on the bottom of the feet.

✔ Smooth feathers with no large patches of bare skin, except in the case of ex-battery hens. Look carefully through the feathers for lice.

✔ Breathe with their beaks shut, unless they've just been chased to be caught or the weather's very hot.

Determining gender

The feathers and colouring in male and female chickens generally differ greatly. In only a few breeds are the colouration and feathering similar. Breeds that are all white or black may be slightly harder to sex, but they still have colour differences, however subtle.

Cockerels generally have longer, arched tail feathers. The hackle feathers on the neck are pointed rather than rounded. Birds that show iridescence in the feathers of the neck and tail are most likely to be cockerels, and this iridescence can be seen in solid-colour birds if you examine them closely.

A cockerel's comb and wattles are larger than those of a hen. Only cockerels crow, and you can often get a cockerel to crow by crowing at him.

Determining age

Telling an adult chicken's age is difficult, but how well a hen lays and how fertile a cockerel is gives you some indication. Hens continue to lay well and cockerels are fertile until about four years old, on average. After that, only a slight chance exists that the birds can reproduce. However, some hens lay sporadically for many years, and a chicken's average life span is about eight years, so you can buy unproductive birds if you aren't careful. That may be okay if you're only looking for birds to scratch about the place, but those birds should be priced lower than younger chickens.

Signs on the birds' bodies can also help you to age them. Hens that are laying have deep-red, glossy, moist-looking combs and wattles. They have widely spaced pubic bones and a moist, large cloaca. When hens are old or not laying, their combs and wattles look dull and dry. Their pubic bones seem close together, and the cloaca looks small and shriveled. Cockerels also tend to have duller combs and wattles as they age.

Aged chickens have thick, scaly skin on the legs. The spur is long and wicked-looking on cockerels. When a hen stops laying, she may develop a spur, too.

Purchasing chickens for show

If you want good show birds, get hold of the proper qualifications for the chicken breeds you're interested in and study them before purchasing birds – the Poultry Club of Great Britain, established in 1877, publishes basic details of breed standards on its website (see www.poultryclub.org), but you need to join the club or buy the book for fuller details. Attending some poultry shows and looking carefully at the winning birds also help. Additionally, you can sign up to clubs for almost every breed of chicken, and they publish information on show qualifications for that breed.

No matter what you do, at first you're going to make some mistakes choosing show birds: picking winning show chickens takes experience and some luck. Good, honest breeders try to get newcomers the best birds possible, and so heed their advice when picking birds.

Transporting Your Birds Safely

If you're going to a breeder to pick up birds, bring a proper carrying crate. You can buy crates specially made for chickens, but any pet carrier works well. Pet shops or the bigger feed merchants sell them at very cheap prices. Before you buy, check them out to make sure that the doors work well and latch securely. Pet carriers can be cleaned easily, and you can stack them so they take up less room without the birds beneath getting soiled.

The old practice of throwing chickens into feed sacks to carry them home isn't very humane or safe – small wire cages like those for rabbits are a better alternative. You can transport baby chicks in cardboard boxes, but be sure to punch through some ventilation holes first. If you have a few chicks choose a small box – they feel safer and stay warmer – but don't try this with older birds – they're bound to escape and you're likely to end up with chickens running all over the place.

Make sure that the carriers you use have good ventilation and secure latches, and don't crowd too many birds into one carrier – allow enough room for the chickens to lie down, stand up and turn around. If the journey is an hour or less the chickens don't need water or food, but if the trip is longer, you need a water container that clips on the cage or carrier. Unless the trip takes longer than 12 hours, don't add feed.

If you're transporting chickens in the back of a pickup or trailer, cover part of the cage or carrier to shade it and protect the birds from the wind. The back of a pickup or trailer can get very hot in the sun, and so pay special attention in warm weather.

Never, ever leave chickens in closed cars in weather warmer than 30 degrees Celsius (86 degrees Fahrenheit). Even ten minutes in a closed car in the summer can be too long. Never leave carriers sitting in the sun – chickens can overheat and die very quickly.

Part II
Housing Your Flock

'Well how was I to know you'd stored all the eggs in the back of the garage?'

In this part . . .

*T*his group of chapters covers the basic care of all types of chickens. If you need to choose housing, Chapter 5 gives you a lot of options to select from, including ready-made chicken houses as well as coops and arks you can build yourself. If you're interested in building your own hen house, Chapter 6 gives you the design and construction tips you need. When you have your housing constructed, you need to set up shop with furnishings as well as keep it clean — Chapter 7 shows you the ropes.

Chapter 5

Going House Hunting for Your Chickens

*U*nless you live in the middle of nowhere and can turn your chickens completely free to fend for themselves, you need to provide them with some kind of housing. Perhaps your property has an old chicken house or a shed or barn that you can convert to chicken housing. If you're lucky enough to have one of these structures, you may not need to do much to get under way, but if you're like many people who've developed the urge to keep chickens, you need to build or buy housing before you can begin. You need to decide what kind of housing suits your needs, and you have many options.

One thing you don't want to do is keep chickens in your house. Chickens simply don't make good house pets. They can't be housetrained, and their droppings are wet, messy and very smelly because they're high in ammonia. Even cute, tiny balls of chick fluff smell bad enough, and you'll soon want to move them. Chicken feathers give off a dust that isn't good for humans to be constantly inhaling and pet chickens can be destructive to houseplants, furnishings and clothing, just like a puppy or kitten. On the more serious side, chicks can carry salmonella and other bacteria that may cause serious human illness.

Chickens do best when they have their own separate housing, preferably with access to the outdoors. In this chapter, as the title implies, we provide you with a handy guide to the issues you need to consider when deciding how to provide housing for your chickens.

Knowing What a Chicken Needs in a Home

Whether you keep chickens for pleasure or to provide your family with eggs and meat, as the keeper of a small flock you need to provide humane, comfortable conditions for your birds.

Chickens, like most animals, are at their best in calm, comfortable surroundings. To create the ultimate 'des res' for your chickens, you need to provide everything they need to be comfortable including:

- ✔ Good feed
- ✔ Clean water
- ✔ Dry, clean surroundings
- ✔ Nest boxes and perches
- ✔ A sandbox to bathe in

They also need enough room to move around comfortably so they can scratch, peck, flap their wings and converse with friends – and they must be able to avoid their enemies.

The needs we discuss in this section aren't just for the comfort and health of your birds but also for your own convenience and comfort. After all, if you as the chicken caretaker have an easy, efficient way to care for your flock and keep your birds healthy and happy, you're less inclined to give up the venture.

Shelter from wind and rain

Being exposed to wind or being wet can make conditions uncomfortable even if the temperature is ideal. Wet chickens are unhealthy and unhappy, and so your birds need a house that keeps them dry in wet weather. If chickens have access to shelter, they go there by themselves when it rains.

Chickens also need protection from winter winds. Wind-chill factors affect chickens the same way they do humans. Choosing a sheltered corner or creating a windbreak to keep housing out of prevailing winter winds is ideal. Trees or shrubs can form natural windbreaks, or you can create a temporary windbreak by stretching a tarpaulin between posts.

Temperature control

Chickens are native to warm, tropical areas, and although breeders have adapted many breeds so that they tolerate cold weather better than their ancestors, chickens do need winter protection. Likewise, weather that's too hot can be deadly for chickens. The ideal spot for chickens is under the dappled shade afforded by tree canopies. Loving this 'jungle-like' environment is hard wired into them and it gives perfect shade, too.

Keeping the cold at bay

The most comfortable temperatures for chickens range between 4 and 30 degrees Celsius (40–86 degrees Fahrenheit), but they can survive the cold if they're dry and out of the wind. If temperatures regularly drop below freezing (0 degrees Celsius/32 degrees Fahrenheit), their egg production slows greatly or ceases altogether.

Some breeds of chicken handle cold better than others, and because some can lose part of their comb or even toes to frostbite when temperatures drop below 0 degrees Celsius (32 degrees Fahrenheit), choosing suitable breeds makes good sense if you're in a very cold region. Breeds with smaller combs close to the head – such as Wyandottes – have less chance of frostbitten combs and heavily feathered breeds such as Cochins and Brahmas are also wise choices. Refer to Chapter 3 for more breed information.

Chicken bodies give off a lot of heat, and so if you provide a well-insulated house that stays above freezing you don't need to heat it. Some basic winter protection, such as covering single-pane windows with plastic, helps heat retention. (We discuss this aspect in a little more detail in Chapter 6.)

Chicken houses need less ventilation in the winter. Reducing ammonia and moisture build-up is still important, though, and so close vents but without making the house stuffy.

Dealing with the heat

Chickens need protection from heat, too. Temperatures in excess of 32 degrees Celsius (90 degrees Fahrenheit) can harm chickens and long periods at these high temperatures may decrease egg production. Just as humans suffer more in heat combined with high humidity, the temperature at which chickens become stressed may depend on the humidity and body bulk of the chicken.

The smaller the housing and the more chickens in it, the more critical cooling becomes in hot weather. If you live in a particularly hot-weather area, you may need larger chicken housing so that chickens can spread out. A good cross-flow of air helps to make life more bearable for your chickens. Don't worry about draughts in their housing: in this situation any breeze is a good breeze. Making higher ceilings, improving ventilation and positioning the housing in natural shade also helps to keep temperatures more bearable.

Insulation can also play a big part in keeping your chickens comfortable. For example, it can prevent a shelter with a metal roof from heating up like an oven inside. However, the gap between insulation and roof material can become a handy hiding place for the dreaded red mite (see Chapter 11) which can multiply like wildfire in hot temperatures. Mineral wool insulation is a perfect medium for red mite to harbour in, as is natural wool. Checking under the insulation regularly to keep an eye on the situation and providing good perches (see Chapter 6) are ways to minimise this problem. Polystyrene sheeting is an effective insulation, but if you don't keep it out of reach of inquisitive chicken beaks, your birds soon pick it to bits. Reflective foil-coated insulation materials do good jobs as long as your birds can't access them and mites can't harbour in them.

A shady outside run where your chickens can sit out of direct sunlight helps them to cope with the heat, too. They especially like dappled shade that mimics sunlight coming through trees – a preference that stems from their ancestry in the jungle. Stretching an army-style camouflage net over the run or the breeziest part of the run on hot days gives that all-important feeling of security from aerial predators while allowing the chickens to take best advantage of any cooling breezes.

If, despite all your attempts, the heat does rise too high, try to minimise stress to the birds. Don't chase or catch them, and don't clean the chicken house while it's that hot. Make sure that you have cool water always available to the birds. Just as they survive cold weather without heat, most chickens survive heat, but their laying may slow due to stress.

The heavy *broiler* breeds (meat birds) are especially prone to dying from overheating. Even broilers housed in pens (a safely fenced area), or given free-range outside suffer in heat, so if you've no choice but to position your chicken housing in a confined yard in full sunshine you may want to rear chickens for meat at times of the year that's less likely to be hot and stuffy. Temperatures below 30 degrees Celsius (86 degrees Fahrenheit) (or even cooler if the humidity is high) are desirable for broilers. Other types of chickens have an easier time handling higher temperatures as long as you ensure that they have shade, good airflow and access to plenty of water.

Protection from predators

Chickens provide a tasty meal for many predators, and chasing chickens is a fun activity for other animals, including some of the two-legged kind. If you keep chickens, you have a duty to keep them safe from harm. An aggressive cockerel may seem to be able to take care of himself, but a car or a big dog can make short work of him.

Chicken-keepers everywhere find predators a big problem – in the city, the suburbs and the country – and everyone who keeps chickens, sooner or later has to deal with a predator. Predators include neighbourhood dogs – probably the number one chicken-killer for all kinds of chicken-keepers. Many other kinds of predators exist as well, and urban areas have their fair share of them. We discuss predators in more detail in Chapter 9.

Strong, well-built housing goes a long way towards keeping your chickens safe from predators. At the very least, give your chickens a predator-proof shelter at night, for when they're most vulnerable. How much additional predator protection they need depends on your location and the predators in the area.

Enough space to move about in comfort

The housing you provide needs to allow chickens the space to engage in normal chicken activities (even if their life is to be short), including flapping their wings, sitting in the sun, taking a dustbath and chasing bugs. The more space you can provide for them, the happier and healthier your chickens are. And because chickens are flock birds and unhappy living alone, you need to keep and provide enough space for at least two birds.

Birds lose the ability to establish a good social structure in housing that's too cramped, resulting in behavioural problems. Crowded conditions increase fighting and disease, prevent normal social behaviours and lead to problems with chickens laying eggs on the floor and pecking each other. Overcrowding also increases moisture and ammonia levels. Housing that contains the right elements, however – perches, nests, plenty of space and so on – promotes good social behaviour.

Allow room for three average adult chickens to 1 square metre (10.75 square feet) of floor space in the shelter plus the same area as an outdoor run as a minimum. With no outdoor run, you need to double the indoor space allocation, but remember that chickens are at their happiest with access to the outdoors. Bantam breeds can get by on a little less floor space. Give your chickens even more space if you can, erring on the side of too much space rather than not enough. And with spacious quarters, you can increase the size of your flock in the future.

You need to ensure that the housing you provide for your birds allows them to stand normally without touching the ceiling. Chickens prefer to sleep on a perch up off the floor, and so taller housing gives them the freedom to do that and feels more natural, but don't count the height of housing as part of the square-metres/feet-per-bird requirement.

In cold winters, an extremely large indoor shelter may be a disadvantage because the birds' body heat can't warm it as well, but being cooped up in the winter inside a small shelter has to be boring. If you protect the outdoor run from strong winds and roof it to keep out deep snow, the birds spend time outside on nice days.

If you keep chickens for showing, provide plenty of room in the housing and around perches so that birds don't rub their feathers on the wire or walls of the housing. Birds with broken or frayed feathers are unsuitable for showing.

Sufficient lighting

Like most birds, chickens are active during the day. Light – whether natural or artificial – is necessary for them to eat and drink by, and the amount of light (day length) influences the hormones that control egg-laying and fertility. Chickens need some natural sunlight (even as little as an hour!) for their bodies to make vitamin D, but if you feed them vitamin D food supplement your chickens can survive well with artificial light only. We discuss light's effect on laying in greater detail in Chapter 15.

Having at least some natural light in a chicken shelter is always a good thing. It saves money and probably feels as nice to chickens as it does to humans. Windows that face south or east work best. If windows face south or west, you need to be able to open them in warm weather to avoid excessive heat build-up in the shelter.

Even when a shelter has natural light, artificial light is often helpful. The benefits of artificial lighting are that it:

- ✔ Enables you to do chores after dark, increasing flexibility and making your job easier.

- ✔ Makes the hens more active on grey winter days, which keeps them eating and interacting normally and helps to stimulate egg-laying. Fourteen to 16 hours of light usually keeps hens laying well all winter. The light needs to be strong enough to read a book (that's you, not the chickens!) in all areas of the indoor space.

If you have a tiny shelter for two hens, having inside lighting or a window may not be practical. The chickens will be fine though as long as they get light in an outside run.

Except for very young chicks, chickens appreciate a time of darkness, too, or at least dim light. For layers, 8 to 10 hours of darkness or very dim light (similar to that provided by a bedroom nightlight) is good; for other types of birds, 12 to 14 hours of darkness doesn't hurt. Dim light inside the house at night is helpful because:

✔ It enables the birds to defend themselves from some predators.

✔ It enables the birds to find their way back onto their roost if something frightens them off it.

Chickens also like their nest-box area to be in a dimly lit location – under windows is usually better than opposite them.

Fresh air

Any enclosed chicken housing needs good ventilation, and the smaller the housing, the more important ventilation becomes, especially in hot and humid weather. Ventilation allows stale, ammonia-saturated air to be replaced with fresh air. Fresh air doesn't mean draughts, though. *Draughts* are leaks of air that pass across living space. *Ventilation* generally refers to clean, cool air that's pulled in naturally through a vent near the bottom of a structure, warms and rises and exits through other vents or windows near the top.

The amount of ventilation your chicken housing needs depends on the number of chickens and the type of shelter you have. The more birds per square metre/foot, the more moisture and ammonia exists. Shelters with high ceilings (2 metres/6.5 feet or more) allow warm moisture and ammonia-saturated air to rise away from the breathing space of the chickens. Shelters with low ceilings may have the same amount of floor space per bird, but the air doesn't have far to rise and needs to exit quickly to keep birds from breathing it.

No magic formula exists to determine how much ventilation you need – too many variables are at play. You should, however, be able to adjust it by opening or closing the doors, windows or vents for different conditions. The key thing to remember is that if you're comfortable breathing in the shelter, if little or no ammonia smell is present and no moisture is building up on walls or ceilings, your ventilation is adequate.

A lack of adequate ventilation can cause a number of problems:

✔ **Build-up of ammonia fumes in walk-in housing leads to an increase in respiratory diseases in chickens.** Ammonia comes from chicken droppings and is a lung irritant.

✔ **Moisture build-up and condensation increases humidity,** which makes an area feel warmer when the weather's hot and colder when it's cold.

✔ **Excessive moisture is conducive to the growth of mould,** and also increases bothersome smells.

✔ **Disease organisms spread more easily in moist, warm environments.**

In a small shelter with only a few birds, the opening the birds use to go outside can serve as the bottom vent and some openings along the upper part of the shelter can exhaust stale air. Larger shelters need to have vents near the floor, windows that open and adjustable vents at the top. If many birds live in the shelter and the weather is warm, you may need an exhaust fan near the ceiling to speed the exchange of air.

Clean surroundings

In both the inside and outside parts of the chicken house, you and your birds appreciate clean surroundings. They usually need some type of litter on the floor inside, and if this litter gets wet, you need to change it immediately. We discuss types of litter in Chapter 7.

Once upon a time wire-bottomed cages were used with cleanliness in mind to allow manure to fall through the wire mesh, but these cages are something to avoid. The wire mesh can trap toes and cause sores on chickens and the design was for more for the convenience of the keeper than the chicken.

Although the ground around your chicken house may start as a grassy patch, it soon deteriorates into a mud pie when the chickens get to work on it. For this reason, and because chickens don't like wet feet and feel uncomfortable in muddy surroundings, you need to have good drainage in outside areas. It also avoids moisture build-up, which increases smells. Some chicken owners put down a concrete apron around the outside areas and sweep it or hose it down. This hose method is great on well-draining ground where the floor dries quickly, but even the tiny feet of chickens can compact an earth floor, causing water to stand on top of it instead of draining through. Some people add sand or gravel to the outside pen area to help if it turns into a mud pit at certain times of the year. Sand or gravel keeps the birds' feet clean and reduces smells.

We discuss chicken-house housekeeping further in Chapter 7, but if you keep cleanliness in mind when designing or choosing accommodation, your housekeeping is going to be much easier.

Surveying Your Housing Options

You have many options as regards housing your chickens. Some types of housing overlap a bit, but in this section we provide a general overview of the more common types of chicken housing available. In the section 'Choosing a Type of Housing', later in this chapter, we help you compare the options according to the most important factors to consider.

Looking around online or at adverts in the back of poultry and smallholding magazines can give you plenty of ideas about what other chicken-keepers already use successfully. You can also buy specialist books about building chicken houses and find step-by-step, do-it-yourself photo guides for some of them. Easier still, flick to Chapter 6 for some good ideas.

Before diving into a discussion about the many variations of chicken housing, it helps to be familiar with some basic chicken-keeping vocabulary. Throughout the following sections and the book as a whole, we use these definitions to identify the various components of a chicken's humble abode:

- **Ark.** An *ark* is a movable structure and is often used as a *chicken tractor* – a contraption for controlling chicken activity for horticultural purposes where the chickens scratch up the ground, weeding, de-bugging and manuring as they move in a restricted area. Arks are usually small and comprise a shelter with attached outdoor run.

- **Coop.** The old-fashioned small house resembling a rabbit hutch is a traditional *coop*. A coop has no access to the outside floor area although the whole thing may stand outdoors. It may become part of an ark when a run is added. The expression 'feeling cooped up' explains what a chicken feels like when confined in one of these.

- **Housing.** Anything you confine your chickens in is *housing*, including their shelter and any outside space you enclose for them.

- **Pen.** *Pen* refers to an enclosure for chickens. A pen is similar to a run, but a run refers only to an outdoor enclosure. Pens can be indoors, sheltered from the weather, or outdoors, unsheltered from the weather. Pens can have a shelter in them. The term 'pen' is also interchangeable with 'housing'.

- **Run.** *Run* refers only to an outside enclosure for chickens – the part without protection from the weather. A run is usually connected to a *shelter* (the area that protects chickens from the weather).

- **Shelter.** *Shelter* refers to the place that protects chickens from the weather; it forms the indoor part of their confined area. Shelters may be part of housing, arks or pens.

Using cages as temporary housing

Housing birds In cages on a temporary basis makes life easy when you need to:

- Travel with your birds
- Clean out the main house
- Prepare and take your birds for showing

> ✔ Look after mother hens with vulnerable chicks (or have a hen who's sitting on eggs)
>
> ✔ Separate a sick or injured bird from the rest of your flock
>
> ✔ Quarantine a new bird
>
> ✔ Deal with emergencies

Choose a cage with solid flooring consisting of a removable pan or a slide-out shelf made of wood or metal. You can remove the solid flooring for cleaning. This type of flooring is natural and comfortable for the chickens and keeps the inside of your vehicle clean if you have to travel with them.

All breeds of chickens kept for showing need solid floors in cages to minimise the chance of foot injury.

Place cages in locations where the ventilation and lighting are good, and unless you're separating chickens to keep them nice for showing, because they're sitting on eggs, or for health reasons, keep two or more chickens in each cage. Make the cages as large as possible, and ensure that the chickens have room to walk around, flap their wings and raise their heads normally. Each confined chicken requires at least 0.2 square metres (2 and a quarter square feet) of floor space. If the chickens are layers, make sure that enough room is available for a nest box too (see Chapter 7 for more on furnishings). A small door above the nest box in the cage wall allows you to gather eggs easily.

Some keepers use cages to house chickens for breeding or for getting them used to conditions they'll experience when caged at shows or travelling. Housing chickens regularly in cages smacks a bit of battery farming, but having a few cages around for special circumstances is good chicken-keeping practice.

Rearing birds inside

If you have a portion of a garage, shed or barn that you can section off, or an enclosed, well-ventilated, well-lit and easy-to-clean building where chickens can use all the floor space, you already have a good place to rear a group of meat birds that you intend to dispatch at 12 to 15 weeks, particularly in cold weather.

If you have no space for outdoor pens or other issues make outdoor pens difficult to use, you can also keep laying hens and pet birds indoors. As long as they're not crowded, birds kept loose in an indoor area can interact socially and behave naturally, and are happier than when confined to small cages (see the 'Sufficient Lighting' section later in this chapter).

You may see eggs in grocery shops with the label 'cage-free' or 'barn-reared'. This label usually means that the hens producing those eggs are kept in large indoor buildings, loose on the floor. (The 'free-range' label on a box of eggs legally means that the hens have access to the outdoors.) These eggs generally cost more to reflect the additional cost of housing birds without cramped cages. Although cage-free conditions may be more humane for the birds, in some operations, producers pack the birds just as tightly as they are in cages.

Pairing a shelter with a run

One of the oldest, most successful and most comfortable methods of housing chickens – both for human caretakers and chickens – is to provide the chickens with a shelter they can retreat to at night or in bad weather and an outside enclosure that protects them from predators, yet allows them access to fresh air and sunshine. Figure 5-1 illustrates this type of housing.

Walk-in shelters allow you to feed, water and collect eggs inside. You can partition-off part of a barn, garage or stand-alone building for the shelter. A smaller door or pop-hole for the chickens usually connects to the outside enclosure, and a larger door enabling human access to the outside enclosure is also a good idea. The best outdoor enclosures allow plenty of room for exercise. You can let out your chickens for a little free-range roaming too, but the enclosed outside area and the shelter give you the option of confining them when that isn't safe or desirable.

You can build walk-in shelters with attached runs as small or as large as you need them to be, and with the intention of expanding in the future as well. Most people who begin with other types of housing end up with this type, unless they have very limited space.

Offering shelter with free-range access

Some people have a shelter where they can feed and water chickens and from which the chickens can come and go at will. To get an idea of this set-up, look again at Figure 5-1, minus the fencing. This arrangement counts as 'extreme' free-range, but we also refer to birds that have a penned outside area as free-range too. Chickens may have their own shelter or be allowed to roost in any barn or shed on the property. This set-up works best when the chickens can roam over large areas without intruding on neighbours and where predators aren't a problem – a farmyard, for example, where the farm dogs (well-trained, of course) keep predators at bay.

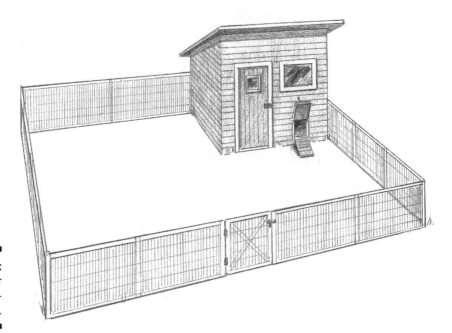

Figure 5-1:
A shelter
with an out-
side run.

Chickens, of course, love a free-range arrangement, as do many of their human caretakers:

- Chickens that roam may get most of their diet free.
- Free-range chickens seem to have less trouble with lice and mites.
- The caretakers of free-range chickens have less cleaning to do.
- Chickens undoubtedly prefer this method above any other.
- Watching the antics of chickens roaming around the yard can be an entertaining and enjoyable experience.

On the other hand, keeping free-range chickens has some noteworthy drawbacks:

- Chickens kept this way are wilder and tend to go off and hide their eggs, particularly when they want to raise a batch of chicks.
- On the range, the chickens are vulnerable to predators from the air and land.
- Chickens can wreak havoc on gardens and landscapes, tasting each tomato as it ripens, eating seeds you've just planted, scratching out a dust bath in the marigolds and picking all the flowers off your roses. They also make a mess on porches and patios.

Cockerel went a-truckin' and he did ride, uh-huh

Free-range birds can cause interesting headaches. A friend of Kim's arrived at her job early one morning, just as it was getting light. A co-worker called to her as she was getting out of her truck to ask whether the cockerel running around the car park was hers. (He'd seen it jump out of the truck.) Sure enough, the woman recognised a cockerel from home. He'd ridden some 50 kilometres (30 miles) in the back of the pickup, where he rested for the night. She spent the next three days' lunch hours and after work chasing the cockerel.

If you're planning to keep chickens for meat, be aware that the flavour of proper free-range meat birds takes some getting used to. Their meat is slightly darker and stronger tasting, and their muscle mass is greater. A decent-sized, free-range meat bird takes several weeks longer to grow than a bird confined to a barn or even allowed restricted outdoor access, but if you allow them to get too old before dispatching, free-range chickens become tough. Modern, fast-growing broiler-types, such as Ross Cobb chickens, don't do well given total free-range. As sedentary birds, they like their feed to be put in front of them, and to grow well they require higher levels of protein than they can get from foraging. You can find more about how to rear your own chickens for meat in Chapter 16.

We don't recommend allowing layers complete free-range until they've established a good laying pattern (we talk about why that's important in Chapter 15), but do allow them some access to the outdoors in an enclosed area.

Looking at small, all-in-one units

You can find chicken housing on the market (as well as plans to build your own) that features a small house, similar to a dog kennel, with a small, outside pen attached or with a pen that surrounds the shelter. The shelter can be on the ground or raised on legs (see Figure 5-2). You gather eggs from outside the shelter through a small door and in some cases feed and water the chickens there, too. In other cases, feed and water go outside the shelter portion.

These small, all-in-one units are a good choice if you plan on keeping two or three chickens as layers or pets, and they're ideal for urban situations. The drawback is lack of space for the chickens, and so allowing chickens kept in these units out for a few hours a day for some supervised foraging helps to make this type of housing a little more pleasant for them.

All-in-one units tend to be expensive, especially if you need to have them shipped to you, and many have features that appeal more to humans than to chickens. Before purchasing one, look it over carefully and think about the provisions that all chicken housing needs to include (see the section 'Knowing What a Chicken Needs in a Home' earlier in this chapter).

If you have the time and skills, you may be able to build a small, compact shelter and run that works better for you and the chickens and costs less than pre-fabricated, all-in-one housing. Even if you don't have any building skills, you may be able to hire a local person to build a nice little set-up for less than the cost of ready-made housing.

Figure 5-2:
An all-in-one chicken house has a combined shelter and run.

Considering an ark or hoop run

Hoop runs (which are particularly common with the modern moulded plastic houses) and A-frame houses (usually called *arks*) are variations of tractor housing (if they're moveable) or all-in-one housing (if they're stationary). Figure 5-3 shows an ark, and Figure 5-4 shows a hoop run. We cover them separately because they're most likely to be used as temporary, warm-weather housing, although you can build them to protect birds year-round.

As the name suggests, the hoop run is like a large tunnel. You can cover it with clear, opaque or black plastic; shade cloth; metal; canvas; or another material. You can use it with or without a more substantial shelter (in the latter case it becomes a hoop coop). A-frames are similarly covered, but the frame comes to a point. To keep them from blowing over in high wind you may have to anchor A-frames and hoops.

These housing forms are quick to set up and economical. They require only minimal building skills, and because of their shape require few supplies. Because they're lighter and easier to move than more conventional houses, many chicken-keepers use them to provide temporary housing or for separating breeding couples and trios (see Chapter 12). Smaller versions can house a few laying hens out on pasture during the summer, or all year-round if the structure is built so that part of it protects your chickens from winter weather.

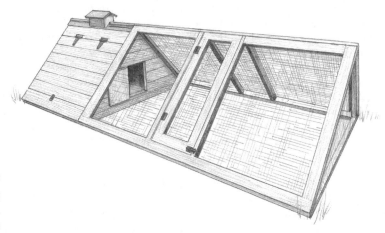

Figure 5-3:
An ark.

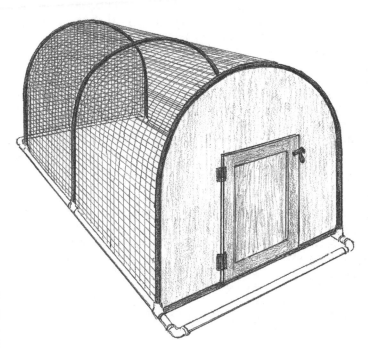

Figure 5-4:
A hoop run.

Meeting some mobile housing methods

Technically, the 'free-range' poultry keeping method (see 'Offering shelter with free-range access' earlier in this chapter for more about 'extreme' free-ranging) doesn't allow chickens to roam completely freely, eating whatever they want. Instead, it's a system of confining chickens on a piece of grass or pasture that you manage or care for so that it provides the maximum benefits for chickens while making the best of grass as free chicken food too, be it on a suburban lawn or a field on a smallholding. Free-range poultry owners also provide a supplement of commercial feed or grain.

Keeping chickens in this way means that you can house them in small, move-able units with built-in shelters or confine them in large, fenced areas that you move, with a free-standing shelter that you move with the chickens. You can also rotate chickens through a series of permanently fenced areas. Each area may have its own shelter, or you can move one shelter with the chickens, among the areas.

Chickens in these units can get a good deal of their food from foraging as long as you put them on good pasture. The birds enjoy this situation and can get a good proportion of their feed free – up to 40 per cent – from eating vegetation. When the birds have eaten down one piece of pasture, you move them to a clean piece. They grow a little more slowly this way, and the meat of free-range poultry tastes different from the meat of birds that are reared indoors on a completely grain-based diet. Laying hens thrive with this method too. Raising replacement pullets on pasture until they're ready to begin egg-laying guarantees that they become more efficient free-range birds.

Some people think of managed free-range poultry as easy poultry-keeping, but it requires some intensive management to be successful. Although the costs for feed may be greatly reduced and the birds get a natural, healthy diet, you have to manage the pasture and move the enclosures. Poultry manure is hard on grasses, as is the scratching of chickens. You need to move free-range poultry on in rotation to a clean pen before they seriously damage the vegetation in their enclosure.

How often you have to rotate the birds is determined by how large the enclosure is, how many birds it holds, the type of pasture grasses and the weather. In mild, wet weather, grass grows vigorously, and you may need to move the enclosures less often unless mud and smells develop that warrant moving housing more frequently. Dry, hot weather may mean shorter intervals between moves or supplementing with other feed. Grass that becomes heavily matted with manure, has turned brown or has disappeared from large areas, means the time has come to move the birds. Over time, you gain a feel for what's right for your conditions and birds.

Chickens love to be outside scratching and foraging and if you manage it well, mobile housing and free-ranging is an excellent way to keep happy, healthy chickens. Using orchard areas or planting up special shelter belts with some taller herbage makes things even better for them by providing extra food and protection, and attracting insects as added free protein.

Many types of housing work well in free-range situations. For small numbers of birds, a chicken tractor-type structure, an ark or hoop housing works fine, but for larger numbers of birds, bigger enclosures with flexible netting and free-standing shelters work better. If the birds on pasture are layers, make sure that the shelters have nest boxes that you can access easily. Hens like to have a shelter high enough to roost off the ground at night. Meat birds are content with a low structure, say, 92- to 122-centimetres (3–4-foot) high, but this arrangement isn't as convenient for you as the keeper.

If you put meat birds in movable pens on pasture, take great care not to crush or run over the birds when moving the pens. The birds can be rather slow and sluggish.

Shelter with an open bottom: The chicken tractor

A *chicken tractor* is a type of housing, usually roofed with an open bottom (at least in the run area), that you move from place to place when you want the soil in various areas turned over and fertilised (see Figure 5-5). You may hear this method being called by its old name of *folding.* Sometimes, owners place these structures over good pasture just long enough for the chickens to eat it down, and then move it to a different spot. Horticulturists favour this kind of chicken-keeping, using the chickens to de-bug, scratch out old plants and manure in one go before moving them on to a fresh patch to work again, all the time keeping the chickens confined well enough to avoid damaging commercial crops.

Chicken tractor-type arks are built in a number of shapes and sizes. You can carry smaller-sized tractors or slide them along the ground, or small wheels may be attached to the bottom frame to make moving the tractor easy. Larger chicken tractors are heavy, and you may need a real tractor to move them!

You can use tractor-type houses to raise meat birds – although the heaviest broilers may do better in more conventional housing – and to house young pullets being reared to laying age, and even laying hens. In the case of hens, include nest boxes in the coop and make sure that the hens have easy access to them. You can also use tractor arks with pet or show chickens.

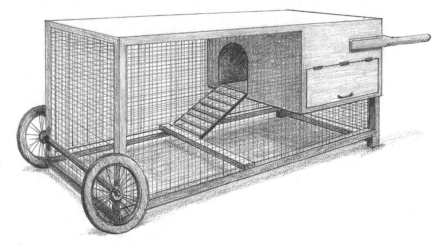

Figure 5-5:
An example
of a chicken
tractor.

The drawbacks to this type of housing include the following:

- ✔ **It works best in mild weather.** Really hot, cold or wet weather causes problems.

- ✔ **You must have space to move the arks regularly to a clean spot.**

- ✔ **The chickens may be vulnerable to predators.** Because of their very nature, owners often place arks far away from human housing; therefore ensure that you use strong wire on the sides or else predators can break in. Predators can also dig under the frame, so you may need to fit extra skirting wire.

- ✔ **In most of these set-ups, catching chickens is hard if the need arises.** You can't get inside some of the smaller units easily, and when you lift a side, chickens can escape.

- ✔ **Feed and water containers may be hard to access.**

Mobile shelter with run

This arrangement is another method of keeping free-range poultry that works much like tractors, hoops and A-frames, except that you provide a separate stand-alone shelter along with some sort of, generally unroofed, enclosure (see Figure 5-6). Each enclosure needs a shelter for the chickens and possibly an additional shaded area – you can provide some shade by raising the shelter up on wheels. The shelter may be a completely covered hoop or A-frame structure, or a small shed. It has to be light enough for you to move it to new enclosures, but you may need to stake or weigh it to keep it from blowing away in high winds.

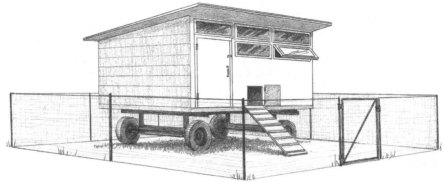

Figure 5-6:
Mobile shel-
ter with run.

You can make the large enclosures with chicken wire, plastic fencing or flexible electrified netting (something we discuss in Chapter 6), placing easy-to-
move posts at intervals of about 3 metres (10 feet). The type of enclosure
material you need largely depends on what type of predators you expect to
encounter. Use heavy, welded wire if loose dogs are in the area or if larger
livestock graze around the chicken enclosures.

Stock your pasture enclosures at a maximum of one bird per 1 square metre
(10.75 square feet), with a shelter size of about three or four birds per square
metre (10.75 square feet). Remember that because the chickens can't use the
spot the shelter sits on for pasture you need to add footage accordingly. The
bigger the area you can provide, the better. In smaller areas you need to move
the shelter and enclosure more frequently. If you have limited netting, making
a square encloses the largest area.

Sussing out specialist designs

You can be as imaginative as you like with chicken housing as long as you
keep all the features that make it comfortable and practical for chicken. The
duplex in Figure 5-7 is a case in point that works if you're raising two small
groups of chickens that you want to keep separated.

The duplex is space efficient, allowing birds to have inside and outside quarters in a relatively small area by stacking the birds' living space. This design
is a great way to keep two different breeds in the same coop while still maintaining the pure breeds, and is a good size for breeding pairs or trios.

It consists of two 1.5-square-metre (16-square-foot) housing areas, elevated to allow more scratching room underneath and accommodating three standard-size chickens or five bantams per side, with 3-square-metre (32.25-square-foot) runs attached to the side. Small doors in the runs allow for alternating free-ranging. Each side contains an auto feeder and two bantam-sized nesting boxes. Barn-style doors in the back give easy access for egg collecting, feeding and cleaning.

Use this duplex idea as inspiration for your own ingenious specialist design for your chicken-keeping venture.

Figure 5-7:
The duplex.

Getting fancy: The chicken garden

One of the greatest benefits to raising chickens is the abundance of fertiliser they provide so freely while eating weeds and bugs. So it makes sense that chickens and gardens go together.

The 3 x 4-metre (10 x 13-foot) building in Figure 5-8 sits between two separate fenced areas – one for a chicken run and one for a garden. You alternate between the two each year, providing labour-free fertilisation. A small composting area is a shovel's throw outside the chicken house door. This large chicken house also features indoor storage space and can accommodate 30 to 40 chickens.

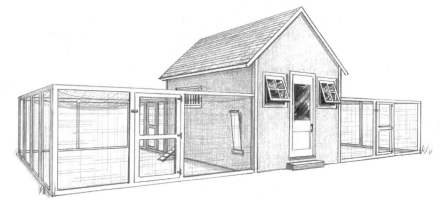

Figure 5-8:
The chicken garden — more than a weekend project.

Choosing a Type of Housing

The many ways to keep chickens successfully depend on the type of chickens you plan to raise, the proximity of your neighbours, weather conditions and potential predators in your area, and how much space you want for yourself. Choosing a type of chicken housing to use goes hand-in-hand with other chicken-keeping considerations:

✔ **Housing pets, layers and meat birds.** If you're keeping a few hens as pets or layers, you're probably best served with a small all-in-one unit or cages or a small A-frame ark or hoop house, with doors close enough to the nest boxes to allow you to reach in easily and gather eggs. If you're keeping more layers, a shelter with a run or free-range is ideal.

For meat birds, indoor housing with access to an outdoor run is perfect; most people want to eat outdoor-reared birds. Because meat birds are here on a temporary basis, the housing can be temporary too, but it needs to protect them from predators and weather.

If you have enough land, using the pastured system may be right for you. Chicken tractors, hoops or A-frame arks can house meat birds in warm weather and you can then store them away over the winter.

✔ **Housing show birds.** When keeping chickens for showing, you need to provide large quarters to prevent feather breakage and fraying. If you keep a breed with really long tail feathers, the roosts must be high off the floor, and bedding or run surfaces need to be easy to clean. Additionally, you may want to attach poles to shelters and runs several centimetres (a few inches) from the wall to prevent the birds from rubbing against the wire or wood walls. Avoiding grass runs with light-coloured birds to prevent feather stains also helps to keep chickens clean, and protecting dark-coloured show birds from direct sunlight prevents their colour from fading or turning brassy.

Many keepers house show birds, especially cockerels, in individual cages so they don't fight. Some show breeders use removable dividers in pens to separate birds they want to mate, but cockerels may attack each other through these barriers if they're not solid.

✔ **Considering neighbours and family.** As a matter of respect for your neighbours and to avoid the hassle of dealing with potential complaints, choose housing that's easy on the eye. Particularly if you live in an urban or suburban area with close neighbours, choose neat, attractive housing that's easy to clean. Instead of building an A-frame ark out of old pallets, consider partitioning off part of the garage and making a neat enclosure behind it, or hiding a small, neat shed in the garden.

If you have small children, you probably want a chicken house that they can't get into without help from you. Also, loose chickens scratching up the garden of a nearby resident or walking down the road may not endear you to your neighbours.

✔ **Keeping your chickens safe and healthy.** Predators are a big concern in some areas, whereas weather extremes are more important in others. The safest and healthiest housing is generally a well-built shelter with a well-built run. Small, combo, shelter-run units may be safe, but they're not a healthy way to house chickens if they're overcrowded or you don't keep them clean. Packing a lot of birds into a small but cute coop makes them stressed and susceptible to disease and behaviour issues.

Chicken tractors, hoops and A-frames can be safe and healthy in good weather, but they're a nightmare in bad weather. Windstorms, heavy rain, lightning, mud and cold all take a toll. If the housing is to be outside, make sure that it's suitable for the weather in your area. The shelter should always have a waterproof top. In cold areas, the shelter should have thick walls or some form of insulation.

Allowing your chickens to range freely does no good if they can wander into heavy traffic, be picked off by dogs or foxes or badly treated by children. They may be getting all the vitamins and minerals they need from foraging only to become a healthy meal for a hawk. For this reason, build or buy housing, whether inside or out, with potential predators in mind. If you have dogs, foxes or badgers in your area, use heavy welded wire rather than chicken wire or plastic fencing on your runs or enclosures. Close and latch doors to shelters for the night – every night.

If two-legged thieves or vandals pose a threat, put a padlock on your shelter doors and consider night lighting or a security camera.

✔ **Factoring yourself into the space and usability equations.** Whether you service the housing from inside or outside depends on the style of housing you choose. Essentially, you need to decide whether you're okay with having a shelter that you must stand outside to service and clean or whether you want a shelter you can walk into. If you're happy to work from outside, you need a door wide enough to allow you to insert and remove the the feed and water. If you want to enter the shelter to work, you probably want ceilings that are taller than you, unless you enjoy stooping. Chasing a bird around in a crouched position isn't fun, and walking on your knees or lying on your belly to clean out a hen house makes you wish you didn't keep chickens.

Small shelters that you can't enter require a removable or drop floor for cleaning. The complete absence of a floor can be a good option because it enables you to move the coop around on grass, set it on cement, or provide a wood base, and then move the coop to clean it. You need to be able to clean these small shelters throughout and that can be a problem if parts of them are hard to reach. If you keep only two or three chickens and know that you'll be diligent about cleaning their quarters, a small, ready-made housing unit or any housing that you service from outside may work for you. These units fit into small areas where other housing may be a problem – that's the biggest advantage they have to offer.

Wire floors with trays under them to collect waste are an option, but they must be smooth and have spaces small enough to prevent a chicken's foot from slipping through and causing injury. Solid floors are the least desirable because cleaning them in small coops is often difficult. Even if these floors are removable, they require more frequent care. Some pre-built shelters have a slide-out pan or door under the perches. Because most of the manure in the shelter collects under the roosting area, this feature is very desirable.

If the housing you choose doesn't have a slide-out pan or door under the roost to facilitate cleaning, lay a flat board under the perches so that you can regularly remove and scrape it off.

The preceding points and Table 5-1 help you to see the pros and cons of each type of chicken housing and decide which type appeals to you and fits your needs. Remember, though, that you can always alter housing for your convenience later on if you find it isn't quite working as you hoped or if you want to add enhancements. For example, in a walk-in shelter, you can add a small door just above the nest boxes so you can reach in from outside and quickly collect eggs when you don't want to go inside.

Ultimately, you need to choose a style to begin chicken-keeping, but don't be afraid to change your housing if you find it no longer works for you. If you discover that trying to clean a small pre-fabricated unit from the outside is tedious work, graduate to a larger, more practical house and keep the smaller one for emergency housing.

Now that you know the basic decision-making factors, consult Table 5-1 to blend your personal considerations and choose a type of housing that's right for you.

Table 5-1 **Choosing Suitable Housing for Your Chickens**

Type of Shelter	2-4 Chickens	4-10 Chickens	10+ Chickens	Layers	Meat	Pet/ Show	Cold Weather	Other Considerations
Cages	Yes	Yes	No	Yes	No	Yes	If inside	More work for care-taker; inhibit natural behaviour
Small, all-in-one unit	Yes	Maybe	No	Yes	No	Yes	Yes, with right material	Can be expensive; good for urban gardens
Indoor, cage-free	Yes, but not practical	Yes	Yes	Yes	Yes	Yes	Yes	Not so humane
Tractor style	Yes	Yes	Yes, if large enough	Yes	Yes	Yes	No	Must be moved often; difficult to access
Ark and hoop	Possible	Yes	Yes, if large	Yes	Only fair; size must be right	Yes	Can work with the right material	Anchor for wind; access may be difficult
Shelter with free-range enclosure	Possible	Yes	Yes	Yes	Yes	Yes	Yes	Best for all
Shelter with total free-range access	Yes	Yes	Yes	Yes	Not best method	Yes	Yes, if shelter is large	Chickens vulnerable to predators; may make neighbours unhappy; eggs may be hidden; birds find a lot of their food
Pastured poultry with mobile housing	Not practical	Not practical	Yes	Yes	Must use the right meat strains	No	No	Pasture must be managed; birds need less grain; birds may be vulnerable to predators

Chapter 6

Designing and Constructing Chicken Housing

*U*nless you're lucky enough to have inherited a nice chicken house when you bought your property, you need to consider creating some type of housing for your birds. In Chapter 5, we talk about the different types of housing options. This chapter goes a little deeper into actual construction.

If you're even remotely handy with a hammer and nails, you can build a chicken shelter from the ground up. If you're totally ham-fisted and no one wants to see you with a saw in your hand, you can buy ready-made housing instead. Or you can combine a little building with a little buying to create a one-of-a-kind chicken Taj Mahal.

We hope you're enthusiastic about getting started, but before you pick up a hammer you need to think ahead – spending lots of time planning is better than lots of time undoing mistakes. A good plan makes for a good chance of success in your chicken-raising adventure. Your first order of business is to make absolutely certain that keeping chickens at your home is legal. You're unlikely to have a problem, but checking is a good idea – especially if you live in a built-up area (see Chapter 1 for more info on consulting the law). You don't want to put forth the time and effort to build a fancy hen house only to be told to remove it.

Deciding Whether to Build, Buy or Recycle? That is the Question

Chicken housing doesn't have to be fancy or expensive, but it needs to allow you easy access and give your birds a place where they:

✔ Are protected from predators

✔ Can stay warm in the winter and cool in the summer

✔ Can lay their eggs

✔ Can prop up their feet and relax after a hard day chickening

We look at different housing styles in Chapter 5. You need to decide whether you want to build that housing, buy it or use a combination of the two methods.

Making do with what you've got — reuse and recycle

No doubt about it – the easiest option for housing is combining pre-built structures with a few of your own personal touches. This method also saves money in most cases. If you have a shed, garage or barn that you can convert to an indoor shelter, you can purchase dog kennel panels to create an outside run. You can also use kennel panels on the inside of the structure to divide the chicken space from areas intended for other uses. Conversely, if you already have a large fenced area, you can buy a small shed and convert it to a shelter. Sheds are generally available at most hardware and home-improvement shops and garden centres, and you don't need planning permission.

If you're looking for something charming, children's playhouses make good indoor shelters. And if you live where rules and regulations are few, an old box trailer – even a horse trailer – can become a chicken shelter, with the addition of an outside run.

A greenhouse frame also makes a good chicken shelter if you remove the clear plastic covering (which makes it too hot inside) and replace it with another, solid covering or use the greenhouse as winter quarters for chickens (be sure to rig up a bit of shade inside for sunny days). We know a woman who turned a two-hole coal bunker into a two-nest chicken coop that was charming. Use your imagination, but remember the neighbours and any restrictions your neighbourhood may have.

Reuse/recycle

Rob says: 'One of the cleverest examples of alternative housing for chickens I ever saw was an old van converted to a small chicken house. The front part of the van was removed, leaving just the back main body with rear axle for easy moving. The windows were painted black for hen privacy and reversed on the van body so they could be opened from the outside. A little ramp to a small, chicken-size door, some nest boxes and perches and the house was good to go.

Tons of old vehicles are out there, waiting to be recycled. Just make sure that your neighbourhood lets you have one sitting on your property, and make it look neat and shiny'.

Building from scratch

Building a full-sized chicken house from scratch is more expensive in most cases than refurbishing an existing structure. It also typically takes more time and demands more in the way of building skills, but it may be the only way to get the hen house of your dreams.

Designing and building a chicken house can be a great family project and everyone can discover lots about keeping chickens as you decide what features you need to include. Whereas one family member may be good with design, another may be good at the construction part.

Chicken house construction is also a good way to hone your latent carpentry skills. We can't cover everything about carpentry in a book about chickens, and so we only touch on the basics here. You can get more information on building by reading home-improvement and do-it-yourself books, including *Building Chicken Coops For Dummies* (by Todd Brock, David Zook and Rob Ludlow), and perusing websites where you can find a wealth of information and some 'how to' videos. Good website information specifically on chicken housing is available at www.backyardchickens.com, www.thecozynest.com and www.poultry.allotment.org.uk.

Assessing your skill base – realistically

Chicken housing can be very simple or very elaborate, but you don't need to be a master carpenter to create a reasonable structure. Only you know whether you have the skills and time to build a chicken house. The good news is that chickens are very forgiving about crooked walls and slanting floors. Just be sure to keep the needs of your chickens first and foremost as you plan and build.

You may be tempted to slap something together from some old wooden pallets and a roll of salvaged wire and, admittedly, some people can do a good job with these types of basic materials. However, you probably want to build something that doesn't make your neighbours unhappy and which has a few years of use in it. Building the right kind of housing in the first place makes sense, instead of constructing something shoddy that you need to replace in a couple of years.

At the very least you need to know how to hammer a nail in, use a tape measure and operate a power saw. Be sure to choose a simple design and preferably the help of a clued-up friend if your skills are few or rusty.

If you already have some building experience under your belt, you can add more complex features to your chicken shelter, such as adjustable vents, manure pits, interior walls over the studs, insulation, exhaust fans, exterior nest boxes and so on. If you have a lot of previous home-renovation projects under your work belt, these jobs should be simple for you.

With more complex projects that require wiring and plumbing, you need to have these skills or acquire them as you go. You may need to have such things inspected by a building authority, and so hire those skills in or be sure that you know what you're doing if inspections are to be made.

If you can afford it, you may be able to hire someone skilled to build the chicken house for you. You can have all the fun of designing anything from a tiny chicken coop to an elaborate hen house while leaving all the work to someone else. Any good handyman can build a simple chicken coop. You can also ask a local college that has a building trades department whether the students would be interested in such a project.

Estimating how long your project will take

Building housing takes time, and so don't wait until the chicks are straining the sides of the brooder, or your spouse is coming home from purchasing the birds, to begin construction. Give yourself time to factor in design, plan changes, trips to the hardware store, bad weather or any other unexpected interruptions in plans.

If you've done building or renovation projects in the past, you probably have an idea in your head about how long your chosen project is going to take. If you need help estimating the time a project's going to take, show your plan to the local building suppliers. They can guide you a bit by looking at your plan to see what building materials you're likely to buy from them. Remember, though, that their estimate is based on what you tell them, and they may assume that you have more building skills than you do. If you have an idea but no concrete plan, larger building-supply shops may help you draw up a plan and make a supply list.

For guidance, even if you're a novice, building a small hoop coop or an A-framed ark or converting a shed or barn into a chicken house, should be possible over the course of a weekend, providing you have all the supplies at hand and you work diligently. With some experience and a good work ethic, a little help from friends and good weather, you can put together a substantial and elaborate hen house in a week. Most people need more time, however, because they work around their day job, the school run and weather interruptions. Add a few additional days to your estimate to be sure that you finish before you need to move the chickens in.

Buying ready-made houses

If you don't have the time, skills or desire to build your own chicken housing, you can buy it. With ever more people raising chickens on a small scale, more and more pre-fabricated chicken housing is coming on the market. Some of this housing is excellent, with great care taken to make the housing comfortable for the chickens and easy for the caretaker to service and clean. However, other not-so-good housing options are also on the market, some of which are too small and cramped even for two hens. Others are hard to clean, and make catching chickens inside difficult. Sometimes the quality of the building materials comes into question and the wire mesh on some ready-made units is very flimsy, making it easy for predators to break in. So the moral of the story is, when you buy ready-made houses, know what your chickens need before you buy, and choose carefully.

Investigating ready-made housing requirements

If you decide to purchase housing, you'll probably look first at chicken coops or small shelter-and-run combined units (we discuss these options in Chapter 5, along with the features you need to check for). You can also purchase a shed for loose indoor housing or as a shelter for free-range chickens.

If a friend or relative has purchased housing for chickens, take a close look at it before you go shopping to see whether something similar would be right for you and ask them what improvements, if any, they'd make to their choice in future.

Just looking at a catalogue or website picture may not give you a good idea of what you're thinking of buying, and so try to get out to see it in person and physically examine it – try opening doors, sliding out floors and so on. Consult Chapter 5 before you go to familiarise yourself with what you and your chickens need from your purchase. If you can't inspect your potential purchase, be sure to look at the description as well as the picture – which tells you the dimensions, the weight and what materials the housing is made from. If you have questions, call sellers and ask them.

Shopping for a ready- made house

If you want to buy housing, check out farm and smallholding magazines, poultry supply catalogues, garden magazines and online sources. If you live in a rural area, a farm-supplies shop may stock poultry housing or have it available for special order, or be able to refer you to a place that does. Don't forget to check your local newspaper, notice boards in farm-supply shops and online options such as Amazon and eBay for bargains on new or second-hand chicken housing. The following are just a few of the other places to look online:

- ✔ www.jimvysearks.co.uk. A wide selection of poultry equipment, sturdy housing and other poultry equipment made to order.
- ✔ www.ebay.co.uk. Poultry housing, new and second-hand, plus all the equipment you may need.
- ✔ www.henhouses.co.uk. Practical, affordable housing.
- ✔ www.essexchickens.co.uk. A great resource for housing, supplies and equipment.
- ✔ www.providencefarm.co.uk. Advice on housing design to minimise pest and predator attacks.
- ✔ www.knowlenets.co.uk. Nets for covering outdoor runs.

Checking Housing Restrictions

Depending on what you want to build, restrictions may exist on the size or location of your building. Finding out about legal restrictions before you begin is better than having to deal with unpleasant consequences later.

Common building restrictions include the distance a structure has to be from existing structures, and from the road. Restrictions can also apply to the size and type of foundation a structure needs. Certainly if you're choosing to build a permanent structure with foundations you need to ask your local planning authority to approve it. Doing so may only take one phone call to your local county council, but can save you a mountain of trouble later on.

Make sure that the official you talk to knows that you plan to use the structure to house animals, even if they're pets. Different rules may apply for animal housing than for other types of structures. Building a mobile shed on a sturdy skid frame usually avoids property taxes and building regulations, but having official acknowledgement of your plans removes any doubts.

Some people think that they can get away with building at a weekend when officials are out of their offices, and no one notices. Satellite big brother, however, can look into the most secluded garden and if you get caught, you may have to move all your chickens, take down the housing and/or pay a big fine. Play by the rules – doing otherwise just isn't worth the hassle.

Choosing the Right Location

After you've got to grips with building restrictions in your area and applied for any necessary planning permission (as we describe in the preceding section), you need to choose the spot where you want to place your chicken housing. If you have a small garden or yard, you may have only one obvious choice. But if you have more room to play around with, the ideal spot is one that's:

- **Close to the house.** You want the chicken house close to your house so you can easily service it and keep an eye out for predators or other unwanted visitors. However, if you have more than a few chickens, you want them far enough away from the house so that you can't smell them on warm, wet days.

- **Close to utilities.** Having electricity in the chicken house is a big plus. Doing chores in the dark is no fun, and lighting makes everything so much easier throughout the winter months. Having water nearby is another plus, making cleaning out and watering less cumbersome.

  Using the wall of another structure (such as a garage or barn) as part of the housing makes running water and electricity to the desired location easier, and saves you money and materials too.

- **Away from the neighbours.** Even if keeping chickens is legal and your neighbours say they love these birds, putting your chickens under their bedroom window or in view of their pool isn't fair. Including the neighbours in the decision about where to locate the hen house may make them happier with your chicken-keeping project.

- **In a well-drained spot.** Chickens don't like to get their feet wet, inside or out, and so don't put your chicken housing where the ground is low or where water drains towards the area. If the area is sometimes damp, you can add gravel to the runs; however, that doesn't always work well because water can stand on top of, or saturate, the gravel spaces. Wet chicken manure isn't a nice smell and is difficult to clean up well.

- **Away from potential environmental issues.** Try not to put chicken housing where manure from the run washes into and pollutes lakes, ponds, streams or other groundwater. If you have a well, place your chicken housing at least 15 metres (50 feet) from it. You may get a visit from the Environment Agency if you're found to be polluting any drains or groundwater.

Oh, the importance of planning

Kim says: 'Here's an example of poor planning. My husband and I wanted to try the chicken tractor/pastured poultry method of raising broilers and replacement layers. We built several 1.3-x-2.6-metre (4-x-8-foot) pens, 1.3 metres (4 feet) high. They were made of plastic pipe and chicken wire with tarpaulins for the ceilings, and were easy to move around the pasture. However, they were a pain to service. At that height, they were hard to lean over to retrieve feed and water containers, and we had to pull back the tarp to do so. If we lifted a side, we risked the chickens escaping, especially the pullets. We inserted some doors in the sides and wired the feeders to the walls. However, when the time came to catch the birds, another problem reared its head. If we lifted the whole thing, birds went everywhere. I had to crawl through the small door, crawl around on my knees to catch the chickens and hand them outside to my husband. Needless to say, we re-thought our housing for the next season. This arrangement may work for you, though, if you have small kids who aren't afraid to catch chickens!'

When your plan is finally on paper, go out to the location where you intend to place the house and chicken run and measure the spot. Make sure that your design actually fits in the space you have in mind. You may not think that this step is necessary if you're building a small, moveable chicken coop, but consider the smallest location to which you may need to move the coop – will it fit? Be sure you check overhead, too. If the hen house is 2.6 metres (8 feet) high, will you need to cut any tree limbs or move any wires and are they yours to move?

Establishing Form and Function: The Basic Blueprint

Chicken houses come in all shapes and sizes. If you've made up your mind to build one, first take a look at Chapter 5 where we discuss different types of housing and their benefits, drawbacks and features. Or check out a friend or neighbour's chicken housing or go to your nearest suppliers and study their ready-made housing.

Whether you're doing this building project by yourself or someone is helping you, start by putting your plans on paper. When you write down ideas and dimensions, you can spot any mistakes in your planning and be better at judging how much timber and other supplies you need. Erasing a line is easier than tearing down a wall, and so take the time to draw out a plan on graph paper. Assign a space measurement to each square, such as 1 square equals 0.1 square metres (1 square foot), and then draw your proposed plan to scale. You can fiddle with the plan until you like what you see and you have something that fits both the site and your needs.

Begin with the major planning issues. Grab a pen and a piece of paper and make a list of all the things you *need* to have in your chicken housing. For example, include the amount of indoor and outdoor floor space needed per bird; lighting, shade and ventilation appropriate for your area; and anything you must do to satisfy building regulations, such as using a certain type of material or colour to blend in with neighbouring building styles.

Just as you have comfortable furnishings in your home, a chicken home needs some basic furnishings to make life easier for the chickens and you; for example, collecting eggs from a nest box is much easier than searching all over the place for the eggs or finding them in the manure on the floor. Make a note of all your 'must haves'.

If you give your birds free-range during the day, remember that perches keep the chickens off the floor while they sleep and make them want to come home at night instead of finding a tree to sleep in.

Most houses for adult birds need nest boxes for egg-laying and perches for sleeping. We discuss these aspects later in this section, but be sure to consider them when planning. You may also want to plan for a manure pit or a dropping board under the perches (see Chapter 5).

Size and shape: Giving your birds some breathing space

You need to allow a minimum of 0.3 square metres (3.2 square feet) of indoor shelter area per bird (although 0.3–0.5 square metres (3.2–5.4 square feet) is better), and you need 0.3–0.6 square metres (3.2–6.5 square feet) of outdoor run area per bird as a minimum – the more the better. To compute the correct amount of space, multiply both the indoor and outdoor minimum spaces by the number of chickens you intend to house and go larger if you can. Always keep a minimum of two birds, and so the smallest shelter is 0.6 square metres (6.4 square feet), but if you have room and resources build bigger.

Ventilation: Allowing fresh air to flow

Chicken shelters need good ventilation year-round. You can accomplish this through windows, doors, roof vents and other means, some of which you can salvage from recycling centres or buy second-hand to save money.

Good ventilation usually consists of some way for cooler air to enter near the floor, be warmed and then exit near the top. If the house is tall enough to walk in, you need high and low ventilation points. You generally use screened vents at the bottom for incoming air, and space between rafters or at the roof peak to allow air to exit. Windows that you can open increase ventilation in warm weather. Small coops that you service from outside can use the entrance point for chickens as the bottom source of air and vents near the top of the shelter for stale air to exit.

The amount of ventilation a shelter needs depends on how many chickens you keep in what amount of space and the weather in your area. Exhaust fans can speed the exchange of air if natural means aren't enough to keep the air free of ammonia fumes and excess moisture, which are bad for chickens and for you. Exhaust fans go near the top of a structure – and sometimes additional fans are used near the bottom – to pull cooler air inside. These fans are most commonly used in commercial settings where many birds are kept in together.

We cover ventilation in more detail, including its importance in chicken health, in Chapter 5.

Having contingency plans

Sometimes, chicken houses have to withstand very hot and very cold temperatures. Building in insulation materials isn't a great idea because it can exacerbate any red mite problem that appears (Chapter 11 goes into this problem in more detail), but keeping some handy to use in emergencies is a wise move.

A sheet of polystyrene placed over a corrugated roof in summer keeps heat out. In winter, tucking polystyrene sheets up against the outer walls and tying them in place, or wrapping the house in bubblewrap, can be effective as a temporary measure. Be careful not to block off the ventilation, though – whatever the weather. Although these solutions may not look elegant, keeping your chickens comfortable is the most important thing.

We know someone who, in the middle of a cold winter, moved her chicken house out of the wind and wrapped it in bubble wrap – leaving access and ventilation, of course. At that time, in freezing temperatures, she had a broody hen with fertile eggs. All the chicks survived!

Providing perfect perches

Perches provide an important function in chicken-house furnishings. Because chickens are vulnerable when they sleep, they prefer to *roost* (perch) as high off the ground as they can when sleeping. The more 'street-savvy' birds also pick a spot with overhead protection from the weather and owls. Perches — poles or boards suspended off the floor – can accommodate these needs.

Some of the heavy breeds, however, including the heavy broiler-type chickens, can't get themselves very high off the floor. Broilers don't need any perches for their short life span, but you do need to provide other heavy breeds with low, wide perches.

Whatever their size, chickens like to roost in the same spot every night – hence the expression about 'problems coming home to roost' – and are like all birds in this respect. As soon as your chickens get used to roosting in your chicken house, they head back home at dusk, even if they've managed to escape that day or you give them the freedom to roam. Most of the time they sit in a crouched position when roosting, but your chickens need enough room to sit upright between the perch and the top of the shelter. Cockerels like to sit on perches to crow, and to be able to extend their head and neck a bit higher than normal when doing so.

The area under the perches accumulates more droppings than any other place in the shelter. To make cleaning the chicken house easier, you can place a separate pan, a board to collect manure, a manure pit or additional bedding under the roosting area. If you plan to add a pit, make sure that you can access it with tools for cleaning.

If your shelter is small, you may not have room for perches, and your chickens have to sleep on the floor, which is fine for meat birds but isn't ideal for adult chickens and makes them feel vulnerable; they may then become stressed and nervous.

Room to perch

You can look at perch size in two ways. Some people think that the chicken should be able to span the perch with its toes in order to grasp it. A round wooden dowel similar in size to a thick broom handle is about right for full-sized chickens, but they don't mind square-edged perches (gently round off any sharp corners or use a natural branch cut to length). Bantam chickens, however, find smaller-diameter perches more comfortable. Other people believe that chickens prefer to sit flat-footed whilst roosting. If you subscribe to this theory provide a 10–15-centimetre (4–6-inch)-wide board to perch on. Boards may be a better option in cold climates because the belly feathers help prevent frostbite by covering the toes while the bird sleeps.

The perch needs to provide about 20 centimetres (8 inches) of length for each bird. Therefore, if you have six chickens, you need 1.2 metres (4 feet) of perch length. If the house isn't wide enough for a single length of perch, you can place several perches 0.6 metres (2 feet) or more apart. Whatever type of perch you provide needs to be strong enough to support the weight of the birds without sagging.

Perch placement

If you can, place the perches at least 0.6 metres (2 feet) off the floor. Even heavy hens can fly up to perches 0.9 metres (3 feet) off the floor. Lightweight hens and cockerels can go even higher – 1.2–1.5 metres (4–5 feet) if you have enough space in the house. A step-type of perch system with each perch being 30 centimetres (1 foot) or so higher than the next in a staggered arrangement used to be the most usual arrangement (see Figure 6-1). A different school of thought says that perches should all be on one level to avoid competition for the highest perches. If you follow this method, allow about 0.6 metres (2 feet) of space between them to allow the birds to jump up easily. Whichever method you choose, don't put one perch directly under another, because the chickens on the lower levels become coated with droppings from above.

Figure 6-1:
Stepped perches may be in a corner, as shown, or up against a wall.

Some people like to put a pan or board to collect droppings under the perches. Manure collects in the greatest concentration under the roosting area, and having a pan or board makes cleaning up easier. An example is shown in Figure 6-2.

Figure 6-2:
Place drop-
ping pans
underneath
a perch
for easy
cleaning.

To avoid contamination, don't place perches over feed and water dishes. If you have the space, don't locate perches over nest boxes for the same reason. Be sure to cover nest boxes if you have to place them under perches. And you may not want to place perches directly over areas where you walk to service the chicken house.

Perches are the danger area for red mite (see Chapter 11). Providing your structure is strong enough, suspending a grid of perches from roof beams by wires minimises the problem of red mites using the perches as a route to infest the walls of the house. Alternatively, construct perches so that they sit in grooves. That way, when you're inside the house, you can remove or lift them for cleaning. Make a spare set of perches so you can swap them over regularly, offering added protection from mites.

Feathering their nests

Nest boxes play an important part if you're keeping chickens for their eggs. Chickens still lay eggs without nest boxes – in fact, some birds never lay their eggs in the nest box – but by providing them you offer a safe and comfortable spot where layers can lay their eggs and where you can easily gather them. You may also want nest boxes for pet and show chickens, and nest boxes are essential if you want your hen to hatch her own eggs.

Young pullets like to 'play house' as their hormones begin to prepare them for laying. If you provide proper nest boxes for them, they try them out: sitting inside, arranging nesting material and practising crooning lullabies. Have nest boxes in place by the time pullets are 17 weeks old.

Chickens like nest boxes that feel secluded, are covered, dimly lit and big enough to turn around in but not so large that they don't feel comfortable (see Figure 6-3). Painting the walls of the boxes with a dark colour or choosing dark materials to construct the boxes helps to get the ambience right. Consider putting boxes under windows rather than opposite them, but if your house doesn't have a lot of leeway for positioning the nest box and you've no choice but to place it in a well-lit place, hang a cover over the front of the nest box to keep the light out. Your hen can then push past it into the dark to investigate and use the nest box.

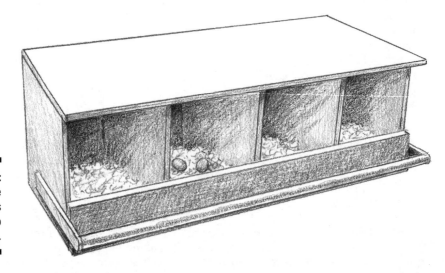

Figure 6-3:
Side-by-side nest boxes are easy to make.

Keep nest boxes lower than the perches. Chickens always try to get to the highest spot in the hen house to roost, and sleep on or in high nest boxes and make a mess. If the top of your nest boxes must be higher than the perches because you're short of space, build the roofs at an angle to discourage chickens from perching on top.

Picking a nest box design

Many commercial nest boxes are available on the market. The boxes may have round openings or flaps for the hens to push through, and the boxes come in various configurations – two nests, four nests and so on. Some have attached backs; others are open backed with an 'egg rollaway' system. The old standard is a set of metal boxes, whereas some of the newer versions are made of plastic resin materials. Plastic and metal are easier to clean and disinfect, but you can make your own nest boxes using wood or use common household items as nest boxes. For example, a covered cat-litter box makes an ideal nest box and deep plastic tubs with a hole cut in the front can sit as nest boxes on the ground or on shelves. At a pinch you can even use sturdy cardboard boxes and replace them when they get dirty.

Another option is the free-standing nest box, which is a bit like a covered cat-litter box. It has a handle on top so you can pick it up and collect eggs without stooping. If you're using cages or small pens to house chickens, you can save room by removing these boxes when hens aren't laying. This type of box also enables you to clean only the dirty nest box rather than a large unit. You can see what a portable nest box looks like in Figure 6-4.

Figure 6-4:
A portable nest box is good for collecting eggs easily and removing when it's not needed.

Accepting that size does matter

For average-sized hens, a 0.1 metre square (1 foot square) nest box is the minimum size to use. Bantam hens can use smaller boxes, and extra-large breeds may need a little more room. Boxes should be big enough for the hen to turn around inside and, if they have roofs, tall enough for the hen to stand up. If you want a hen to sit on eggs to hatch them, you may want larger nest boxes so she can be comfortable while she sits. Boxes can be square or rectangular and even round.

Boxes need a lower side in the front that encourages the hen to step into the box rather than jump in over a side. Sides can be anywhere from 15 centimetres (6 inches) high (if not roofed) to 30 centimetres (12 inches) or more high (if roofed).

Having roommates – a box for every two chickens

Hens do share or wait their turn to use nest boxes. When designing your chicken house, provide enough nests so that the hens don't have to wait too long to use one and the nests invite your hens to come in and sit a while. One nest box for every two hens is ideal. Fewer boxes are okay if space is limited, but one box for more than four hens leads to fighting and laying eggs outside of the box. You can have one box for each hen if you have the room, but you may find that some boxes aren't used at all.

If a box is large enough, two or more hens may crowd in together and share a nest box. Usually one lays and departs quickly. To avoid a situation where all the chickens prefer the same nest, make all the nests the same size and colour and provide them with the same degree of darkness. Even under these circumstances, however, hens often favour a certain nest box, and you can do little about it. The hens line up to use that box and may even squabble over it.

Provide extra nests if you intend to let some of your hens sit on eggs to hatch them, because a nest is out of commission when a hen is sitting.

Making it a suite

Hens enjoy company while laying, and so don't scatter the boxes around the coop. Place nest boxes in a sheltered, dimly lit area of the hen house and keep them out of draughts. If you have numerous nest boxes, place some on or near the floor and some higher. Hen houses often have nest boxes stacked one above each other up to just over 1 metre (3.2 feet) off the ground. All nests that are more than 30 centimetres (1 foot) off the ground need a perch or alighting board in front of them, though – the hen needs a spot to fly up to and land on before entering the nest. Our experience is that hens don't use the highest nest boxes very often, and heavier hens prefer low nests, so we keep nest boxes lower than 90 centimetres (3 feet) off the ground.

Hens willingly accept floor boxes if you don't have enough height in the shelter to raise boxes off the floor. Communal nest boxes are quite common and providing, as with all nest boxes, you place them in a dimly lit area of the house, your chickens are happy to nest together.

Designing an outdoor space for your birds

When designing the outdoor run, use your imagination. Chickens are curious, inquisitive birds and appreciate interest in their lives, and so build in some areas up off the floor – without giving them a platform for escape! In this way you can make good use of a small area. If you have a vegetable patch, polytunnel or greenhouse, try to think of ways that you can incorporate that space too, even if just for short periods of the year. A constantly used run soon becomes a bare patch if it can never be rested, and other parts of your garden may benefit from being open to your chickens for some time.

Miscellany: What more can you possibly want?

As well as the essentials, list the things you *want* in your chicken house, such as room for additional birds in the future, a quarantine space if you're going to show birds, windows that open, a structure that's nice to look at and fits your human housing style, and a manure pit under the roosting area. In this list, include anything that you'd have if money and space were no obstacles. Including in your plans a place to store feed and extra bedding separate from the chickens' living space is a good idea. You may also want to plan for a place to store manure, especially if you live in an urban area.

As you plan further, you can consult your list to make sure that you're covering the necessities as well as giving yourself as many of the wants as possible.

Being Mindful of Materials

Building regulations may place some restrictions on the types of materials that you can use for building your chicken house, and so check first before deciding on materials. You need materials for the framing and roofing. And then there's the flooring. And fencing. And posts. And don't forget the chickens! Oh, wait. That's another chapter.

After you complete your paper plan and ensure that your planned house fits the site, start making your supply list. Count the number of lengths of wood, fence posts, rolls of wire, wood or metal panels, and so on that you need by looking at your plan. Don't forget to add nails, screws or other fastenings to your list, as well as windows, vents, paint and roofing material. You may also want to add materials to build the nest boxes and perches with. Most people build these when constructing the basic house.

 Do some comparison shopping to help you find the best prices on the materials you need. After doing a bit of price-checking, you may want to alter your plans. If money is a factor, consider building a simple coop that you can add to or modify as funds permit.

Getting to the bottom of flooring

Wood or cement flooring is preferable to earth floors in chicken shelters because it helps to prevent pests and predators and makes cleaning easier. You can buy rubber and other types of roll-out, waterproof materials designed for barn floors that make excellent coverings over cement and wood, if you can afford them. Materials such as rolls of vinyl made for human homes and metal sheeting work less well – they may be easy to clean, but can be slippery underfoot. Metal is never good in conjunction with chicken manure because the acids in the manure soon eat their way through and metal's trickier to repair than wood or concrete.

Earth floors are acceptable for outdoor runs. Adding a layer of sand or fine gravel to the runs makes them more absorbent. In all but the biggest runs, the earth floor soon becomes hard-packed.

Cages may have solid or wire floors, and slatted wood flooring is used occasionally. Wire flooring needs to be smooth as well as heavy, and strong enough to support the birds without sagging. Keeping birds clean is easy with wire floors, but they prevent some of the natural behaviour of chickens, such as scratching at the ground. Also, toes can get caught in wire floors, causing injuries, and so reserve that idea for temporary or transport housing.

You need to be able to clean solid floors in cages and small shelters easily. These floors may be pull-out-tray-type floors, or floors that drop down on a hinge to remove droppings and litter.

Constructing the frame

You don't need top grades of framing materials and panels for building chicken housing, but you do need materials that are straight and sound enough to make your job relatively easy. Most chicken housing is built of wood because wood is probably the easiest building material to work with. Plastic lumber works well for chicken housing, but is quite expensive unless you have a good source of recycled plastic around, and you can also use lightweight aluminium panels like those used in pole barns.

Treated lumber is fine to use when building chicken housing because the chickens can't chew it. For your shelter, use exterior grades of wood unless you intend to cover the wood walls, in which case a lighter grade is fine inside. (Remember, though, that adding coverings may create gaps that bring their own problems to chicken-keeping – particularly in the way that they provide hiding places for mites.) Use a good-quality, exterior-grade timber such as plywood for walling – scrimping on something cheaper only creates problems in the long term that mean returning to the drawing board and rethinking the materials you've used.

Painting or staining the walls with exterior-grade products gives your chicken house further protection against the elements. If you're unsure about what to buy, ask a building store salesperson for recommendations based on your area and what you can afford. Make sure that the person's aware that you want it for animal housing, because some of the more old-fashioned wood treatments, although effective, give off fumes that can kill or injure chickens.

If your plan involves a wooden-framed roof for your chicken house you need shingle or similar to protect it. For many years a felt roofing material was used extensively on chicken houses, but this has gone out of favour because of its potential for harbouring mites. Good corrugated plastic roofs that incorporate ventilation with waterproofing provide a better, modern alternative. Metal roofs don't need any kind of finishing.

In areas where hawks and owls can be a big problem, or when you must keep chickens inside the run, you may need a roof for your outside run in addition to the roof on your shelter. First, decide whether you want to be able to walk upright in the outside run. You then need to plan how to support the roof. If you have a wide pen, you may need additional posts inside it to hold roof supports that keep roof material from sagging in the middle. Fruit cage netting for excluding birds and flying predators from gardens is a tried and tested – and lightweight but effective – idea that can work equally well for chicken runs.

Ice and snow can build up on any kind of netting or wire that you use on the top of an enclosure, causing the roof to sag or break if the load gets too heavy. You can stretch netting or wire tightly to keep ice or snow from clinging, but some roof supports may still be needed.

Building materials are usually sold in specific lengths and areas, and so try to draw your plan with those sizes in mind. For example, most wood panels are pre-cut; ask your timber merchant what its standard is and plan your design to avoid a lot of excess cutting and waste.

Understanding fencing

When designing the perimeter of an enclosure, the kind of wire you need for any fencing exposed to the outdoors depends on the predators in your area. You need fencing strong enough to keep out the strongest predators, with openings small enough to prevent the chickens from squeezing through.

Chicken wire

Inexpensive chicken wire is perhaps the obvious choice and is almost synonymous with chickens. It has hexagonal-shaped openings in 1 to 4 centimetre sizes (0.5 to 2 inches) and comes in several thicknesses. However, although chicken wire keeps chickens in, it doesn't keep strong predators such as dogs and badgers out. The wire rusts and becomes weaker after a few years, and can be difficult to work with.

One place where chicken wire is very useful, because of its flexibility, is as a predator-proof 'skirt'. Digging a trench to bury chicken netting as deep as 0.6 metres (2 feet) into the ground can deter predators such as foxes and badgers from digging under a fence. As an alternative, using chicken wire 0.6–1 metre (2–3 feet) wide, attaching the long edge to your existing wire on the outside of your enclosure at a height of 0.3 metres (1 foot) and bending the chicken wire at ground level so that it extends outwards from the fencing for a minimum of 0.3 metres (1 foot), creates a skirt to stop any digging predator from gaining access. Check the chicken wire regularly, though – it rusts eventually. You can adapt this method for movable runs, too. It works very well.

Nylon and plastic fencing

The many new types of nylon and plastic fencing on the market offer an alternative to chicken wire. They can be good choices for chicken fencing if they're strong enough because chickens don't chew their way out of enclosures, but some predators may be able to chew their way in. Whereas the fine black nylon netting used to protect crops from birds and deer works well on the tops of pens to keep the escape artists at home or protect them from hawks, we don't recommend it for the sides of enclosures.

Pheasant-fencing panels

Another fencing option is the panels used for pheasant rearing that you see advertised in game and shooting magazines. Although designed for pheasants you can adapt these heavy-duty wire, fence-like panels for chicken by fitting them together to make a pen that keeps predators out. Some have a solid part-panel round the bottom that you can cover with a lighter weight net or with a heavy-duty tarp to make a shelter. These panels only work where your ground is level and with no gaps at the bottom to serve as exit points for chickens or entry points for predators.

Weld mesh

Weld mesh, a strong welded wire, is a very good alternative to chicken wire. It comes with a choice of sized mesh, from very small (1 centimetre/0.5 inch) to something that restrains an ostrich-sized bird. Choose a mesh size appropriate for your flock or mix and match both weld mesh and chicken wire. If you need to confine tiny chicks you can add smaller meshed netting onto the existing mesh at a later date. Look also at the range of heights available. 1.2 to 1.5-metre (4 or 5)-foot high fencing works for most breeds and keeps most predators out. You may want to arrive at your builder's merchant's yard with a definite size in mind and have it cut by professionals, unless you have strong wire cutters at home.

One of the most common mistakes that new chicken owners make is to underestimate the strength and cunning of predators. Predators lurk in urban areas too, and most of them love chickens and eggs. A fox can pull chicken wire apart. Big dogs make short work of flimsy wire; badgers are stubborn and work on it until it rips. Because of its strength, we highly recommend using weld mesh on outside runs.

Electrified fencing

If you plan to allow your chickens to free-range, you may want to consider using flexible electrified netting. You buy this netting in lightweight, easy to move 50-metre (160-foot) lengths available at a standard height of about 1.2 metres (4 feet) and an enhanced height of about 1.5 metres (5 feet). You can add as many or few lengths as you want, and you need to use special insulated posts, included in the netting kit, which you push into the ground to support it.

The vertical strands of the netting aren't electrified; nor is the bottom strand, but you can peg it to the ground to stop predators from pushing under the wire. The rest of the horizontal plastic wires have fine metal wire threaded through that carries a charge when you attach the fence to an energiser and electricity source. The lowest horizontal wires are close together to stop chickens from walking through and the gaps become larger the higher up the netting you go. These wires can be powered by solar, battery or mains electric and need an energiser too, which converts the power from the electricity supply to a pulse. Batteries need charging from time to time, at least in midwinter, even where they have a solar supply.

Although the netting itself isn't very strong, the electric shock it delivers to anything that touches it keeps chickens in and predators out. It does have some limitations, however. Overgrown vegetation or other objects that touch the fence weaken the electric pulse running through it, and so it needs constant maintenance. Having said that, this netting is a very flexible piece of kit and in the right situation is the perfect way to use all the nooks and crannies available in a large or irregularly sized garden. If you have lots of space for your chicken-keeping you can use this netting as part of a really efficient r otational system as well.

If you choose to go with flexible electrified netting, ask a farm or smallholder supplier for advice about energisers, because you need to consider how far your birds will be from an electricity source and how often you can service it.

Supporting fencing with posts

On the sides of the outdoor run, chicken wire requires a top and bottom board to keep it stretched tightly; welded wire probably doesn't need this extra support. All wire needs posts to support it, however, and taking a trip to a garden centre to see the various types available is a good idea, as well as asking at a builder's yard. Some posts come as a kit where you bury the bottom part in the ground, leaving an opening into which you slot the post. Wooden posts rot and break over time; however, many people prefer the look of wooden posts to metal ones, and they're easy to work with. Treated timbers can make long-lasting posts.

In recent years many people have taken to using PVC pipe (the kind designed for water or electricity) to make supports for various types of chicken homes. This type of piping makes a lightweight, easy-to-move structure.

Some types of PVC piping become brittle in cold weather, and the types that don't become brittle are much more expensive.

Place posts every 3–5 metres (10–16 feet) and at corners for outdoor runs. If the soil is sandy, you may need to use cement to keep the posts in place. You need about half a bag of ready-mix, quick-set cement per post.

Installing Wiring and Fixtures

Although not totally essential for successful chicken-keeping, having electricity in any kind of chicken housing is a big plus. With electrical power running to your housing, you can provide light and heat with ease.

Make sure that you understand how to install electricity safely. If you're not familiar with electricity, leave this part of the job to someone who is. If you have to get planning permission for your housing, you're obliged to use a certificated electrician anyway. We cover the nuts and bolts (in a manner of speaking) in the following sections, but for detailed guidance, please consult a registered electrician. Your building supply shop may be able to help, but remember that working with electricity can be a shocking experience.

Providing the hook-ups (electrical, that is)

Although rodents are usually the ones responsible for electrical fires in barns and sheds, keeping all electric outlets and fixtures out of the birds' reach or enclosed in metal boxes or wire cages prevents curious chickens from being electrocuted or starting a fire. (We look at rodents and how to spot them in Chapter 9.)

Lighting up

Egg-laying hens need a regular lighting schedule, with 14 to 16 hours of bright light and 8 to 10 hours of darkness, in order to lay well year-round. If you're not worried about egg production, and most small flock owners are happy to accept seasonality in the flow of eggs, good natural light through windows meets all the chickens' needs unless your shelter area is in a very dark corner. (For more about the light needs of laying hens, see Chapter 15.)

Providing artificial light has its benefits, however, for you and your chickens. When chickens sleep, they really sleep. Total darkness makes them go into a kind of stupor. They make an easy target for predators at this point because they don't defend themselves or try to escape. They also sit still through rain or snow, even if they go to sleep in an unprotected place. A dim light in the shelter enables them to defend themselves and helps them find their way back to a roost if disturbed. It also prevents them from flapping around in a panic and from possible injury when strange noises wake them.

When installing electric light in your chicken house, locate the fixtures near the ceiling. If your chickens can reach the light bulbs, cover them with wire mesh or plastic covers designed for light bulbs so they don't get broken. The best place for a light switch is near where you enter the house, just inside or outside the door.

If you're not able to turn lights off and on at regular times, you may want to put the lights on an inexpensive timer or buy something more sophisticated. If your switch is outside, be sure to fit a special weatherproof design. You can find a good range at equine supply shops – you can trust stable equipment to be robust.

If you intend to visit your chickens in the dark, provide a way to dim the lights as you leave. Light stimulates chickens and if you disturb them after they've gone to roost for the night on their perches they probably wake up and jump down to investigate what you're up to, believing it's another day. When you turn out the lights to leave, they're stranded on the floor, which can be quite stressful for them. When you leave, take a few minutes to dim the lights and allow the chickens to return to the safety of their perches.

Try installing a blue bulb in the house for when you have a job to do with the chickens after dark. Blue light doesn't stimulate a chicken in the same way as other light and they remain calm and stay on their perches, making it easier for you to work. When you leave the house, just turn off the light. If electricity isn't available at the site of your chicken housing, consider one of the new battery-powered lights, use solar lighting or just get all the chores done in daylight hours!

Chapter 7

Furnishing and Housekeeping

In This Chapter

▶ Making cleaning out easy with good bedding

▶ Luring hens into nest boxes with comfy nesting material

▶ Organising food and water dishes

▶ Keeping your chicken housing spick-and-span

*W*ith all the hustle and bustle of everyday life you may barely have time to keep your own home clean and yet, as a chicken-keeper, you have to find a way to keep your chickens' home clean too. Cleanliness is important to the health of your flock, and it may also be important to your own health, no matter where you live. If you're an urban chicken-keeper or a chicken lover with close neighbours, keeping things neat and tidy is especially important. If you become frustrated and feel overworked caring for your chickens, or your neighbours complain of smells and unsightly piles of muck, your chicken-keeping days may be numbered.

In this chapter, we talk about some creature comforts that every chicken house requires. You can add personal decorating touches such as curtains for the windows and pictures on the walls if you want, but the chickens certainly don't expect it and doing so only adds to your problems when trying to keep the house pest free. We also describe what you need to do to keep everything clean enough to meet your own and your chickens' needs.

Bedding Down

Most people like to have some type of bedding material on the floor of chicken shelters. *Bedding* (or *litter*) is essentially floor covering and isn't something chickens need for sleeping. It makes the floor of the hen house easier to clean, absorbing moisture and keeping down smells and making it a better place for you and the birds.

Several types of bedding work well, and so consider the cost of the various options and how they best suit you and your chickens. Some examples of bedding are:

✔ **Sand.** Sand is a great flooring material and provides a natural dustbath at the same time, but it's heavy to carry and shovel when cleaning out, so be sure to take its weight into consideration. Buy clean, salt-free, kiln-dried sand from a builder's merchant or be really generous and give your chickens sand that has been prepared for children's play areas.

✔ **Sawdust.** You may be able to get sawdust for free if you live near a sawmill. In rural areas you can sometimes get truckloads of sawdust delivered. The one drawback to sawdust is the large amount of dust it produces – breathing it in can be unpleasant and although chickens love dry, dusty conditions, sawdust can irritate your eyes. Avoid sawdust from wood that's *tanalised* (impregnated with preservative under pressure) or has been treated with a preservative of any sort – it's poisonous for both humans and poultry.

✔ **Straw or hay.** You can use straw if you get it at a good price, but other materials work much better. Straw isn't the best material for chicken housing but it may be the only practical option if you have a large number of chickens. Avoid it for chicks under four weeks' old and always choose barley straw over wheat or oat straw, which are rather stick-like, not very absorbent and can be dusty as well as messy – be sure you ask what type of straw is on offer before committing yourself to a shed load just because it seems good value. Chopped straw makes for better bedding if you can find it or have some way to chop it yourself.

Hay is definitely a poor choice, even though chickens love to pick through it out in the farmyard. Hay is full of mould spores that can set up an infection in the chicken's lungs if you force them to spend any time confined on it.

✔ **Wood chips or shredded bark.** These materials are commonly used as garden mulch. If the cost is right, mulch with smaller pieces can work as bedding; however it isn't very absorbent. Stay away from artificially coloured mulch because it may harm the chickens or stain their feathers.

✔ **Wood pellets sold as horse bedding.** Farm stores sell large bags of these compressed sawdust pellets. The pellets are very absorbent and less dusty than normal sawdust. Just pour them out, and then lightly moisten to fluff them up. Make sure that you moisten the pellets before putting the chickens back into the house, however, because the birds may eat the pellets. They aren't poisonous, but they can swell in the crop and cause problems. Don't use wood pellets sold for fireplaces; they may contain chemicals.

✔ **Wood shavings.** Shavings are a common material available in pet shops as well as farm stores. Pine shavings (but *not* pine needles) in particular are a good option, being economical, containing a natural disinfectant and smelling fresh and clean. The best wood shavings for both you and your chickens are 'dust extracted', which are sometimes called 'first class' or 'premium'; cheaper shavings have the same drawbacks as sawdust. You may be able to get a discount on the shavings if you buy several bales at one time. Remember that the bales are compressed and cover a lot of area when opened, and you have to find somewhere to store them.

✔ **Other materials.** If you have access to products such as coir, shredded leaves, shredded paper, fine gravel, ground oyster shell and so on, you can evaluate how they work in your chicken housing and use them if they're more economical than other types of bedding. The general rule to follow is that bedding needs to be dry, dust free (for your benefit), absorbent, non-poisonous and easy to move and dispose of – bedding only gets heavier when it's soiled.

Avoid using certain materials as bedding. These include:

✔ **Cat litter.** Unscented, cheap clay litter is dusty and may swell in the birds' crops when they pick it up, thinking it's grit. Other types of pet litter may be treated with chemicals for scent, which can harm chickens if they consume it.

✔ **Pine needles.** This material is too slippery and isn't absorbent.

✔ **Treated mulch.** Any garden mulch treated with weedkillers, insecticides or dyes can harm your chickens.

✔ **Vermiculite.** The dust from vermiculite can harm the lungs of both the chickens and their caretaker. It tends to blow around when dry and is relatively expensive.

✔ **Whole leaves.** Most leaves pack down when wet, and eventually they turn into a mouldy mess.

Making Nests Comfy and Cosy

Hens love to rearrange and fluff their nesting material as they sit. It makes them feel more comfy. Nesting material also helps cushion eggs and keeps them from rolling around and breaking. Unlike some birds, hens in the wild don't carry bedding material to a nest, but may scratch out a hollow in loose vegetation or soil and then sit in it, turning around and around to form the perfect nest. They then reach out with their beaks and tuck bedding material in around themselves. These hens like to have the nest deep enough so that when they sit in the nest, the tops of their backs are the same height as the nest wall.

Whereas bedding material doesn't do much for chickens themselves aside from providing them with the chicken-house equivalent to a nice carpet and something to scratch around in on rainy days, nesting material is more important for their comfort. Chopped straw makes the ideal nest bedding, and shredded paper or wood shavings also work well. Put 7–10 centimetres (3–4 inches) of nesting material in each nest box. Some people simply put a piece of rubber matting in nests – easy to clean, but not very natural for the hens.

Nest bedding gets kicked out, broken up and packed down over time, and so replace it frequently and add some louse powder occasionally (make it a non-toxic one because it's going to be close to your eggs), just as a precaution against mites. You need to replace nest and floor bedding if it gets dirty, which it will do over time. (See the section 'Good Housekeeping' on coop cleaning later in this chapter.) Clean nesting material keeps eggs clean.

If you're worried about how your chickens fare in colder weather, give them more bedding – not actually for a bed, because they sleep on perches, but they spend more time indoors if it's cold and rainy and bedding is a form of entertainment for scratching about in. Add some sand or wood ash to it and it turns into an indoor dustbath as well. (See Chapter 6 for more about keeping the house warm in dead of winter.)

Setting the Table: Equipping Your Birds' Dining Room

In years gone by, you may have seen Grandma throwing chicken feed from a bucket onto the ground. Or, in charming movie scenarios, she throws it from her turned-up apron. This technique may work for half-wild chickens, but your nicely confined birds need a better method.

Proper feed and water containers help you avoid wasted feed and save you money. A wide range of chicken feeders and water containers is available from farm supply stores and online shops. These range from inexpensive, small red plastic dishes with clear plastic tops (which serve as a reservoir) to huge – and pricey – metal dispensers. You can buy containers – or build them from materials you already have to hand – that hold a lot of feed or water and dispense it slowly to your chickens. These automatic containers help you to control waste because you use just a small amount of feed or water at one time, and what isn't needed at that moment isn't being wasted or contaminated. Automatic containers also save time because some are big enough to hold feed for several days, and you can use them for small groups of chickens. Whatever you choose needs to suit your chicken-keeping set up, be easy to clean and make feeding and watering convenient for you as well as the chickens.

The following sections show you options that require daily filling as well as those that you fill less frequently because they dispense food and water as needed.

Self-feeding and watering systems save time and labour, but you need to check them daily to ensure that they're full enough and functioning properly. Chickens can suffer or die if valves become clogged or feeders are empty and you're unaware of it. Similarly, nipple self-watering systems may freeze up.

Feeding containers

Feeding equipment for chickens need not be elaborate, but it must provide the following functions:

- ✔ Hold at least a day's supply of feed for the chickens.
- ✔ Allow all the chickens to eat at the same time.
- ✔ Keep chickens from scratching out or otherwise wasting feed.
- ✔ Be easy to keep clean and sanitary.

Most chicken feeders today are made of galvanised metal or plastic so you can clean them easily. Occasionally, home-made feeders are made of wood. Figure 7-1 shows a few of the feeders that are available for purchase; if you prefer to make your own feeder, you can build a small wooden trough; use a small piece of metal or plastic gutter; or use a narrow pan, such as a loaf tin.

If you tend to have many hard frosts or freezes, remember that plastic becomes brittle and cracks easily in freezing weather, and so you may want to avoid plastic dishes for everyday use. (However, keeping a spare plastic dish can be handy if you ever want to use cider vinegar as a remedy for poorly chickens, something we discuss in Chapter 11. In metal containers, the acidity of the vinegar dissolves the coating and releases zinc into the water, rendering it poisonous.)

Some feeders hold more than a day's worth of feed. These are called *self-feeders*. They usually work by gravity – when the dish is empty, feed flows out of a container attached to the dish to fill it. Self-feeders are a great choice, but you still need to check every day to see whether the feeder needs filling and is properly dispensing the feed – if the dish is full of scratched-up bedding, the feed may not flow out.

You need to ensure that feed dishes remain clean, and so don't place them under perches or roosts where manure can fall into them. For the same reason, avoid feeders that have tops the birds can perch on.

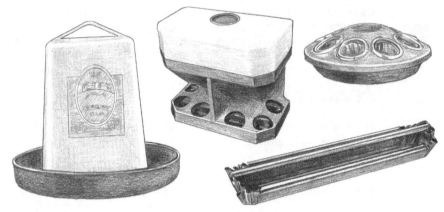

Figure 7-1:
Some feed-
ers, both
automatic
and manual
filling, that
you can
purchase for
chickens.

Sizing things up

When deciding which feeder to buy, consider how many birds are going to eat from it. Each adult chicken needs about 15 centimetres (6 inches) of space at the feeder, and the best arrangement is to have enough feeder space so that all your chickens can eat at the same time. Of course, dishes must be commensurate with the size of the chicken. Baby chicks require smaller, shallower dishes than adults, and each chick in a brooder needs about 10 centimetres (4 inches) of space at the feeder.

Chickens eat according to their pecking order, and so if feeding space is limited the highest-ranked chickens eat first and get all the choice feeds, or pick up the bigger particles of feed, and the lowest-ranked chickens may not get the feed they need. To maximise the space available, most feeders allow access to both sides of the dish or are round. If sufficient space isn't available for all your chickens to eat together, provide a second feeder. Doing so is especially important when the feeders aren't self-filling or when you offer an assortment of feeds for the birds to choose from.

If you have a bullying hen in your flock, provide an extra feeder preferably out of sight of the others – she can't guard them all at the same time.

Keeping feed in and chickens out

Feed dishes should be designed to keep chickens from getting into them and scratching feed out, so are generally long and narrow. Most chicken feeders come as long, rectangular, narrow troughs or as a circle. Narrow troughs should be deep enough to stop chickens from easily throwing feed out.

Dishes can be covered, with holes to let the birds access the feed, or just too narrow for them to climb inside. Some feeders (called *roll-top feeders*) have a bar over the centre of a long dish, and the bar rotates if a chicken tries to perch on it. Other feeders go on the outside of pens, and the birds stick their heads through the wire of the pen to reach the feed. Narrow edges discourage chickens from trying to perch on the dishes, and the feeder should be heavy enough or attached to something so that it can't be tipped over easily.

If chickens persist in tipping their dish over or scratching the feed out, you can suspend a feeder by a chain or wire so that it's about level with the birds' backs. If you hang the feeder on a wall, ensure that it has a steep slanted top or cover to discourage the birds from perching on it.

Chickens not only scratch in open dishes if they can get inside, flinging feed out, but they may sit in mash or crumbs and take a dustbath, showering feed everywhere. Chickens also use their beaks to fling food around, especially if the diet contains whole grains and they're trying to sort what they like. Think of the tale of Cinderella when the birds help her to sort out seeds from ash – this is what chickens are clever at doing. They can pick some of this feed out of the bedding, but they waste a lot. To minimise the feed being lost, sit the feeders in large plastic or metal trays or use feeders with narrow openings so the birds can only get their heads in. A common type is the 'toast rack' – a small wire structure that fits on the top of a long feeder.

Watering containers

All sorts of watering devices are on the market, from simple dishes you fill daily to automatic drinkers you fill less often and nipple devices that chickens must push to dispense a drop of water (see Figure 7-2 for some examples). Whatever water container you choose, it needs to hold enough water to satisfy all the birds for a reasonable period of time. However, unlike food dishes, water containers don't need space for all birds to drink at the same time.

Water containers need to be clean at all times, so avoid placing them under perches. The rims must be narrow enough to discourage the chickens from perching on them. You can leave them outside in the day for free-range birds.

Choosing water dishes

Choose water containers made of a material that doesn't crack and break if they get iced up. Glass and many kinds of thin plastic, for example, crack in cold weather. Stainless steel dog dishes work well to hold water for small groups of birds in cold weather. If ice forms in the dishes, simply turn them over and pour a small quantity of warm water on the bottom. You hear a pop as the ice releases from the pan.

So that you don't have to spend ages defrosting an iced-up drinker, empty it the night before and replenish it with tepid water first thing in the morning. Your birds don't drink in the dark, so as long as you're ready to provide for them as soon as the sun comes up on winter mornings, they'll be fine.

Opting for automatic watering systems

To minimise your daily chicken-keeping chores, you may want to consider using an automatic watering system (see Figure 7-2 for examples), which continuously offers water to the birds. This type of system needs to be connected to the household water system and works with a valve that shuts off water when it reaches a certain level in the dish that the birds drink from. You need to check every day to make sure that the system is working properly and is clean, but you don't need to fill the water dishes daily.

If you keep more than ten chickens, you may want to buy or build an automatic nipple watering system. This system contains a plastic pipe connected to a water source and fitted with nipple valves that drip water when touched. Chickens quickly discover how to use the valves, touching their beaks to the nipple to get a drop of water. Remember that while the birds are getting used to this system, dishes of water must be available to them. You may have to train them to use the nipples by frequently touching them and leaving beads of water hanging. Because these systems ensure that fresh water is always available, they're especially useful in warm weather.

Nipple watering systems save time and are clean, but have some drawbacks:

✔ You need to attach them to the household water system.

✔ You need to keep an eye on them, because they don't work if they freeze or when hard-water deposits or dirt clog the valves.

✔ You need to flush the system through with a disinfectant from time to time, to avoid a build-up of algae in the system in summer.

✔ You may have difficulty at first seeing whether all the chickens know how to use the system.

Other types of automatic drinkers simply hold a large quantity of water in a reservoir and release it to a dish as birds drink. The smaller plastic drinkers are efficient and easy to fill; for larger quantities the metal double-walled drinkers are a better design. If you opt for this kind of drinker you still need to check it each day to see whether the reservoir requires topping up or whether the hole that releases the water is clogged up from the inside. How often you need to fill the reservoir depends on how many chickens you have and how much water the reservoir holds, but choose a size to suit your numbers of birds. Give chickens fresh water daily if your system doesn't already do so; they're reluctant to drink from stale or tainted water.

Self-watering containers where water flows from a reservoir into a dish often become clogged with bedding from the outside, preventing birds from getting water. Check these types of container often and clean out any junk from the dish area. Elevating the container off the floor or placing a plastic tray under it with no bedding in it helps avoid this problem.

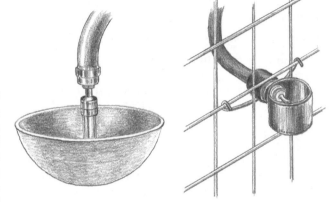

Figure 7-2:
Some examples of automatic watering devices.

Storing feed properly

All types of chicken feed are perishable and start losing nutrients the moment they're processed. The older feed gets the more likely it is to become infested with insects, which further degrade the quality of the feed. So the first lesson is not to buy more feed than you can properly store and use within a reasonable time, generally about a month.

Storing feed can be as simple as purchasing a metal dustbin, placing your bags of feed in it and keeping the lid closed. If you have only a few chickens and buy one bag of feed at a time, a robust dustbin is the best method. If a clever dog or other animal discovers how to take off the lid, you can drill a small hole in both sides of the can and run a bungee cord or chain through the handle of the lid to the holes on both sides, or weight the lid down.

If you buy larger quantities of feed to save money or because you have a lot of birds to feed, you need a secure way to store it. Metal containers work best, because they're lighter than wood or resin-type plastic and rodents can't chew through them. But a storage unit of heavy wood may also be used if you don't mind the extra weight. Heavy-duty, resin-type plastic is fine for storing feed, but thinner plastic may be chewed through by rats and other pests.

An excellent way to recycle an old freezer or refrigerator is to use it for feed storage. They have close-fitting seals on the doors and may keep feed fresh longer than a can. If you have an upright model, place it on its back. Remove the door latch or jam it so that if a child or pet climbs inside and the door closes, it can be pushed open. The weight of the door should keep pests from lifting it, but if they do, putting some sort of lock on it is relatively easy.

Other feed storage options include:

- ✔ Metal storage boxes
- ✔ Metal bins that are sold to hold rubbish bins
- ✔ Recycled metal drums with lids

Don't use barrels or tubs that held pesticides or other toxic substances. Some residue may remain and contaminate the feed.

Tight or heavy lids are important for all recycled containers. Also keep in mind that grain products can be dusty and bad for the lungs; be careful when handling them, keep a dust mask handy if you're concerned and make a habit of wearing it when scooping out feed. Feed should always be kept dry. Immediately discard any feed that gets mouldy.

You can pour feed directly into the container or leave it in the bag. We prefer to leave it in the bag because this keeps the fine particles on the bottom of the bag from building up in the storage unit. If you purchase more than one bag at a time, the unused bag remains fresh for longer sealed in its original bag.

If you have no option but to store feed in bags in the open, store it off the floor and about 30 centimetres (1 foot) away from walls. Many bags of feed can be stacked on a pallet – the higher off the floor the better. Cover the bottom of the pallet with a piece of metal or thick wood before laying the bags on it to prevent mice and rats from going underneath and between the pallet slats to get to the feed. Keeping some clear space all around the pallet also discourages pests. (Traditionally, this space was called the 'rat run' and gave enough space for the farm cat or rat catcher's terriers to do their job!)

Feed bags stored near to pesticides or other chemicals can absorb dangerous products from a spill, or even from gasses in the air, and so store feed away from all potential contaminants, including petrol, antifreeze, oil, pesticides and pool chemicals. Keep medicine cupboards out of the feed room.

Good Housekeeping

In the same way that teenagers are content to live in a room with empty pizza boxes under the bed and clothes all over the floor, chickens don't mind a little disorder and dirt either. But dirty, wet conditions smell and attract pests. Plus, for their health and yours, chickens need dry litter, clean nesting material in the nest box, clean feeders and drinkers and fresh air to breathe.

You want to accomplish the task of cleaning your chicken house as quickly and easily as you can. Undoubtedly, as a chicken-keeper you adjust your housekeeping methods as you gain experience and decide what works best in your unique situation. The level of cleanliness you maintain is dependent on your preferences or needs. (Luckily, chickens themselves aren't too fussy about their accommodation, but then, they don't tip well either, unless you count the eggs!) Knowing what your options are in terms of how often and how thoroughly you need to clean is helpful for establishing a routine. As experienced chicken-keepers, we give you some insights in the following sections, and try to keep you from making the same mistakes that we did.

Gathering your tools

Before you start cleaning, you need to gather some tools, most of which you probably have lying around already. At the very least, you need the following:

- A rake
- A shovel
- Something to carry manure and debris away in (most people want a large 'muck' bucket for small amounts or a wheelbarrow for larger amounts)

Other useful items you may like to have around include the following:

- **A broom and a stiff brush.** Don't use brooms or other items from the human household in the chicken coop.

- **A dust mask.** The dust from chicken droppings and their feathers is bad for human lungs, and when you clean, you stir it up, and so we highly recommend using dust masks when cleaning out. Ignore the cheapest models and go for the more effective masks that have two straps and an exhale valve. Draw the mask tightly to your face and check the fit by placing your palm lightly over the mask and inhaling. Masks aren't fun to wear, but if you get into the habit for the very dusty chicken tasks such as cleaning out and feed times your lungs benefit.

✔ **A hose.** If you have cement or other solid-surface floors, a hose is handy for washing them after you remove most of the litter and manure.

✔ **A long-handled dustpan or a large, flat shovel.** These items can help you scoop up mess easily without straining your back.

✔ **A vacuum cleaner.** Some people use these on small coops, but be aware that they raise a lot of dust and are really only for the very fastidious.

✔ **Window cleaner and some paper towels or rags.** You need these supplies only for housing with windows (obviously).

✔ **Work gloves.** A pair of gloves keeps your hands clean and protects them from splinters and blisters.

Seeing what you need to do, and when

What you do when cleaning the coop and how often you do it are fairly subjective decisions that vary based on how many chickens you have and how you're raising the birds. Some circumstances – such as free-range – make houses easier to clean than other situations.

Some people believe that the right way to go is to be immaculate. They clean droppings up at least once a week, hang flower boxes full of petunias under the windows and install automatic air fresheners (we're not joking!). This may be the way to go if you live in an upscale residential area with close neighbours, although the air freshener is going a bit too far!

Other people clean once a month or so and make sure that they keep pests under control, minimise smells and keep things looking neat and tidy. Many people fall into this category of chicken-keepers. If you live in an urban or suburban area, this method is the least you should consider.

A third method of chicken-keeping, often called the *deep litter method,* works well for many people, although it's really appropriate only if you live in a rural area. With this method of chicken-keeping, you clean the housing and remove litter only once or twice a year, and carry out minimal routine maintenance. This method can work pretty well if the bedding is dry and loose. Adding wood ash encourages the chickens to take dustbaths and stops the bedding material from compacting. In fact, several studies have shown that chickens raised this way may be more resistant to many diseases and parasites. As long as the chickens aren't too crowded and you keep them dry – and we mean really dry – smells are minimal.

If you decide to go with the deep litter method, you must add fresh litter when manure piles up, and you may need to remove and replace areas under the water container and round the doors of free-range systems if they get wet. You need to ensure that the housing is well ventilated so that moisture and ammonia don't build up, and don't allow the litter to pack and crust over.

If it does, the litter doesn't absorb moisture and smells, and the chickens can't keep maggots and other insects under control, resulting in an insect problem. You may need to stir up the bedding with a rake or pitchfork if it gets compacted and hard. Even throwing a little whole grain amongst the bedding and letting the chickens dig for it can work.

You can adopt compromises and customisations of these general guidelines for cleaning. For example, many people put pits or dropping boards under the perches because a large amount of droppings concentrate there. They clean the pits or boards frequently, and the rest of the housing less often.

Regular, essential cleaning

At the very least, your chickens need a dry space, clean litter and nest boxes, and clean food and water dishes. Keeping feed and water dishes clean may mean attending to them more frequently than the rest of the coop. Brush out any caked feed, wash and rinse them and then spray them with an anti-bacterial spray, or use a commercial poultry-safe disinfectant, following the directions on the packaging. Rinse and dry in the sun if possible, and ensure that the feed containers are totally dry before refilling them.

You need to keep algae, slime and scum from accumulating in water dishes. You may need a bottlebrush to clean some of these items. Check the nipples of automatic water devices for rust or hard-water scale build-up; if needed, soak them in a lime and scale remover liquid. An old toothbrush is handy for cleaning nipples and other small surfaces.

Keep nest boxes clean at all times as well, and frequently replace any dirty or lost bedding. Clean nests make clean eggs and happy hens, and clean eggs are healthier for both eating and hatching.

Don't clean the nest of a hen you've left to sit on eggs. If you notice smashed or leaking eggs, remove them and any soiled nest material. If the area around the nest becomes filled with droppings, you may want to pick them up. After the eggs have hatched, immediately clean out that nest box completely.

Avoid using water for cleaning unless the floor drains well, the day is warm and sunny and you can use ventilation to dry the house quickly. Try to avoid getting anything wet that won't dry before nightfall.

You don't need a lot of cleaning products for chicken housing. A general purpose cleaner and a cleaner for windows (if you have them) are fine. Most of the common human household cleaners available at feed stores and your local shop are safe for general use. You may choose to use a product that breaks down easily in the environment. That way, you know that it doesn't leave unsafe residues behind – particularly important if you have meat chickens or laying hens. Steer clear of ammonia, which isn't good for the lungs. You're looking for cleaning products, not necessarily disinfectants.

Unless you've had a disease problem, don't worry about disinfecting the general quarters. If you have had this problem, ask a veterinarian what products you need to use to eradicate traces of disease, and always read and follow the label directions exactly. Remember to clean all surfaces before sanitising. For sanitising, a solution of one part common household bleach and three parts water makes a good general disinfectant if you feel you need it, but be sure to rinse surfaces after use. A tried and tested alternative for a really good clean and disinfecting of poultry premises is dry lime sprinkled on the floor or a limewash on the walls. Lime counteracts the acid environment that chickens create, but it's a strong alkaline so take care when using it.

If you've had a problem with lice or mites, don't spray the hen house with creosote or other old-fashioned remedies, and avoid using household pesticides or products for other pets until you've read the label completely. Some pesticides are extremely toxic to birds and any revised doses need to be worked out by a qualified person; the label should mention use on or around poultry if the product is safe. Animal feed suppliers generally stock a range of suitable and effective products that are safe to use. To get rid of any mite eggs that are lodged in cracks and crevices, run a lighted blowtorch over them.

Deep cleaning

Whether you're an urban or rural chicken-keeper, clean everything out once or twice a year, when it smells or gets wet, or when the bedding gets too high. Start your indoor cleaning by shooing out the chickens. Then take the following steps:

1. **Scrape off the roosts.**

2. **Dust out the cobwebs.**

3. **Brush down the walls.**

4. **Remove all the dirty litter.**

   Some people lightly dampen the litter to lessen the amount of dust that gets stirred up, but don't overdo the wetting. Removing litter is easiest when the litter is dry, and so wear your dust mask.

5. **Sweep the floor with a damp broom.**

6. **Wipe light bulbs carefully after they've had a chance to cool.**

   The bulbs get coated with dust, which reduces light.

7. **Clean any windows.**

8. **Wipe off any screening that protects windows or ventilation flaps.**

9. **Place fresh litter in the house.**

Disposing of manure and old bedding

Sometimes the hardest part of keeping animals, including chickens, is finding a way to dispose of their waste, especially in urban areas. Before you have your first wheelbarrow load of manure, you need a plan for what to do with it. You may want to check with your county council officials to see whether any laws exist about the disposal of animal waste.

You have a few options for getting rid of manure. Although your first thought may be to use it as fertiliser, doing so on a small urban garden may be more of a hassle than it's worth to you. Even if you have a good place to store it as it ages, if you have a large flock a time may come when you simply have too much to use in your own garden. If you have more room, composting manure and bedding is an excellent green way of disposing of the waste.

Composting and using as fertiliser

Composting is an excellent way to use manure and chicken bedding, breaking down the materials into a soil-like substance. You can buy compost tumblers and bins to work with a small amount of material, but all you really need to get going is a circle or square of heavy wire to contain the pile – and an area where a compost pile doesn't offend neighbours or the rest of the family.

You can keep a pile of manure and bedding in an out-of-the-way place and allow it to break down slowly – a *passive compost pile* – but by managing the compost pile well you help the material to break down faster. If what you want to add to the pile is heavy on manure and light on bedding, you need to add coarse, carbon-filled material such as straw, dry leaves, shredded paper or coarse sawdust to allow air into the mix to help it break down.

Chicken manure and bedding are usually dry when you remove them from the housing, but compost needs to be reasonably wet to begin working – about as moist as a sponge that has been wetted and then wrung out – and so you may need to moisten it. Add too much moisture, though, and you cause the pile to stop decomposing correctly. Anaerobic bacteria are the result, causing the pile to smell. Turning the pile and adding more dry material helps you to reduce the smell. If you have free-range chickens, you can employ them to turn your compost pile for you – they love it, but they can make a mess. *Controlled composting* needs a bit more space and attention, but can soon turn your chicken manure into something very useful.

Chicken manure mixed with bedding can be a good fertiliser, but it burns plants if applied to them when fresh. You have to let it age first, from four to six weeks in warm weather to several months in cold weather. When your manure has aged and composted well, applying it to tree fruit crops, flower beds and lawns is safe so your neighbours or local allotmenteers may be keen to take some off your hands. In your own garden you can apply compost directly to your lawn and beds (lawns where children play may be the exception, because some disease organisms can remain). On bare soil, work in as much as you can, dispensing it lightly on planted areas, a few centimetres (or inches) at a time, to keep from smothering plants. You can put 7–10 centimetres (3–4 inches) of compost between established plants as mulch.

To avoid food pathogens such as salmonella, which can be present in manure, don't use manure on food crops within several months of harvesting. You can incorporate it into vegetable plots or use it on small fruit crops, such as strawberries and grapes, in the autumn after everything is harvested, giving it time to break down before you plant the next crop.

The problem with using manure as a fertiliser is that after you apply it to the soil it can become a major water pollutant if its nutrients run off in heavy rain and pollute any surface water, streams or rivers. For this reason, don't spread manure on slopes near any water, including roadside ditches that may drain into streams and wells. City storm sewers often drop untreated water into surface water such as streams or lakes, and the excess nitrogen from manure can pollute that water. Thus, where you can spread manure may be regulated as a nitrogen-sensitive area. Call your local council's environmental health department to find out if spreading manure on your property is allowed. Putting any form of manure into storm sewers is usually illegal. Nitrogen, phosphorus and other components of manure can also leach down through the soil and pollute groundwater if applied heavily to the surface.

The best way to use chicken manure as a soil fertiliser is to work it into the soil by hand or by using a rotavator, which causes the waste products to bind with soil particles, making them less likely to pollute the water.

Don't add dead chickens and associated waste to your compost. These things attract animals and insects and create smells in a compost pile. Chapter 17 gives you information about how to dispose of animal matter in a safe way.

Other disposal methods

If you live in an urban area, you may not have room to compost manure, or doing so may be prohibited. Some places may allow you to bag waste for collection with green waste and garden trimmings. Many councils have good municipal compost schemes that work well for park and garden waste. Find out if one operates in your area.

If local rules prohibit you from using your normal rubbish collection for chicken waste and you're overwhelmed by compost, check with gardening and allotment clubs; some may want the waste for composting and are happy picking it up from you. Alternatively, you may have to pay to take it to a landfill. Or perhaps Farmer Joe will let you dump it on his farm.

Part III
Caring for Your Flock: General Management

'It says quite clearly on the pack –
WARNING! – DO NOT exceed
the stated measure!'

In this part . . .

*I*f chickens could read, they'd definitely come straight to Chapter 8, which stays with the basics and looks at feeding chickens! In Chapter 9, you get the info you need to keep your flock safe from pests and predators.

Knowing how to handle your chickens and how to prevent problems with disease keeps both you and your chickens happy and healthy, and we tell you how to do just that in Chapter 10. And for those times when your chickens aren't feeling well, we discuss in Chapter 11 how to work out what's going on, as well as how to treat the problem.

Chapter 8

Feeding Your Flock

. . . .

. . . .

*P*eople probably disagree more over feeding chickens than over any other aspect of keeping these fascinating birds – *everybody* has an opinion. Some people believe that you don't need to feed chickens at all; they say you just turn the birds loose to find their own feed. Others devise elaborate diets that chickens love but aren't really necessary. Chickens aren't fussy eaters; they eat just about anything. However, they're like small children: when faced with an abundance of options, they don't always make the wisest choice.

In this chapter, we explain why thoughtful diet planning is an important part of chicken-keeping, and we cover the spectrum of what chickens need nutrient-wise, what options you have for providing those nutrients and what special considerations you need to take into account depending on your chickens' ages and purposes (laying, providing meat and so on). We also tell you when you need to supplement your flock's diet and with what, talk about treats as a means of keeping your chickens happy and explain how much water your chickens need in order to stay properly hydrated.

Feeding Basics

Chickens aren't vegetarians by nature. They're *omnivorous,* meaning that they eat both animal protein and plant-produced foods such as seeds, fruits and leaves. Wild chickens get their protein in two ways:

> ✔ **They actively hunt for it,** chasing butterflies and grasshoppers and gobbling up tiny frogs and baby mice.
>
> ✔ **They scavenge it,** with no qualms about picking apart a dead carcass, including that of another chicken.

In a domesticated setting, chickens still love to range freely looking for food, but the places people choose to keep them don't always support their nutritional needs. Most chickens need to be fed by you, the chicken-keeper.

The sections that follow spill the beans on why you need to feed your flock, what they need from their food and what your options are for feeding them.

Understanding why you need to manage your birds' diet

People often tell us that their grandparents kept chickens well enough by letting them run around loose in the farmyard and throwing them a little grain every day. Nowadays, though, things are different. Most gardens aren't like old-fashioned farmyards with large animals close by spilling feed and passing undigested grain in their manure for chickens to find. Modern chicken-keepers don't generally have large piles of manure breeding a tasty crop of maggot protein for their chickens, either. And when chickens are confined to a pen for most of the time, they certainly can't find the food they need.

In bygone days, your grandmother expected to get few eggs during the winter, whereas modern breeds of laying hens lay many more eggs each year – including through the winter months – especially if you have young hens or artificially light your henhouse (see Chapter 15). Granny expected to rear a handful of chickens each year for meat, unlike current broilers that are ready to eat in as little as 12 weeks. She anticipated that lots of birds would die over the year, many from poor nutrition or predator attacks. And unlike today, good, nutritionally balanced chicken feed was seldom available, even if Grandma had the funds to buy it in the first place.

You can try to keep chickens in the same way as your grandparents or great-grandparents, but we don't advise it. A far better bet is to read this chapter and see how modern chicken owners feed their birds.

The land is unreliable

Ideally, you'd turn your chickens loose, and they'd find all the food they need themselves. You can do this, however, only if the environment contains the right mixture of foodstuff in sufficient quantity throughout all the seasons of the year. A suburban garden doesn't support many chickens very well and your neighbours may not want them visiting their gardens. Even a large rural area with room for chickens to roam may not have enough of the correct nutrients to maintain chickens year-round, especially in cold weather.

Chicken digestive system overview

Food only takes about 2½ hours to pass completely through a chicken's digestive system. The food a chicken picks up in its beak is first sent to the *crop,* which is a pouch-like area in the neck for storage. The crop is stretchy and allows the chicken to quickly grab opportune food finds and store them for a slower ride through the rest of the digestive system.

When the food leaves the crop, it's then slowly moved along to the *proventriculus,* or true stomach, where digestive enzymes are added. Then it moves to the *gizzard,* which is an oval sac composed of two pairs of strong muscles. The chicken picks up bits of gravel and grit and stores them in its gizzard. With a squeezing action of the muscles, the grit helps grind down food particles much as your teeth break up food. The food passes next to the intestines, where the intestine walls absorb nutrients.

Liquid and solid waste combine in the end part of bird digestion and pass through the *vent,* or *cloaca.* The vent is a multipurpose organ through which waste passes, the chicken mates and through which it passes its eggs.

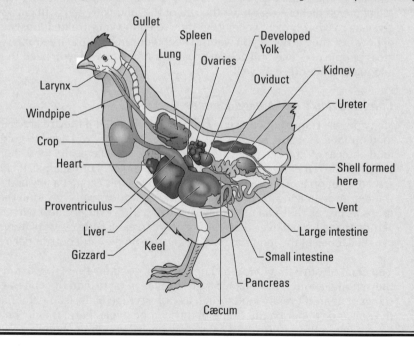

Your chickens need to be fed if:

- ✔ You don't want them to go too far or become dinner for something else. Whether in town or country, hungry foxes are never far away.
- ✔ You want eggs all winter or tender meat grown quickly and reliably.
- ✔ You want your chickens to survive when winter brings frost and snow.

If chickens have unrestricted access to a large amount of relatively wild land, they may be able to find enough of a properly balanced diet to satisfy both their hunger and their nutritional needs. But a lot of productive land is necessary to fully support chickens, and the more chickens you have, the greater the chances are that your free-range birds aren't meeting their nutritional needs, even in the best weather.

Avoiding total free-range conditions, in favour of managed pasture (see the section 'Letting birds feed on pasture: Free-range chickens' later in this chapter), makes particular sense if you want to rear birds for meat on grass. Free-range meat chickens that have the run of the land with no particular care given as to what type of vegetation they're eating vary tremendously in terms of how fast they grow. If you want to keep your meat birds free-range – and most people do – and still achieve a good rate of growth, you need to provide adequate broiler feed as well.

Chickens don't make good choices

Many people think that chickens choose the right diet if they're given many food options. Not so. Chickens are a lot like children – you need to guide their food choices. A chicken that eats an abundance of a favourite food simply fills up on that, even if it isn't the best nutritional choice. For example, if chickens come upon a bonanza of raspberries, they don't eat just a few and then wander away looking for protein to supplement the meal. They stuff their little crops as fast as they can until they're bulging. However, a chicken in a wild setting that's forced over time and territory to eat a wide range of items to feel satisfied, does in the long run tend to balance its diet and nutritional needs.

You can't be sure that chickens know what's good for them to eat and what they should avoid. Chickens have little sense of taste, and they often eat things that aren't really foods, such as polystyrene balls, lead paint flakes in the soil and rubber bands. Some of these 'foods' may harm them. Some may get caught in the gizzard, and some may pass through the body.

Chickens also can ingest pesticides from eating grass and other vegetation. Small urban and suburban gardens are mowed and trimmed and treated with pesticides to reduce the insect population. Even rural fields can be treated with insecticides and weedkillers. If you do let your chickens roam for part of their dinner, make sure that they aren't eating from 'poisoned plates'.

Poultry cannibalism

You may be disturbed to find your chickens greedily picking the meat off the fast-food chicken bones you've discarded. But chickens have no idea that they're cannibalising one of their own, and they wouldn't care if they did.

Rob says: 'Before I had chickens I visited a friend who was raising a small flock. The friend said, "Watch this!" as he tossed a recently laid egg onto the ground. Within the blink of an eye, the whole flock was fighting over the delicious treat smashed on the ground. Of course, doing this isn't advisable for a host of reasons, but I remember being shocked and amazed that chickens love eggs so much. It seems most people are pretty amazed by this as well.'

Knowing what nutrients chickens need

Because chickens, like children, don't always make the best food choices, you need to be in control of your birds' nutritional needs. The sections that follow show you the general nutritional needs of all chickens and mention the variations depending on the birds' age; we explain the specific dietary nutrient proportions various chickens need (called *rations*) in the section 'Demystifying commercial rations' later in this chapter.

Protein

Protein is an essential part of a chicken's diet. It consists of a variety of amino acids, and each type of protein provides a slightly different balance of these amino acids. In nature, chickens obtain their protein from insects and invertebrates such as bugs and worms. Both contain a wide range of amino acids but generally form a small part of the diet in terms of volume.

Protein is the most expensive part of the diet, and so a little good-quality protein containing all the right amounts of amino acids is preferable to large amounts of poor-quality protein. Choosing the right sources of protein is important, as is balancing the protein content of chicken feed between too much and too little.

Chickens need 16 to 20 per cent protein in their diet, depending on their age and intended use. Too much protein in a chicken's diet is a waste because it's secreted in the manure as nitrogen and ammonia, which are pollutants of both water and the air. Another downside to feeding too much protein is that a chicken's body has to work harder to metabolise protein and break it down into urea to be eliminated. This effort creates body heat, and birds on high-protein diets may suffer more from heat stress. Many people think that chickens need more protein in the winter because it creates body heat, but this assumption is faulty. Too much protein places stress on the bird's organs, and the energy it uses in breaking down the protein can't then be used to produce meat or eggs.

Diets that are low in protein or in some amino acids cause problems such as slow growth, poor feathering and fewer and smaller eggs. Moult is stressful to birds and requires good nutrition to see the birds through it. Feathers need protein to develop properly, and increasing the protein level at this time doesn't hurt. A little supplementation with a high-protein feed such as sunflower seed, peas or beans can be helpful. Some chicken-keepers like to give alfalfa as a supplement at this time.

If laying hens don't get good-quality protein in their diet, they struggle to produce quality eggs for long periods of time. If broiler-type birds don't get enough protein, their legs and wings may become deformed and they may become unable to walk, be in pain and you then have to dispose of them. The law considers this to be a welfare issue and you can be reported to the RSPCA (Royal Society for the Prevention of Cruelty to Animals) if you let your birds get into this state.

Fat and carbohydrates

Chickens don't require a lot of fat. High-energy broiler feeds need 6 to 7 per cent fat and laying and chick starter feeds about 3.5 to 4 per cent fat. Hens do need some fat for the production of egg yolk, but chickens can use carbohydrates for most of their energy requirements. Unneeded calories are converted to fat and stored. Excess fat is deposited around the abdominal organs and under the skin and can cause many problems such as heart failure.

Carbohydrates – which get burned in all animals' bodies as fuel for life processes such as breathing and growth – need to furnish the bulk of a chicken's diet, and a commercial diet based on grain easily supplies them. Most diets have enough carbohydrates for all types of chickens, but diets based primarily on pasture or vegetation need grain supplements to provide sufficient carbohydrates.

Vitamins and minerals

Chickens require all the vitamins and minerals that people do except vitamin C, which their bodies can make. Table 8-1 gives you an idea of what each vitamin and mineral does and what benefits they provide. Chickens need many other vitamins and minerals too, but the requirements for some still aren't fully known, and so in the table we cover just the most important ones. If these vitamins and nutrients are adequate, the others are almost always available in healthy quantities. Chickens get these vitamins and minerals from a properly formulated feed.

Some types of chickens may need more of some nutrients. Layers, for example, need more calcium than other birds in order to continue laying well. That's why making use of chicken nutrition experts' expertise in buying specially formulated rations is so valuable.

Table 8-1	The Benefits of Vitamins and Minerals
Vitamin or Mineral	*Benefits It Provides*
Calcium	Promotes strong, smooth eggshells; prevents rickets; aids hatching
Choline	Promotes growth, egg production and a healthy liver
Cobalt	Promotes growth and general good health; helps prevent hatching problems
Folic acid	Promotes healthy feathers, growth and egg production; prevents anaemia
Iron	Prevents anaemia
Magnesium	Helps prevent sudden death in broilers
Manganese	Helps prevent *perosis* (twisted legs) and hatching problems
Niacin	Helps prevent leg deformities; keeps mouth and tongue healthy
Pantothenic acid	Promotes healthy skin, mouth and feet
Phosphorus	Promotes strong, smooth eggshells; aids hatching; prevents rickets
Riboflavin (vitamin B2)	Prevents curly toe disease; increases egg production; promotes growth
Thiamine (vitamin B1)	Promotes a healthy appetite; sustains life
Vitamin A	Promotes growth, egg production and good general health
Vitamin B12	Promotes growth; helps prevent anaemia and hatching problems
Vitamin D	Promotes growth, egg production and strong eggshells; prevents rickets
Vitamin E	Decreases leg joint swelling; strengthens immune system; helps prevent mental problems
Vitamin K	Promotes good blood clotting; helps prevent bleeding in the muscles
Zinc	Helps prevent shortened bones and feather abnormalities

Comparing your feed options

After you know what constitutes a balanced diet for your chickens (as we describe in the preceding section), your next step is to work out how to provide it. Do you rely on the feed merchant, design your own custom blend or send your chickens out to the pasture? Before you decide, weigh up the pros and cons of the choices available to you.

You have four main options for feeding your chickens:

- ✔ **Commercial feed, specially formulated for your chickens' needs.** Animal feed nutritionalists spend hours working out these rations so you don't have to. If you have layers, you want to optimise their diet for egg production, and so we highly recommend feeding a commercial ration that's properly balanced for laying chickens (instead of feeding them only whole grains). These commercial mixes contain the proper ratios of calcium and other minerals, which are added for egg production. Even outdoor layers confined to a pen need access to a good laying ration.

  We prefer to use commercial feed for meat birds too, because it blends all the ingredients and prevents birds from picking and choosing what they like and wasting the rest. Feeding only whole grains (also called 'straights') isn't a good idea for meat birds. Pets and show chickens also do better on a balanced commercial diet that doesn't allow them to pick and choose.

- ✔ **Individually mixed feed for which you supply the recipe to a feed mill.** Some people have an abundance of home-grown grain or other products available to them and want to use them for feed. These people often enjoy experimenting and tinkering with formulas and have a specific idea about what they want to achieve or avoid when feeding their birds.

- ✔ **Individually mixed feed you prepare in your own home.** Another way to make home-made mixes is to buy the ground ingredients separately (or purchase a small grain grinder) and mix small batches of feed at home. This option isn't easy, however, and has room for mistakes. The problem is that most places require you to buy the ingredients in large quantities, such as in 25-kilo (55-pound) sacks. You don't use even amounts of the ingredients in mixing feed, and ground grains begin to lose nutrition quickly, with the result that the grains can go stale before you get to them.

- ✔ **Pasture.** Chickens confined to arks or other moveable pens are sometimes referred to as 'free-range' when in fact they're confined on the range. Pasture alone doesn't provide enough goodness for growing or laying chickens but, combined with another source of feed, it can save you money at the same time as satisfying your chickens' needs and keeping them healthy.

Some people advocate feeding kitchen scraps to poultry. Back in the Second World War, the government actively encouraged people to produce their own food cheaply and feeding scraps to chickens was one way to achieve this aim. Things are very different now and feeding kitchen scraps to your birds is actually against the law. All is not lost though – you can still treat your birds to garden trimmings (see 'Keeping the Diet Interesting by Offering Treats' later in this chapter). Just don't take them into the kitchen or they're then automatically classed as kitchen scraps! (For more information phone the DEFRA help line 08459 33 55 77 or visit www.defra.gov.uk.) Garden trimmings can make a useful supplement to commercial *compound food* (a pelleted whole diet especially made for chickens) and add entertainment and enjoyment to your chickens' day.

Milk and other dairy products are also forbidden by law as chicken food, being considered animal by-products by 'the powers that be'. However, a small, one-off feed of yogurt as a probiotic can cure a mild tummy upset in chicks. Stale bread, biscuits, cake, cereal and so forth are also out of bounds, however much your chickens may adore them.

Don't feed pet chickens a diet composed of dog or cat food. These pet foods are too high in protein and fat for regular chicken consumption, and they don't contain the right blend of vitamins and minerals for your birds. Likewise, meat and bone scraps are an absolute 'no no'. Of course, your chickens are omnivores and happily eat meat, but unless it's an unfortunate mouse or frog that strays into their vicinity, meat's out of bounds.

Other than grit, which you need to add to your chickens' food as an aid to digestion, regular supplements of vitamin concoctions or other things advertised as 'immune boosters' or 'energy creators' are unnecessary if the feed is balanced. Most supplements are just a waste of your money, and some can harm your chickens. For example, too much calcium and phosphorus can cause kidney stones, and too much vitamin K can cause bleeding problems.

All kinds of problems can develop if a supplement:

✔ Alters the taste of the birds' water, causing them to avoid drinking it.

✔ Makes the water taste salty, causing the birds to drink too much.

✔ Gives the birds' feed a flavour that they don't like.

Don't waste your money buying supplements. Instead, spend it wisely on a good, balanced feed.

The virtues of commercial feed

The main benefit of commercial feed to be aware of is that you don't need to spend time agonising over what amounts of protein, fat, carbohydrates, vitamins and minerals to give your chickens. With commercial feed, all this hard work is done for you. The nutrients available in grains and other products vary from batch to batch, which means that the ingredients must be tested to see how much protein and other nutrients are in the feedstuff. Large companies employ animal nutritionists who test feeds and decide on the recipe or blend of ingredients necessary to meet the standards of the feed product to be made. These specially trained nutritionists also test the feed for mould and other toxic substances and make sure that the feed is safe and nutritious after mixing.

All sources of protein vary from batch to batch as to which and how many amino acids they contain. In the past, fish and blood and bone were included in chicken feed to provide the protein, but most commercial chicken feeds now use only plant sources for protein. These feeds use synthetic amino acid supplements to cover the amino acids that plant sources are low in, such as methionine and lysine, enabling the chickens to get what they need without having to consume more crude protein.

Commercial poultry feed ranges in protein content from about 12 to 22 per cent, with the higher protein feeds made specifically for growing meat birds. To feed your meat birds, you can purchase commercial meat bird, game bird or broiler feed (the terms vary).

Almost all commercial feeds use synthetic and natural vitamins and minerals to balance the feed, including all the vitamins and minerals we refer to in Table 8-1, plus some additional ones. Therefore, if you keep your chickens on a commercial diet, they're unlikely to have vitamin or mineral deficiencies.

Chicken feed producers choose their carbohydrate sources carefully, too, and have lots of experience in getting the carbohydrate content just right for your birds. A diet high in fibre may make the birds feel full before they consume enough feed to get the protein, vitamins and minerals they need. Birds on pasture that's become old and fibrous need regular access to grit, and birds that don't have access to some high-calorie grains and protein typically lose weight, grow slowly, lay poorly and suffer from vitamin and mineral deficiencies.

Chickens that receive lots of treats or additions to a commercial diet or are fed a home-made diet may suffer from deficiencies from time to time. When birds get full on things they like, they may not eat the properly balanced diet in sufficient quantity to get the nutrients they need. And without some animal nutrition education, few chicken owners can devise properly formulated diets.

Flavouring or colour is rarely used in chicken feed nowadays although some sneaky yellow dye used to be added to colour up the egg yolks. Some feed manufacturers add preservatives to keep the feed fresh longer. Growth hormones aren't used in chicken feed, and if a medicine is added to the feed, it must be on the label.

Local feed for local chickens

Most feed mills rely on imported cereals and use soya and soya by-products that can't be grown in this country. However, you can buy British – try asking your feed suppliers if they support local cereal growers in the UK. Doing so helps them and benefits you in the long run as world markets come under ever-growing pressure from increased transport costs.

Medicine is added to some chicken feeds, the most common being *coccidiostat* – a medicine that controls the disease *coccidiosis,* which can be quite deadly in young birds. Don't assume that you need to buy medicated feed, though – you can keep chickens in good health easily, without resorting to routine medication. Although antibiotics may be in the feed that large commercial poultry operations use, they're rarely put in feeds that consumers buy off the counter for their chickens.

The plain truth about home-made feed

Despite what most people assume, the home-made recipe approach has more drawbacks than positives. In this section, we give you some advice on making your own feed in case you feel strongly about doing so, but first, consider these cold, hard facts:

- ✔ **Home-made *isn't* cheaper than a well-formulated commercial compound feed.** Some people think that making up their own recipe for chicken feed is cheaper than buying commercial feed but if you only have a small flock of chickens, it would be extremely rare to get a cost saving from devising your own chicken feed when you have to buy all or most of the ingredients.

- ✔ **Making your own recipes requires a complex understanding of chicken nutrition.** Keep in mind that feed companies hire experts to formulate and test their feeds. If you intend to devise a recipe of your own, you need to study chicken nutrition thoroughly.

- ✔ **Most mills don't custom mix less than 1 tonne of feed.** This amount of feed is difficult for most people to transport and store – much less use before it goes stale. In many cases, a person making an order who's using relatively small amounts of grain (and 1 tonne is a small order) is charged a milling fee, too. If a group of chicken owners in an area band together and order a large amount of a custom feed, some cost savings may be realised.

- ✔ **Costs for each ingredient vary as the cost of grain goes up and down.**

- ✔ **You may have to pay for bags or supply your own.**

If you *really* want to make your own feed, and if you have a feed mill in the area that grinds and mixes feeds on a custom basis, you may be able to devise your own feed formula using locally grown grains. Some mills have vitamin and mineral mixes and protein supplements on hand to mix complete feeds. They may even have an experienced animal nutritionist on staff.

These mills generally just grind grains together to a certain size, and then mix in a powdered protein mix and a vitamin-mineral supplement. If you want soya beans in the recipe, they often have processed soya bean meal on hand. But usually these recipes aren't cooked or bound together, and the ingredients may separate or be picked through by chickens. If the operation has a pelleting machine, you must pay extra to have the feed made into pellets. Most of the time, the mix is a fine grind, and some of the separation of ingredients can be overcome by moistening the mix before feeding.

If you have a moderately large flock – 100 layers or 200 or more meat birds – and want to devise your own rations, you can consult with a poultry expert from an agricultural college or an animal nutritionist employed at a large feed mill that caters to special mixing. These experts may help you with smaller flocks, but they generally work with larger quantities of feed and their formulas may be difficult to reduce to small-flock quantities. The cost for mixing small quantities of feed usually outweighs any benefit. In fact, some mills don't even consider it. So unless you have access to a marvellous supply of home-grown grain and can share feed with other people, this option just isn't suited to small-scale chicken-keeping.

My home-made happening

Kim says: 'One summer when we were raising three pigs for meat as well as some broilers, I decided feed would be cheaper if I had my own recipe mixed. Pigs and chickens are fed a similar diet, and I thought I could customise one mix that would suit both. After lots of researching formulas, consulting with the chap at the mill, and pricing the various feedstuffs, we ordered 1.5 tonnes of feed. We loaded that feed in the back of an old van and started the ten miles home.

Each batch of the finely ground and dusty feed had to be mixed with water at every feeding in quantities small enough to be eaten in a few hours, because it would mould quickly in hot weather. The huge stack of feed bags attracted mice by the dozens. It was messy and time consuming, and after paying for the feed bags and a milling fee, our wonderful recipe cost us at least as much as commercial feed would have. Our meat tasted wonderful, but I vowed never to go the home-made route again.'

Letting birds feed on pasture: Free-range chickens

Nature didn't design chickens to subsist on vegetation – wild chickens eat lots of insects and seeds to balance their diets. Because you expect more eggs from domestic chickens than wild birds lay, and you want your meat birds to grow heavier and faster than wild birds that subsist on a natural diet, domestic chickens need extra concentrated protein and calories to thrive.

Free-range chickens may get ⅓ to ½ of their food from foraging at certain times of the year, but even chickens that have unrestricted access to large pieces of land still need feed. They may be able to find bugs and seeds to round out their diet in some seasons, and because chickens can only eat a certain amount of food each day, filling up on yummy outdoor treats like these means that they don't empty their feed hopper very fast! In other seasons, however, the chickens may find those vital ingredients missing.

If you manage your pasture properly and the weather is such that the vegetation is growing rapidly and is young and succulent, more of the birds' dietary needs can be met. A good nutritionist once explained that you can grade the quality of your grass by scoring the best short, sweet spring and summer pasture at a rate of '10' and the tough, stringy grass of mid-winter at '1'. The 30 per cent protein level in grass at the top end of the scale drops to low single figures at the other, giving you a decent 'guesstimate' of how much goodness your range is contributing to the general feeding of your hens at different times of the year.

If you have well-managed pasture and not just a grassy spot in the yard, pasture can make up a large part of the diet of layers, breeding birds, show birds and pet chickens. In addition to grass, managed pasture includes plants that provide quality protein and carbohydrates. Clover, alfalfa and some other legumes, for example, raise the protein content of the pasture. You can use your imagination to provide exciting belts of herbs and vegetation that provide cover as well as food for your outdoor birds and which encourages them to be adventurous on the range. Even a garden with scrubby vegetation can provide the sort of environment chickens just love to be in. The birds usually need some high-protein feed on the side, but in reduced quantities.

With adult laying hens you also need to supplement the grass-fed poultry diet with a daily scattering of whole-grain feed for a few extra carbohydrates, as well as grit for aiding digestion. Your layers soon get used to the routine and you can use it to your advantage if you want to manipulate them in any way. Feeding your chickens well has the added advantage of keeping them closer to home, where you can find the eggs they lay and keep predators away.

Vegetation alone can't keep broiler-type (particularly Ross Cobb) chickens growing properly; you continually need to supplement their diets with a good concentrated feed. Pasture can, however, eliminate the need for methionine and other synthetic supplements. The feed for these birds needs to be a commercial feed of about 16 per cent protein or a mixture of whole grains and a high-protein supplement. High protein feeds can be sunflower seeds, safflower seeds, peas or beans and alfalfa. Allow the birds free access to these foods and they eat enough to supplement their pasture diet.

Free-range birds may take a little longer to get to a good eating size. Because they move around a lot more, you find more dark muscle meat and less fatty breast meat in their carcasses. Also the skin of the birds may be a bit more yellow due to pigments in the grass.

You need to move chickens to clean pastures frequently. How often depends on the weather, rate of vegetation growth and number of birds. Move them before all the grass has been eaten to the roots or the pen is too dirty. However, do take care when moving outdoor chickens. Organise things so that you move them in their house at dusk or early in the morning, and keep water available at all times. You don't want to harm them or stress them too much. Also ensure that your birds have a shaded place to go to when the sun is too hot. Their ideal situation is dappled shade that mimics the jungle canopy where their ancestors came from.

In addition to feed values, outdoor chickens are happier, better amused and produce tastier eggs with golden yolks than birds kept in more confined circumstances. Most meat birds get a healthy yellow tinge to their skin, and their beaks and legs colour up beautifully compared to the same breed that's 'confined to barracks'.

Avoiding feeding chickens the wrong things

In this section we list some things that you absolutely shouldn't feed to chickens. Experts often disagree on what's poisonous to poultry, and because whether a given plant is poisonous may depend on the circumstances in which it's grown, we leave out many disputed plants. Certain plants that are poisonous to other forms of livestock aren't toxic to birds, and some books don't differentiate between species of livestock when they list poisonous plants.

Some plants or plant parts, such as apple seeds, may be poisonous to chickens if eaten in large quantities, but in all practical and normal uses, a chicken never gets enough to harm it. If you have a question about the safety of a plant growing in the area where you keep your chickens, consult a veterinarian.

Never feed your chickens the following:

✔ **Alcohol.** Never give your chickens the stale beer that went with last night's chips. Birds' bodies can't tolerate much alcohol.

✔ **Anything mouldy.** Some moulds are harmless, but many have toxins that can damage organs or cause neurological damage.

✔ **Avocado.** Birds shouldn't eat avocado because it contains a fatty acid called *persin,* which can be fatal to birds; no guacamole dip for the chickens!

✔ **Green areas of potatoes or potato sprouts.** Both are poisonous, and you too should avoid eating them.

✔ **Leaves from tomato, pepper, potato, aubergine or nightshade plants.** Chickens normally don't eat these, but they may if really hungry.

✔ **Morning glory, sweet pea and datura (angel's trumpet, jimson weed) seeds.** All these are common garden plants.

✔ **Raw dry beans.** Fresh beans (except soya beans), like green beans, are safe, as are bean plants. Some types of raw dried beans have toxins, and so cook all beans.

✔ **Raw peanuts.** Peanuts can have a fungus called *aflatoxin.* Cooking helps destroy the fungus but may not totally eliminate it. Peanuts for human consumption are generally safe if roasted or boiled.

✔ **Rhubarb leaves.** They contain oxalic acid, which is a poison.

✔ **Tobacco.** From the leaves to cigarette butts picked up off the pavement, avoid tobacco in all its forms. The nicotine in tobacco is a poison, and birds' small bodies can't handle it.

For a completely different reason, also avoid feeding kitchen scraps to your chickens – it's illegal (see the 'Comparing your feed options' section earlier in the chapter).

Choosing the Right Commercial Feed

Just as with other animals' feeds, chicken feed runs the gamut from top-quality, brand-name feeds to poor-quality, cheap feed. Many feed mills produce one feed that's sold under several different brand names, with the result that the same feed can cost different amounts just because it has a different brand name! Equally, some brand-name feeds may be made by different mills and contain different ingredients in different areas of the country. Chickens don't have a high degree of taste, but at times we've seen chickens prefer one brand of feed to another.

When choosing a feed to buy, base your decision on the ingredients and the guaranteed protein and other nutrient levels instead of on cost or brand name. Also make sure that you're getting the quantity of feed you expect; check the weight on the bag because it can be deceptive. The smaller the bag, the higher the price per tonne; you have to balance your budget against the amount you're likely to use. Remember to check the 'best before' date on the bag, too, because minerals and vitamins degenerate over time, though this may not be much of a concern in high summer when your chickens have plenty of healthy outdoor supplements. If you have chicken-keeping friends, try working out whether sharing a larger bag of feed would help you with both costs and freshness of feed.

To find the feed that's best suited to your chickens, you need to be familiar with the purpose of different types of feed, the forms in which feed is available and the info you can expect to find on package labels. The following sections address these topics.

Demystifying commercial rations

Feed labels indicate the age and/or type of chicken they're designed for, such as 'chick crumb', 'layer' or 'all adult poultry', and so when choosing commercial feed, first look for feed that suits your chickens' age and type. (For more information, ask your local feed merchants to explain rations to you; they usually have an animal nutritionist on hand to answer questions.)

The following sections provide a rundown of what you're likely to encounter.

Starter rations (for chicks)

The ration for layer-breed chicks needs to be 20 per cent protein, and is usually called 'starter feed' or 'chick crumb'. Fast-growing meat chicks need a high-protein feed of about 22 to 24 per cent protein from the time they start eating for the first four weeks. Their feed is called 'meat bird starter' or 'broiler starter'. Meat chicks that you've chosen as 'suitable for free-range' are a slower-growing variety and the starter ration for laying bird chicks with 18 to 20 per cent protein is just fine for them.

If you're raising more than one type of bird, best practice is to brood them separately or brood one type of bird at a time. Mixing chicks for a laying flock and chicks for meat isn't a good idea because they have different feed requirements and it can lead to problems – meat chicks grow much faster and personality clashes can occur.

Giving medicated feed to chicks was pretty standard practice not long ago, but with good levels of management and hygiene it's no longer necessary, even if your birds are destined for a free-range lifestyle. If your chicks do develop a health problem (with coccidiosis, for example), you can buy medication to put into their water or use a homeopathic coccidiosis remedy, which some experienced chicken rearers swear by.

Layer rations

If you're raising chickens for egg production, you may need to buy one of three types of feed, depending on the birds' ages. Taking care to feed laying chickens correctly results in healthy birds that give you optimum egg production. Because they're around for longer than meat birds, you need to take an additional feeding step when raising them from chicks.

Don't feed adult layer rations to other types of chickens, because the higher mineral content can damage the kidneys of birds that aren't laying. The exception is for a cockerel housed with a laying flock; he's fine consuming laying rations.

Here are the types of layer rations to use:

- ✔ **Chick starter ration.** A chick starter ration is formulated at about 18–20 per cent protein and the pellet size is reduced by kibbling the pellets to make a smaller, finer texture without reducing them to dust. This ration is suitable for the first 4–6 weeks while your chicks are under a *brooder* (a heat lamp). They need an ad-lib supply.

- ✔ **Grower pullet ration.** If you're raising young *pullets* (females who haven't started laying) to become layers, you want them to grow slowly enough to develop good strong bones and to reach a normal body weight before they begin producing eggs. High-protein diets tend to hurry the birds into production before their bodies are quite ready. Therefore, the ration for growing pullets, from leaving the brooder at 4–6 weeks to about 18 weeks, needs to be about 18 per cent protein.

- ✔ **Adult layer rations.** After the hens reach the age of 18–20 weeks or begin laying, and throughout their laying careers, they need a protein level of 16–18 per cent. The calcium and minerals need to be formulated for laying hens.

Don't force extra calcium and minerals on hens by adding things to a properly formulated feed. Too much calcium can cause kidney failure. If you're getting a lot of thin eggshells or soft-shelled eggs, give your hens some calcium in the form of crushed oyster shells in a feeder where they can choose the amount.

Broiler rations

Home-flock owners may encounter two types of meat birds, both with different needs: the Ross Cobb broilers that grow extremely quickly and require precise diets, and the slower-growing strains of meat birds suited to free-range, traditional dual-purpose breeds or commercial strains, which require slightly different management.

Here are the details:

- ✔ **Ross Cobb broiler hybrids.** This breed of bird is the one you see on supermarket shelves at just five weeks' old. They grow extremely quickly and need expert management. Rations need to be constant and at the higher end of the protein spectrum and a dispatch date adhered to because they easily succumb to heart failure that's immediate and final. For these reasons, you're better to leave Ross Cobb broiler hybrids to the commercial world. Avoid them if you can.

- ✔ **Traditional and free-range meat birds.** These types of meat birds grow more slowly, add less muscle meat than the broiler hybrids and take longer to reach a satisfactory dispatching size. Traditional breeds are likely to be any of the Sussex breeds, Rhode Island Red, Maran and Welsummer. Cockerels of these breeds all produce a satisfactory carcass at about 18 to 20 weeks. The more common commercial slow-growing breeds are Sasso and Hubbards.

  For these birds, feed chick starter crumb for the first 4 to 6 weeks, after which you can lower the protein to 18 per cent for the next 6 weeks with a 'grower' ration, and after that, protein content can be 16 per cent. These birds enjoy a bit of whole grain as well, as long as they have non-soluble grit to complement it.

Whole-grain mixtures

Sometimes corn is cracked, and other grains may be hulled or rolled in these mixtures. Chickens do love these mixes, but they're almost impossible to mix accurately to provide all the nutrients a chicken needs. Many of the cheaper mixes are filled with pieces of corn cob, seeds and other junk that chickens don't seem to care for. Chickens pick through these mixes, and the dominant birds in the pecking order get first choice, often concentrating on the corn or another part of the mix and leaving the lower-pecking-order birds with little variety.

Whole-grain mixtures are best used to supplement the diets of chickens that have good free-range or pasture conditions or to relieve boredom in confined birds. A little whole grain is thrown on the floor for the birds to search for, and, in the process, they stir up the litter. However, remember that if you only use whole grains on confined chickens, the diet may be lacking or deficient in important amino acids found in the protein of animal origins.

Whole grains tend to attract mice and pest birds such as sparrows more than processed feeds. They're also more likely to be infested with weevils and other insects.

Remember that one grain type doesn't supply all your chickens' needs. The most common 'grain' used for the protein part of chicken feed is a soya bean, and these have to be cooked and processed a certain way before use. Soya isn't grown in the UK, and so is imported.

Variations for breeders and show birds

If you're breeding birds, you may want to increase the protein content for cockerels and make sure that hens have a laying ration. If you keep any traditional/free-range birds for breeding, switch their diets to a layer diet.

Show-bird-keepers often tinker with their birds' diets based on elaborate formulas thought to grow feathers or improve colour. Really, these birds just need a good, balanced diet that contains a relatively low protein content of 14 to 16 per cent. Birds caged in small areas need a more fibrous-based feed to make them feel full without getting fat. You can add small amounts of whole wheat, triticale or barley to the diet for this purpose.

Selecting a form of feed

Most chicken feed is ground, mixed, cooked or steam-treated and then turned into mash, crumb or pellets. When products are mixed and bound together, each piece contains a balanced proportion of the mix. The vitamins and minerals don't sift out to the bottom, and chickens aren't able to pick out favourite pieces and waste others.

Feed comes in three forms:

- **Crumb.** Crumb consists of pellets that have been kibbled down to a smaller size. It's used mainly in chick starter feed but is also an appropriate texture for adult birds. Adult birds happily pinch chicks' feed if they can find a way to get to it!
- **Pellets.** Pellets are the most common method of feeding and the second most popular form of feed with chickens. They're long, narrow cylinder-shaped pieces of compressed feed. Pellets are usually used for adult birds.

✔ **Mash.** Mash is the least preferred form of feed. It's finely ground feed; the texture resembles cornmeal. Home-ground feeds are usually mash texture.

If mash is the only type of feed available, you can add a little warm water to the feed just before serving it, which gives it the consistency of thick oatmeal. Chickens generally gobble this porridge-like meal down and it's a good way to use up the fine pieces of crumb or pellets left in the bottom of a bag or the feed dish. However, don't let this wet mixture sit too long; it spoils and becomes mouldy, which can harm the chickens.

In the past, farmhouses had a pan of mash on the stove constantly. Potato peelings and other scraps were boiled up with it and fed warm to the chickens. It had a distinctive homely, comforting smell. Unfortunately, in today's world of global trade and with fears about transmission of diseases through inappropriate feeding, those days are gone. Feeding kitchen scraps to chickens is no longer allowed by law.

In some areas you can buy a pellet and whole-grain mix, usually under the name of 'mixed grain' or 'molassed whole feed'. This mix is covered with molasses or another sweetener to hold it all together. Although these rations may list poultry on them (or more often, feature a picture of a chicken on the sack), they really aren't formulated for poultry. You can use these feeds on your other farm animals, and you don't need to worry if the chickens steal a bite, but definitely don't use them as your sole chicken feed.

Double-checking the label

To be sure that your feed contains what your birds need and to ensure that you don't eat eggs or meat contaminated with medication from feed, be certain to check the label . . . twice!

Here's what the label tells you:

✔ The ingredients, including percentages of basic nutrients such as protein and fat and the percentages of recommended dietary vitamins and minerals.

✔ The manufacturer's address in case you have a question or complaint.

✔ Whether any medications are used in the feed, what they are and how long the chicken must be off the feed before eggs or meat from the bird can be eaten.

✔ A 'manufactured on' date or an expiry date. (Feed loses nutrients the older it gets, and after six months, most feed is stale. Stale feed may not harm the birds immediately, especially if the access to range and fresh insect and worm protein complements it, but long-term use of old feed can lead to vitamin and mineral deficiencies.)

Another reason to check the label is to make sure that you don't accidentally buy medicated feed. If you choose to buy medicated feed, you can spot it easily – sacks are clearly labelled as such – but remember to check the label even if you're not planning to buy medicated feed. Some of these feeds are non-returnable even if you buy them as a genuine mistake, and so look carefully before you leave the feed merchants.

Supplementing Diets with Grit

Because they have no teeth, wild chickens pick up small rocks, pieces of bone and shells and store them in their *gizzard,* a muscular pouch that's part of the digestive system, to help break down and digest food particles. Similarly, chickens in captivity need access to non-soluble grit, especially if they're free-ranging or eating any kind of home-made or fibrous diet, such as whole grains, grass and other vegetation. If you're feeding a whole-grain diet or pasture to laying hens, you may also want to offer them a dish of crushed oyster shells or a soluble calcium and mineral mix designed for hens. You can buy both at feed stores. They supply the calcium and minerals that high-producing hens need.

If you're feeding only a commercial mash, crumble or pellet, your chickens don't need additional grit to grind their food because they can quite easily digest these feeds. However, some chicken owners feel the birds are happier when they have grit and oyster shells because one bird's dietary needs are different from the next, and what works for one chicken on a commercial diet may leave another lacking in a particular nutrient. Offering grit and oyster shells to all your birds does no harm, as long as they're free to eat it at will. They don't fill up on it.

Birds that range freely part of the day pick up enough grit in certain parts of the country, but provide it for them anyway. Any kind of small stones ranging from the size of sand to 5 millimetres (a quarter of an inch) are greedily pecked up by your birds. They know when they need it and don't overeat.

You can purchase grit composed of crushed limestone, granite and even pieces of flint in feed stores, with different sizes for chicks and adult birds. If you have just a few chickens, you can purchase canary or parakeet grit in pet stores. It's finely ground, but is fine for chicks or, in a pinch, for older birds.

You can supply grit from about the fifth day of life. Chicks should be eating their normal feed well before you add grit. The chickens are quite happy if you simply place a pile on the floor of the house and if it's fine enough it may well double up as a dustbath too.

Make sure that the grit you're purchasing is for birds. Feed stores sell a coarse salt, mineral mix for large livestock that people mistake for grit. This mix can cause serious harm to your chickens.

Deciding When to Put Out Feed

Most people fill their chickens' feed dishes so that food is available much of the day, or they use feeders that hold several days' worth of feed. You can use these feeding methods for all types of chickens; they eat this way – small amounts frequently – in nature.

Other people still feed their chickens at certain times of the day, generally morning and evening. That way they can control the amount of feed that may attract pests and, if the chickens are too heavy, restrict the amount they can eat. Usually however, you can go with your personal preference; some people like to observe and tend to their chickens more often than others. This method works well for all but meat birds.

Free-range chickens may be more inclined to come to the hen house to lay if you feed them there in the early morning. And if you need to lock the chickens up every night to protect them from predators, feeding them inside in the evening can be a handy trick to entice them in.

Because of their excessively fast rate of growth, the heavy meat-type broiler chickens need to have food available to them at all times, day and night. Chickens don't eat in the dark, and so the lights need to be on for these birds all night. In commercial situations, the very fast-growing broilers have a regulated artificial light regime to ensure maximum eating time. Most home chicken-keepers find that difficult to regulate and prefer to adopt a more natural way to keep their birds, and so choose a slower-growing meat bird strain and make sure that they have feed ad lib. That way you don't need any fancy lighting schedule. Let the birds clear up any spilt food regularly before topping up again, otherwise it attracts vermin and can be an expensive way to feed chickens. Laying hens, pets and show birds are fine with restricted times of feeding and don't require feed at night.

Free-range birds are going to be subjected to natural daylight and darkness, which is one reason they grow a little slower. When days are long, this situation may not make a big difference, but when you're raising meat chickens on pasture in early spring and late autumn, it can have a significant effect.

Be very careful not to feed mouldy food, which can kill or harm your chickens, and ensure that food is stored so that it doesn't attract rats, mice and other pests (check out Chapter 7 for more on food storage).

Determining How Much to Feed

Telling someone how much feed their chickens are going to eat per day or week is difficult simply because it involves so many variables. The type of chickens, whether they're growing or laying, and how active they are, affects the amount of feed each bird needs. How tidy you are, the type of feeders you have and the number of scavenging pests you support also change the amount of feed you require. The weather is also a factor: chickens eat more in cold weather and less in hot weather.

Modern, high-production egg breeds convert feed to eggs very efficiently, especially if you feed them a ration formulated for laying hens. After they're laying well, they need about 2 kilos (4.4 pounds) of a quality feed of 16 to 18 per cent protein to produce a dozen eggs. The breeds kept for dual purposes (eggs and meat) generally have heavier body masses to support and need more feed to produce a dozen eggs than a lighter production breed.

About 1 kilo (2.2 pounds) of feed is necessary to produce 0.5 kilo (1.1 pounds) of body weight on a growing meat-type bird. So if a broiler weighs about 3 kilos (6.6 pounds) at 10 weeks, it has eaten about 6 kilos (13.2 pounds) of feed. Remember that it ate less when it was small, and the amount of feed consumed increased each week. A medium-weight laying hen eats about 115 grams (4 ounces) of feed per day when she begins producing. These amounts are rough estimates, but they give you an idea of what to expect.

When you pick up your birds, ideally they feel well fleshed but without rolls of fat. If the breastbone feels very sharp and prominent, they're probably too thin. Chickens that don't get enough feed stop laying. The thinner birds in the flock and the ones that vitamin and mineral deficiencies hit first are the ones lowest in the pecking order, because dominant birds tend to eat first and eat the best of what's available.

If you're using a lot more feed than you think is correct, pests such as rats may be eating it overnight. You may want to empty feeders at night, hang them out of reach or put them inside a pest-proof container for all birds other than the broiler-type meat birds.

Keeping the Diet Interesting by Offering Treats

Chickens' diets need to be well balanced, but an occasional treat can be good for the birds. Treats can help relieve boredom in confined chickens, including those that are being kept inside because of bad weather. Treats may deter chickens from pecking at each other or eating things they shouldn't, such as the paint off the walls.

If you do feed treats to your flock, try to keep them nutritious and not more than 1 to 2 per cent of the overall diet. That's a small amount, usually less than one cup per bird per week, and you need to divide it up over several days. Make sure that you only feed as many treats as the chickens can eat in a small amount of time; treats left out attract pests or smell.

Chickens don't care much for sweet foods, and you should avoid foods that consist primarily of sugar and fat. Also ensure that none of the treats are mouldy, because mouldy food can cause a wide range of problems in animals. Too much of certain foods, such as cabbage, onions, garlic and flaxseed, may cause your chickens' eggs – and even the meat – to have an 'off' taste if you feed these foods for long periods of time.

Remember to feed treats in small quantities and clean up any the chickens don't eat straight away. The following treats are good and safe for chickens:

- ✔ **Dark, leafy greens.** Hanging a cabbage up above head height for chickens to jump and peck at is an old tried and tested trick, providing food, amusement and exercise all at the same time.

- ✔ **Eggs and eggshells.** Before the advent of chick crumb, people would commonly hard boil an egg and mash it up as chick food. Eggshells baked and crushed were returned to laying birds to supplement their calcium needs. These practices are considered old fashioned now. Feeding eggshells to layers, no matter how well disguised, can encourage egg eating, a nasty habit that's difficult to break and miserable to experience. Modern feeds and oyster shell are a better solution. Put the eggshells on the compost heap.

- ✔ **Fruits.** Apples, pears and other fruit raked up off the ground provide excellent treats, especially if wormy, unless they've been sprayed with pesticides. Most fruits can be fed to chickens, although they probably refuse to eat citrus fruit. Fruit can be soft or damaged but not mouldy.

- ✔ **Home-grown grains.** Growing your own dedicated chicken feed is perfectly possible. Just scatter your chickens' whole grain in a patch of the garden and harvest it by stringing up bundles; the birds do the rest. Sweetcorn, maize and sunflower heads can contribute to chickens' diets too. Just remember the golden rule: don't take them into the kitchen.

- **Other green, orange and red vegetables.** As long as they come straight from the garden, allotment or greengrocers, and not via the kitchen, these treats are fine to feed to your chickens.

- **Potatoes and potato peelings.** Don't feed raw potatoes or peelings to chickens. The sprouts and green areas of skin can be poisonous.

- **Pumpkins and squashes.** The 'guts' from a hollowed out pumpkin are quite popular with chickens. You can even feed the rind after Halloween if it isn't mouldy. Chickens also adore those monstrous marrows and gone-to-seed cucumbers that no one else wants.

- **Weeds from the lawn and garden.** Most weeds are quite nutritious – just make sure that they haven't been sprayed with pesticides. Every area has weeds that are poisonous, and so consult a book or authority before feeding your birds unfamiliar weeds. Never feed yew (a soft-needled evergreen common in churchyards) trimmings to any animal, and don't include any mushrooms or fungi in your offerings. Dandelions, goosegrass chickweed and thistles are all safe. A little cut grass is okay, but don't overdo it with this snack.

- **Miscellaneous.** Cooked nuts are fine, as are raw crushed acorns and walnuts. Wild bird seed and sunflower seed are okay, and leaving the hulls on is fine. A little dry pet food or a few pet treats occasionally are okay, but don't feed too often or too much. Rabbit pellets can be an occasional treat as well.

Keeping the Water Flowing

Having a source of clean water is vitally important to your chickens. People often don't realise how important water is to their chickens until they go from pouring water in a dish once a day to a system that allows birds to always have fresh water available. The birds with unrestricted access to clean water grow better, are healthier and lay more eggs.

In moderate weather a hen may drink a pint of water a day. In hot weather, that amount nearly doubles. Broilers may drink even more because their metabolism works much harder, producing more heat and using more water. Birds roaming freely may drink more or less than confined birds, depending on the moisture content of the food they consume and how active they are.

Drinking can be restricted because water isn't available or because the water available is unappealing. Chickens don't like water that's too warm; in very hot weather, providing an unlimited quantity of cool, clean water may mean the difference between life and death for your birds. Move water containers away from brooder lamps and out of sunny areas. You may want to change water or flush the pipes of automatic systems more frequently so that the water is cooler. If you keep only a few hens, have a small drinker that you fill frequently with fresh water rather than having a large drinker containing possibly stagnant water that the birds are reluctant to drink. (For more information on water containers and systems, see Chapter 7.)

Chickens also drink less if the water has an 'off' taste from medications. Make sure that any medications are truly needed in hot weather and avoid all those fancy additives; doing ensures that your chickens drink enough to avoid heat stress. Chickens also avoid dirty water full of algae, litter, dirt and droppings, and so scrub out those water containers. Even if you see your chickens drinking from muddy puddles, provide fresh water for them too – they use more than one source.

In winter, if temperatures are below freezing, offer water at least twice a day in sufficient quantity that all birds can drink until they're full. A good tip is to empty water containers at night to avoid them freezing solid; that way you don't need to deal with iced-up drinkers that you left ready to fill for the morning drink.

Any time birds stop eating as much as usual, check the water supply. Nipple systems need to be manually checked frequently to make sure that they're working. We've heard horror stories of nipple valve systems becoming clogged by mineral deposits or other things and failing to work. These devices freeze up easily in cold weather, too. If a chicken caretaker fails to notice that the chickens can't get water, the birds die of thirst.

Chapter 9

Controlling Pests and Predators

• •

• •

*Y*ou may adore your chickens so much and think them so cute that the idea of anything wanting to harm them seems unlikely. Unfortunately, though, chickens are attractive for different reasons to a good many baddies – enemies that want to kill and eat your chickens or steal their feed or eggs. Most chicken owners have to deal with pests or predators at some point in their chicken-rearing lives, and it can be extremely frustrating and, at times, heartbreaking. A predator in the hen house not only ruins your day but also ends your chickens' days, and so understanding how to deal with the predators that patrol your area is important. Your best defence in keeping your chickens safe from harm is to be prepared and informed, which is where this chapter comes in.

Keeping Pests at Bay

Pests are creatures that don't directly kill and eat chickens or feed off them like parasites (see Chapter 11); instead, they eat chicken feed or eggs. Too many scary disruptions and too many unwelcome snouts in the feed can be the start of major problems.

Pests cost you money and may make your neighbours unappreciative (to say the least!) of your chicken-keeping. Mice and rats, for example, can be found anywhere that humans and domestic animals reside. When only a few are hanging around, you may not even know that they're there. However, they can multiply rapidly to become a serious problem, seemingly overnight. The same goes for most other pests.

Preventing pests

Preventing pests is always better than dealing with an established population. Be diligent, clean and tidy, and pest problems may never bother you or your birds. The following list shows you several things you can do to prevent pest problems:

- **Store feed properly.** You can go a long way in preventing pest problems by storing feed products correctly – in insect- and mouse-proof containers. Insects destroy the nutritional value of feed, and mice can munch through the bottom of a feed bag in seconds to eat and soil the feed inside. After people discover how to store feed correctly and keep these scavengers away from it, they're often amazed at how little their chickens actually eat. (Turn to Chapter 7 to find out about all things feed-related, including information on dishes and storage.)

- **Keep feeding areas clean and dry.** Clean up any spilled feed immediately. Keep things dry, because wet areas are most likely to support maggot growth.

- **Deny rodents a place to hide.** Pick all rubbish up and remove piles of junk that can shelter rats and mice. Keep grass and weeds trimmed around buildings. Rats are more likely than mice to come from a neighbouring building or garden and make regular 'rat runs' between feeding at your place and sleeping at theirs, and so you may have to enlist the help of your neighbour to control these pests.

- **Cover what you can.** Cover feeding stations and water containers outside to prevent wild bird droppings from getting into them and spreading disease. You may need to cover your free-range pens with fine bird netting (like the netting for protecting fruit from birds) if the problem is bad.

- **Prevent wild birds from nesting inside buildings where you keep chickens.** This advice applies to all types of birds, from sparrows to swallows. Besides disease, wild birds can bring parasites such as lice to your chickens. The best ways to handle these problems is to exclude the birds from the building and remove any nests you find promptly.

Identifying and eliminating common culprits

The most common pests are insects and rodents, which occur in both rural and urban areas. Your best defences against them are to keep conditions clean, store feed properly and have an action plan to deploy when you start seeing signs of pests.

In the following sections, we discuss common pests and ways to control them.

Mice

Believe it or not, chickens eat mice, and so mice seldom set up house in the chickens' home. Instead they work on the fringe, getting into stored feed, chewing up building insulation and wires and running under your feet when you least expect them. Mice can't directly harm you or the chickens, and they don't eat eggs, but they can eat and soil a lot of feed because they're incontinent and spread their urine as they go. They also cause allergy problems for some people and spread certain diseases; and so you need to try and get rid of them, or at least control their numbers.

Mice have small territories. They build nests close to a food source, don't travel far and are likely to move indoors to a feed shed in winter. You may have a mouse problem if you notice:

✔ Small, dark, oval droppings.

✔ Little round ball nests and nibbled paper or polystyrene debris in concealed places.

✔ Tiny holes in walls, floors and feed bags (mice have a skull that can collapse, which allows them to squeeze through openings as small as 1 centimetre (0.5 inch) wide).

✔ Shallow surface tunnels in loose soil, litter or snow.

Mice and rats rarely exist in large numbers together, and so the one good thing you can say about a mouse problem is that you probably don't have rats.

Using traps is a good way to control mice when the population is high, or when you don't want to use poisons. Because mice are curious about things in their environment, trapping them is rather easy. You can buy numerous styles of mousetraps, some of which work better than others. The best ones are those that you can squeeze to open and don't require you to touch the dead mouse inside. To entice a mouse into a trap, a dollop of peanut butter usually does the job – mice only go for cheese in cartoons!

Emptying traps and disposing of dead mice afterwards can be unpleasant, but if you're kind-hearted and choose a trap that catches mice alive, you then have the problem of doing something with them. Whatever you do, don't just take mice outside the door and set them free; they'll be back inside before you know it. You can kill them, but why set a live trap to do that? You can feed them to the cats or chickens, but that's not such a wise or humane option either. Your best bet is to take them far away to the woods or a field, far from other homes, and set them free again.

An alternative to trapping is to use poison bait. You can control mice fairly easily with poison bait, but you must be sure to place the bait where children, pets and curious chickens can't get to it. You can buy bait stations that safely hold bait from hardware shops and farm supply stores; you need to remember to replace the bait as soon as it's eaten.

Don't let chickens or pets eat mice killed by poison, because most poisons remain potent in the dead mice and can harm the chicken or pet.

If you don't have a cat yourself, try inviting a neighbour's cat into your feed store every now and then. The cat's behaviour tells you whether you have a hidden problem – even the smell of a cat can be enough to put mice off taking up residence with you.

Rats

Rats are more secretive, larger and meaner than mice – chickens and even most cats leave rats alone. They can eat eggs and baby chicks and even feed on larger birds as they sit on perches or nests in the dark. The chance of rats eating live chickens is low, but it generally happens when other food is scarce, the shelter is unlit at night and the chickens don't roost off the floor. Chickens go into a kind of stupor in the dark and don't defend themselves very well. Rats come up under them where they sit on a nest or the floor and begin eating them alive. Just having a small nightlight on after dark allows the chickens to move around and defend themselves.

Rats take and eat a lot of feed and destroy more by soiling it. They are, however, very sophisticated and carry feed back to a store near their nests. They can move a surprising amount of food to this store over time, and return again and again until the food source is gone. Rats also do considerable damage by chewing on the structure of your chicken house and things such as wiring and plastic pipes. Rodent damage to wires is a common cause of fires.

Rats travel farther than mice to get food, and their nests may be outside the chicken coop. You may have rats if you notice the following:

- ✔ **Narrow, worn paths on the surface of the ground or shallow tunnels through loose earth.** These routes are known as *rat runs*.

- ✔ **Large holes chewed through heavy wood, plastic or even cement.** Rat holes are larger than mouse holes and often have greasy, dirty smears and tiny paw prints around them.

- ✔ **Droppings that are much larger than mouse droppings, but shaped similarly.**

Unlike mice, rats need liquid water, and you sometimes find them drowned in water buckets with steep sides – a sure sign they're visiting. They can swim pretty well, however, and are adept at climbing too.

You can eliminate rats by using traps and poison. Rats are very suspicious of new things, however, and so it may take days to get them to eat poison bait or fall victim to a trap. If you decide to set a rat trap, you need to place it close to pathways, tunnels or 'rat runs' that rats habitually use.

Traps, however, aren't as effective at controlling rat populations as poison. If you decide to go for poison bait, buy a poison specifically for rats and change the type of poison you use from time to time so resistant populations don't build up. Remember to use bait stations to protect pets, children and chickens, or place the bait where they can't find it. Place chunks of bait in tunnels if the chickens don't have access to the area, otherwise they may scratch them up.

Don't throw any dead rats killed by poison into open fields or woods. They may be eaten by birds of prey or other animals, and the poison left in the rats can kill them. Bury them instead or wrap them tightly and dispose of them in the rubbish.

Weevils, grain moths, meal worms and other insects

Weevils, grain moths, meal worms and other insects attack stored feed. They may provide a little extra protein for the chickens, but their feeding strips away the nutritional value of the feed, leaving only husks behind. Grain mills and feed stores attempt to control these grain pests with pesticides (organic feed mills use non-chemical pesticides), but some level of infestation is almost always present.

Your feed has been attacked by one of the many grain pests if you notice:

- ✔ Fine webbing in your feed
- ✔ Tiny worms in your feed

These pests are more likely to attack whole grains than processed feed, but you can find them in both.

To minimise the risk of getting these pests, buy only the amount of feed that you can use in about two months and keep it in tightly closed containers, not open bags. Even with these precautions, insect eggs may be present in the feed when you buy it and so a problem can develop from there. One option for dealing with it is to try sticky traps that lure grain insects with phero-mones (you can buy these at garden centres). Read all the label directions carefully and follow them.

If you have a way to freeze feed for a few days, doing so effectively kills most grain pests. In the winter, just leave the feed in your car for a few days when the temperature is below freezing. If, after that, you still have insects in the feed, use that feed up promptly if it isn't too badly infested, or discard it if it is. Then clean the feed container thoroughly with hot, soapy water and allow it to dry in the sun. (Use metal and plastic containers rather than wood, because they don't absorb moisture. Also, insects and other pests can't chew through them as easily.) Be sure to clean up any spilled feed, and empty and clean the feed dishes.

Flies are another insect that may become a problem. Chickens love maggots, which are baby flies, and if chickens can reach them few maggots make it to adulthood. If the maggots are outside the chickens' reach, however, such as in a manure pile outside the pen, large quantities of flies may be produced. If you have a manure pile, compost it in such a manner as to keep it hot and cooking. If you have a passive compost pile that just sits there, keep it far away from your house and the neighbours.

A good way to catch flies in a chicken house is to use sticky paper, but hang it where the chickens can't grab it and where your hair doesn't get stuck to it. Don't use pesticide sprays unless the labels say they're safe for use in poultry housing, and if you do, follow the directions exactly.

Never treat feed with pesticide sprays or powders intended for garden or home use. Some of them are highly toxic to birds. Similarly, if you have a problem with lice or mites, don't spray chicken housing with creosote or use other old-fashioned remedies. Smallholders and farm supply shops sell modern disinfectants and anti-pest remedies that are safer and more effective.

Nuisance birds

You have good reason to keep wild birds away from your chickens if you can. Wild birds are carriers of many diseases, including some (such as avian flu) that are harmful to human health. In addition, hawks, crows, owls and seagulls all prey on chicks if given the chance, and even on small adult birds in some cases. Chickens have an instinctive wariness of aerial predators and run for cover if the skies darken. Helicopters and aeroplanes can also send them dashing for shelter!

Wild birds also can make your feed bill jump in a hurry, and so feed your chickens inside a building if possible. Cover windows to inside shelters with screens, close doors and plug holes in eaves and under rafters. Most wild birds don't enter buildings from openings close to the ground (such as most chicken pop holes), so that's seldom a concern.

In pasture situations, keeping wild birds away is difficult. They're less likely to be a problem if you use concentrated feeds rather than whole grains, but occasionally they become a problem even with this kind of feed. Some times of year are more problematic than others. In spring, when pest birds have young, they're more daring about stealing both chicken food and small birds – beware bantams and young chicks – as a food source. Watch out for nesting crows in your area and remove nests if you can, before the problem begins.

It's illegal – to poison pest birds such as hawks and owls. You can sometimes buy live traps, but then you have to dispose of the birds you catch in them. The best defence against nuisance birds is simply to exclude birds and chase them away.

Ultrasonic devices said to scare off birds without you hearing them are useless. If they worked, they'd seriously offend your chickens. But they don't work, so don't waste your money. Fake owls or hawks may work temporarily, but you have to move them around regularly to fool the wild birds. One option is bird scarer ropes – you can buy them from smallholders and farm suppliers; farmers use them to scare birds away from fields of growing corn, but you need to use them sparingly and safely for them to be of use. Birds are quick to discover that something isn't real, and any benefit is then over.

Fending Off Predators

Man isn't the only species that likes eating chicken. A predator in the hen house can be a huge frustration – particularly for the chickens it kills. *Predators* are animals that eat other animals or kill them just for fun. Chickens are high on the preferred food list for many predators. Some predators destroy all the birds they can when they gain access to a hen house; others take one every so often when they're hungry.

Some cockerels can be aggressive, but most chickens do little but squawk and run to defend themselves, and so knowing how to deal with predators yourself is important. Case in point: one loose dog can completely destroy a prized flock of chickens in just a few minutes. And though chickens don't seem to be bothered much by cats and farm animals such as goats, sheep and horses, that doesn't mean letting them hang around your chickens is always safe. Pigs, especially, catch and eat chickens.

Predators are more likely to begin attacking poultry when the birds hatch their young in the late spring or early summer or in winter when food is scarce. When predators find a good source of food, they often return. Remember that predators exist in both urban and rural areas. In fact, some urban areas may have higher numbers of predators than rural areas.

The following sections help you keep your flock safe from harm and introduce you to the most common predators that chicken-keepers need to beware of. In case you do find yourself with a predator on your hands, we also tell you how to work out which one you're facing and how to deter it.

Providing safe surroundings

A little planning when building chicken housing goes a long way towards keeping your chickens healthy and safe. You may be surprised at what's after your chickens, even in an urban area. Of course, free-range chickens are at the greatest risk of encountering a predator, but predators can be remarkably resourceful about breaking into chicken housing. Build the strongest, safest housing and pens you can afford. (Chapter 6 tells you all about constructing your chickens' home.)

The following tips can help you further protect your chickens from predators:

✔ **Be careful about letting your chickens roam.** Chickens that range freely may disappear without a trace. In areas with a heavy predator presence it may be impractical to let chickens roam completely freely. Keeping chickens penned until later in the morning and bringing them in early in the evening can help.

✔ **Prevent night-time attacks.** You can close up good, predator-proof shelters at night. Leaving a small light source in the hen house at night further ensures the safety of your chickens.

✔ **Cover the coop windows.** Cover any open shelter windows with strong wire.

✔ **Keep an eye out for dogs.** Keep your own dogs from chasing chickens, even in play. You can train your own dogs to leave your chickens alone, but don't trust any dogs you're not familiar with.

✔ **Fence your chickens in.** Pens of sturdy wire are the best protection your chickens can have. To keep out larger predators, we advise you to lay 30 centimetres (1 foot) of heavy wire bent outwards at the bottom of the run fencing and buried or weighted down with large rocks.

Electric fencing (which we discuss in Chapter 6) is effective at keeping foxes, dogs and badgers out of chicken pens. A single strand of electric wire near the top of outside runs and near the bottom of any flexible fencing deters them.

✔ **Keep clear of trees.** Make sure that predators can't use nearby trees as a way into the pen.

✔ **Beware of aerial attacks.** If wild birds are taking your chickens out of fenced runs, you need to cover the runs with nylon netting or fencing.

Recognising common chicken predators

The most common predator of chickens is the dog; this animal may be man's best friend, but it certainly isn't a chicken's. However, many other predators occur in both urban and rural areas as well. We discuss the most common ones in the following sections.

Luckily, some animals you just don't need to worry about. Domestic cats and even semi-feral cats seldom attack adult chickens, and they don't eat unbroken eggs. Some areas are believed to have a problem with big cats – the 'Beast of Bodmin Moor', for example – but unless you live in a remote place these creatures are very rare, and so don't suspect big cats unless you actually see them.

Domestic dogs

Where you find humans, you always find dogs. Even the tamest dog may enjoy chasing chickens, and even if it doesn't intend to kill them, it may chase them into harm's way, cause them to pile on each other in pens or just run them to death.

When domestic dogs get into a chicken pen or find chickens ranging freely, they generally kill all they can catch. They kill birds in a variety of ways but don't eat them. Dogs usually leave behind a mess, with blood and feathers everywhere. Birds left alive may have deep puncture wounds or large pieces of skin pulled off and most likely you need to destroy them. Dogs that kill frequently may get more 'efficient' and cause less mess. When dogs have killed chickens – and even small dogs can be deadly chicken-killers – they generally kill again and so you must always control them. When a dog gets the taste of freedom and hunting, it wants to find other dogs and form a pack, which can be really dangerous for your chickens. This behaviour is natural for dogs whose owners let them go out of control.

If you know the owner of the dogs (and it isn't you!), report the situation to the police and tell them where the owner lives. The police officer handles the process or explains what to do. In most places dog owners are responsible for paying for the damage their dogs cause.

On the other hand, a well-trained dog can be your first and best defence against mice, rats and bigger predators. Their regular presence and scent marking can be so off-putting to foxes that they may never bother you, which would be a blessing. If you have a good dog, take it with you every time you visit your chickens. Some terrier breeds, although killers by nature, can happily run among chickens without paying any attention to them while on the trail of a rat.

Foxes

Foxes live in both urban and rural areas, and plenty of them are around. They have one or two litters of cubs each year and are always on the lookout for food to feed their growing families. These cunning animals soon get to know your routine, and visit your hen house every night just waiting for the first time you forget to shut the chicken house doors or are a bit late in doing so.

You can tell when you have foxes in the vicinity by the clues they leave. The eerie human-like scream you sometimes hear in the night is a *vixen* (a female fox) calling for a mate to make even more foxes! In winter you may see their tracks in the snow – dog-like paw marks, often with another mark where their long bushy tails (called a *brush*) dragged behind. You may also notice the very distinctive pungent and musty scent they leave when marking their territory. Their dung is similar to a dog's but in the autumn it may contain berries and have a purple hue. Although they're nocturnal animals, don't be surprised to see foxes in broad daylight.

Foxes can dig under fences, squeeze through small gaps, rip at chicken netting with their teeth and chew their way through any flimsy structure that stands in between them and your chickens. They use ploys such as frightening the hens into a frenzy, flapping against the wire and then biting at them. If they can drag a head or a leg through a gap in your housing or through large wire mesh they chew it off.

When they gain access to a house or chicken run they kill all the birds. Given enough time without being disturbed they eventually take all the dead birds away and make a store. Usually, though, the flock owner returns before the fox has time to do this. Discovering the carnage is very upsetting for chicken-keepers, and has given foxes the reputation of being wanton killers. If you're unfortunate enough to have this happen, remember to check for survivors. Chickens instinctively run and hide in these panic situations and some birds may have escaped. If you find any escapees, they'll be in shock and need careful treatment. (Chapter 11 looks at how to cope if this situation arises.)

To keep foxes at bay and your chickens safe from harm, use strong wire netting buried 30 centimetres (1 foot) into the ground around a chicken pen, or bent to form a skirt, and weighed or pegged down. This stops foxes from digging their way in. Additionally, build roofs of wire on runs or install electric wires at the top and bottom of runs. Well-maintained flexible electric netting keeps all but the most persistent, or high jumping, of foxes out. Take care not to leave wheelbarrows or any other equipment next to the fences – foxes are smart, and use these as stepping-stones into a run. They aren't called 'crafty' for nothing.

Keeping foxes at bay is a full time occupation, especially for free-range flock owners. Foxes are territorial creatures, and so you need to play them at their own game of scent marking. A good deterrent is for you or your dogs to patrol your area frequently. The more often you can visit your chickens, the less likely you are to get problems.

If the problem is severe, shooting foxes is allowed because they're considered to be vermin. A rifle, or a shotgun fired at close range with the correct weight lead shot, is effective. Don't ever attempt this action yourself –

especially not with an air gun, which doesn't do enough damage to kill a fox outright. Wounding a wild animal is not just inhumane; they become more dangerous to your chickens if you compromise their hunting skills – your chickens then become their target meal. A professional pest control officer, however, has all the correct gun licences and is trained to use guns with care.

Badgers

Badgers live in sets on the edges of woodlands and in the countryside. Although not so common in towns and cities, badgers do travel around visiting other sets so some may pass your way. Fortunately badger raids are pretty rare and they only attack chickens when their normal food source is too difficult to find. Badgers are usually happy to dig for their main foods, which are worms and insects, but after a really dry spell, when the ground is hard and the worms have retreated deeper underground, the badgers' hunger may drive them to look for other sources of food – potentially, your chickens.

When you have badgers nearby, you don't see their droppings – they bury them in latrines – but patches of earth may be scooped out where they've been looking for worms. In the winter you can spot their tracks in the snow by the claw marks.

Badgers are big and very strong animals with long claws and a reputation for being stubborn. When in the mood to do so, they rip open a sturdy chicken coop, kill the chickens they find inside and frighten the others into a panic, causing them to injure themselves or die of a heart attack. Make sure that your housing is as strong as it can be if you know you have badgers in your area. Better still, keep badgers away from the housing with an electric fence to deny them the use of their strength against your chickens.

Hawks, crows, seagulls and owls

Chickens seem to have an instinctive fear of large birds overhead. When they spot one, they squawk loudly and distinctly and run for cover or flatten themselves on the ground. Hawks, crows (including rooks and magpies) and seagulls are a threat and are active only in the daytime. Owls, on the other hand, generally hunt at night but when times are lean, as in the winter or when they have young to feed, some owls also hunt in daylight.

Hawks are efficient killing machines that live pretty much throughout the British Isles. Hawks are most likely to prey on chickens in rural areas, but it takes quite a large hawk (such as a buzzard) to prey on chickens. Most small hawks (such as smaller kestrels and sparrow hawks) don't bother adult chickens, but may pick off baby chicks. All tend to avoid taking birds out of smaller fenced enclosures because of the danger posed by anything they may get caught up in, but it sometimes happens.

Some hawks carry away their prey; others pick it apart where they catch it. If a hawk kills one of your chickens, you won't find it inside the shelter. Hawks pull the meat and feathers off the chicken in clumps, often starting at the breast and pulling out organs. If left alone, hawks can pick a carcass apart in a short time.

Crows and seagulls are also problematic. They smash and eat eggs and carry away small chicks or pick them clean where they've caught them. These birds often kill with a sharp stab of the beak to the back of the neck. Your best approach is to keep them away in the first place – as soon as one bird discovers that your chicken house has a food source, all its mates want to join the party.

Owls can be a problem, and have been known to fly through windows or down from lofts to pluck chickens off perches while they sleep. They usually take the chicken with them from a building, but outside they may sit on a post or on the ground to consume part of the bird. If your chickens roost outside, they're a prime meal for owls. However, like some hawks, not all owls are large enough to kill chickens.

Owls that get into a pen or house with lots of chickens may get a little kill-happy and kill several birds, usually by pulling off the heads. They may take a single bite out of each breast, or they may eat just one bird.

Hawks and owls are protected birds and so you can't kill or harm them. The best way to avoid these predators killing your chickens is to prevent them from getting to your birds. Screen chicken house windows and put some kind of roof over runs (check out Chapter 6).

Minks and weasels

Minks and weasels aren't as common as they used to be. They generally live in rural areas near some source of water. They're active day and night but are very secretive; human activity generally keeps them away during the day.

When minks or weasels enter a chicken coop, they're not looking for chicken feed; they eat only meat and eggs. The meat may come from rats, which are a favourite food. But if, in the pursuit of rats, they find chickens, they're quite happy. And if your chickens range freely by waterways, they're fair game.

Weasels typically kill one or two chickens by biting the heads, and then they primarily drink the blood, without eating a lot of flesh. Minks may kill many more chickens than they need, also by biting the heads and then piling them up. They may eat small amounts of flesh from one or more birds. Minks seldom eat eggs when chickens are available, but they sometimes eat eggs in nests.

The case of the crafty buzzard

Some friends of Pammy's owned a field in which a free-range flock grazed. Occasionally, a buzzard would fly down and settle in the field about 20 metres (64 feet) from the chickens. He sat quietly waiting but the hens, of course, were naturally curious and eventually just had to go and investigate this new addition to the flock. Straight away, 'pow!' – the buzzard jumped on one and killed and ate it there and then. This behaviour is unusual and particularly clever, but shows just how canny wild birds can be.

Minks often enter chicken coops through the opening a rat has made or by following a rat tunnel. They can squeeze through quite small holes, and so to prevent their entry, close all openings in housing that measure more than 2.5 centimetres (1 inch). Keep chickens out of brushy areas near water and maintain a clear area around pens and shelters. Noise, such as that from a radio, and bright lights are more likely to discourage minks and weasels than some other predators.

Working out who's causing trouble

Determining which predator killed or injured your chickens can sometimes be difficult. If a chicken simply disappears, it's probably impossible. In many cases, however, you can find some clues. When damage or killing has been done, examine the victims and chicken housing closely and consider the following questions (Table 9-1 can help you answer them):

- ✔ Is there an obvious way the predator entered, such as a large hole in the fence?

- ✔ Has more than one bird been killed?

- ✔ Were the birds eaten, or just killed?

- ✔ What part of the body did the animal feed on, and where are the wounds?

- ✔ Were eggs eaten?

- ✔ Are there animal tracks in snow or mud?

Consider the area you live in. Minks and weasels are seldom far from water. Badgers aren't often found in the middle of built-up areas but foxes and loose dogs are. Table 9-1 is a quick guide to help you identify which predator you're dealing with.

Table 9-1		Working Out Whodunit	
Predator	**What It Goes For**	**Hunting Hours**	**Signs**
Dog	Chasing and catching chickens	Day or night	Chickens killed but not eaten; survivors may have deep puncture wounds or large pieces of skin pulled off; scattered blood and feathers
Fox	Chickens	Early morning and late evening, day time if they're hungry; cold weather and times when they have cubs to feed	Chickens are free-ranging or pastured away from humans in a rural area; scattered blood and feathers; many birds dead or dying; other birds in shock or hiding nearby
Badger	Chickens and sometimes eggs; breaks into chicken housing	Especially after a spell of dry weather	Wrecked housing and many birds killed, some eaten
Hawk, owl, crow and seagull	Baby chicks; only large hawks and owls can kill adult chickens. Crow family also steal eggs	Hawks and seagulls during day; owls mostly nights, though some days during winter	Hawks: birds carried away or picked apart on the spot where killed; empty egg shells, often with one large hole Owls: birds plucked from roost and carried outside; in coop, heads may be pulled with a single bite taken out of each breast Crows: empty egg shells; chicks with heads removed or bones picked clean; seen returning again and again Seagulls: similar to crows
Mink or weasel	Chickens, eggs and rats	Day or night, but they shy away from human activity	Chickens kept in rural areas near water sources: heads bitten off; small amounts of flesh eaten; birds may be piled up

Even if you've seen certain animals in the vicinity of your chickens, don't assume they're necessarily the killers. And even though certain animals have been around a long time without bothering the chickens, don't assume they aren't the killers. Determine the killers based on the facts, not assumptions, so you can better protect your chickens.

Catching the troublemaker

Catching predators can vary in difficulty depending on the troublemaker.

If you catch a dog you're not familiar with in with your chickens and you can handle it, take it to a vet. The vet looks to see if the dog is micro-chipped and – if it is – contacts the owners, who are responsible for it. If the dog isn't marked and has been causing a nuisance, the vet will probably put it down.

In the UK, foxes are considered to be vermin. Pest control firms or professionals with game and gun licences can deal with them, but remember that as soon as you get rid of one fox, other foxes move into their territory, and so dealing with them is an ongoing job. You can also trap nuisance predators.

When we say trap, we mean *live* trap. Other types of traps can leave animals in pain for long periods of time (and you then have to kill them anyway) and can kill or maim animals you didn't intend to kill, such as your chickens or the neighbour's cat. But before you set a trap, be sure that you're willing to deal with whatever you catch. You must check live traps twice a day and deal with what you trap promptly.

Live trapping isn't as humane as many people think. We offer information on trapping in this section because many people feel as if it's the only option they have, but in reality, excluding predators with the proper fencing and shelters is preferable to trapping or killing them. (We provide some suggestions for excluding individual predators in 'Providing safe surroundings' earlier in this chapter.) You may have to avoid letting your chickens roam as freely as you'd like in areas where you know of many daytime predators. Shutting up your chickens in a secure shelter at night protects them from night predators.

In many cases you have to kill the animal you live trap (remember though that badgers, hawks and owls are protected species and killing them is unlawful). If you want to take the animal somewhere to release it, you need to take it miles from your home, or it comes straight back. But when you take it out of its territory and into some other animal's territory, it may be killed shortly after you release it anyway. Get permission before releasing any animal on property you don't own, including parks and nature reserves – don't leave your problem on someone else's doorstep.

If you don't want to shoot an animal you trap, ask local pest control officers if they can handle the disposal for you. Think ahead, and before you set the trap, decide what you're going to do.

You can purchase various kinds of live traps. They're all set in different ways, and so follow the directions that come with the trap. Practise setting and opening the trap to release an animal before using it. Live traps come in a variety of sizes, so make sure that the trap you pick says that it's suitable for the predator you want to catch.

For all four-legged predators of chickens, a can of cat food makes great bait. Bacon fat on bread also works. Place the trap outside of your chicken run so you don't catch chickens, but keep it close enough and disguise it with branches or grass. It may take a few days for the predator to enter the trap, so be patient.

Handling an angry animal in a trap can be quite an experience. If you do move an animal in a trap, wear heavy leather gloves to protect your hands and cover the trap with tarpaulins or blankets. This cover calms them and protects you. If you get bitten by the animal, you may need rabies prevention injections. You can sometimes hire licensed professionals to shoot or trap predators and remove them for you.

Chapter 10

Keeping Your Flock Happy and Healthy

Caring for animals means going beyond simply knowing how to house and feed them – it's also about knowing how to manage them so that they're healthy and content. The very best animal managers do this intuitively; they've observed the animals' behaviour so minutely that keeping a happy and healthy balance really is second nature to them. You can't expect to have this level of empathy with your animals to start with but this ideal is something to aim for. Essentially, if you maintain clean surroundings, feed a balanced diet, keep your birds occupied, inoculate when necessary, don't move birds in and out of the flock and reduce stress, you'll have few health and injury concerns.

In this chapter, we look at ways to keep your flock healthy with preventative care and how to handle your chickens humanely.

Taking Basic Precautions to Protect Your Flock

Maintaining good environmental conditions is very important in maintaining the health of your flock – poor conditions stress your birds' immune systems, which can lead to disease problems. Indeed, many more chickens die each year from poor environmental conditions than from disease. Chapters 5–7 look at designing housing to protect chickens from environmental problems such as dampness, heat and cold. This section deals with more 'personal' factors and relationships, including the way you interact with your chickens.

Dealing with heat, cold and dampness

The temperature and the amount of moisture in your chickens' environment have a significant impact on their wellbeing. Chickens need protection from extreme heat and cold, and you must keep their environment dry. The following sections address these concerns.

Handling the heat

Being subjected to heat stress for too long kills chickens. For heavy meat birds, who suffer greatly in the heat, death can occur within a few hours in heat above 32 degrees Celsius (90 degrees Fahrenheit), especially with high humidity. So, when the temperature rises above 30 degrees Celsius (86 degrees Fahrenheit), especially when the humidity is high, check on your chickens.

Make sure that your chickens have plenty of the following:

- ✔ Shade
- ✔ Ventilation
- ✔ Cool water

Chickens don't like to drink warm water, and so move their water dishes to the coolest areas, change the water frequently and make sure that water is always available and easy to get to.

Free-ranging chickens survive heat much better than penned birds because they instinctively find the coolest places. If you keep your birds in a pen, improve their conditions by giving them access to extra shade, or in extreme heat play a sprinkler on the shelter roof, which cools the house as water evaporates from it.

Chickens that are too hot become inactive and breathe with their mouths open (called *gaping*). If your birds appear heat-stressed, move slowly and calmly when you're among them, for example when you're feeding and watering them. Any more stress from fright may cause death.

Heavier meat bird breeds may not eat enough in hot weather and therefore not gain weight as quickly as cooler birds.

Fighting off the cold

Chickens handle cold better than heat, as long as their shelter is dry and out of the wind. But frostbite of the combs, wattles and toes may occur when temperatures get down near 0 degrees Celsius (32 degrees Fahrenheit) for any length of time, causing blackened areas that may eventually fall off. If you live in a cold weather area, try to pick breeds of chickens that have small combs and wattles because these lose less body heat and don't get frostbitten as easily.

Having wide perches is a good preventative measure because they allow chickens to sit with their feet flat instead of curling their toes around the perch. When chickens sit with their feet flat, their feathers cover the toes, which makes the toes less susceptible to frostbite.

In addition to causing frostbite, cold temperatures can cause hens to stop laying, or in hens that continue to lay, cold weather may contribute to egg binding (see Chapter 11 for more info on egg binding and other health problems).

To help your chickens' bodies to function properly in cold weather, you need to give them more feed. The type of feed is of little importance, just as long as they have enough of it and it's always available all the time.

Water is vitally important at all times – most chickens survive freezing conditions as long as their water supply is unrestricted. In cold weather, however, drinking water can freeze up, leaving your chickens thirsty; keeping the drinker inside the hen house may stop this from happening so quickly. Your chickens need to be able to drink freely at least twice daily. If they don't drink, they don't eat as much, and if they don't eat enough, they're more susceptible to dying from the cold. Emptying the drinker last thing at night when your chickens go to bed stops it from becoming frozen solid by morning and is a good habit to get into. You can then refill the drinker with fresh water first thing in the morning.

Doing away with dampness

Damp or wet environmental conditions are big contributors to illness in chickens. Damp conditions can come from moisture produced by birds' respiration and droppings in poorly ventilated housing, or from rain or snow. Poor ventilation and overcrowded houses often result in the formation of condensation, which causes wet, unhealthy conditions; the warmer the air becomes, the more moisture it holds.

Wet areas favour the growth of moulds and fungus, causing additional discomfort for your chickens in hot weather, and high humidity makes ideal conditions for many respiratory diseases. Moulds on feed and bedding cause a wide variety of illnesses, and they aren't good for the human caretaker either. To avoid this nastiness:

✔ Always keep your chickens' bedding dry.

✔ Muck out the house regularly.

✔ Don't give your birds mouldy feed.

✔ Make sure that the house is well ventilated.

✔ Drain outside runs well to prevent chickens from tracking in a lot of moisture on their feet.

Handling your birds safely

Home flock owners often wonder how their chickens became ill, especially if they don't take them to shows and sales. Well, brace yourself – studies done on commercial chickens show that 90 per cent of diseases are carried to chickens by their human caretakers. Disease organisms can survive on shoes, clothing, unwashed hands, car tyres and equipment, and so keep that sobering thought in mind when you next visit a poultry show or your friend's flock.

In addition to diseases that affect your chickens' health, the media reports some bird diseases as being harmful to humans; unfortunately, these news stories scare people into believing that keeping chickens is dangerous. We're here to assure you, though, that your risk of catching any disease from chickens is minimal.

Of course, you always need to follow safe, sensible methods of handling and caring for chickens; take heed of the following precautions:

✔ Wash your hands, at the very least, before caring for your own birds after you've been visiting other flocks. To go the extra mile, also change your shoes and clothing.

✔ Avoid going to places where poultry live if poultry disease breaks out in your area. If you do go, take a shower and put on clean clothes and shoes before you take care of your own flock.

✔ Wash your hands carefully after handling and caring for chickens.

✔ Keep chickens away from your face – no kissing or nuzzling.

✔ Wear disposable gloves when handling sick or dead chickens.

✔ Cook chicken meat and eggs thoroughly and store them properly. Always clean countertops and sinks where raw poultry has been. Wash dishes and utensils that come into contact with raw poultry with soap and hot water before using them for anything else.

✔ Wash vegetables and fruits thoroughly that may have become contaminated by chicken droppings if you let your chickens roam freely around your garden. You must also clean patio and picnic tables and railings that chickens have soiled before you use them.

✔ Quarantine sick chickens. See the later section 'Quarantining new birds or those that have been off your property' for more info.

✔ Be sure to supervise children when they're handling chickens, eggs or poultry equipment, and make them wash their hands immediately and thoroughly – not just splash them with water – after they're done. Toddlers have that irresistible urge to put everything they find into their mouths, and so keep a watchful eye on them when they're near your chickens or any chicken equipment.

Keeping your chickens safe from poisons

As with all livestock, the best approach with chickens is to try and anticipate problems before they occur and prevent them from happening. Chickens aren't particularly discriminating about what they eat and although you wouldn't purposely feed your chickens poisonous substances you may not realise the kinds of things that chickens view as 'food'. The following is a list of some commonly overlooked dangers:

✔ **Garden seeds treated with fungicides or other pesticides.** If you use these types of seeds, make sure that your chickens can't scratch them up and eat them or get to them in storage.

✔ **Granular fertilisers.** If a chicken comes across some widely scattered bits of fertiliser it may sample one or two and then move on. But if chickens see you scattering fertiliser they may come running and, in the spirit of competition, gobble up quite a bit before they stop or you chase them off. And if they come upon a bucket of fertiliser granules they can become excited and eat more than is good for them.

Keep all fertilisers and pesticides in places where chickens can't get to them and where children can't find them and feed them to your chickens. Lock up the chickens before spreading fertiliser, or make sure that they aren't nearby to watch you do it. Better still, try making compost from their droppings and use that instead – it works well and is harmless.

✔ **Paint.** Watch out if you're scraping paint off old houses or buildings and the flakes fall down into the soil. If the paint is lead-based and the chickens eat the pieces – whether purposefully or when eating other things – you can then have a problem. Lead poisoning is generally slow, and sometimes connecting the symptoms with the lead paint chips is difficult.

Speaking of paint, when the time comes to apply a new coat, don't go away and leave that can of paint open next to the chicken shelter.

✔ **Pesticides.** Keep chickens out of any areas to which you apply pesticides to kill weeds or bugs. Just getting some pesticides on their feet may be enough to harm your birds. Snail and slug baits often come in pellet form and are extremely toxic, as are the insects that have been killed by these pesticides. Unfortunately, chickens still view these as food.

Keep a part of your garden for wildlife; the creatures living there help you to control garden pests without resorting to pesticides. When you know that your garden is safe for your chickens you may find that they too can help with pest control.

✓ **Rodent poisons.** Mouse and rat poisons are deadly to chickens if they ingest them, as are the poisoned rodents. Be quick to remove any dead rodents – they may still have active toxins in them – and be careful when placing mouse and rat baits in the first place. Chickens are curious birds and they may reach their heads under things and in holes to sample what they see.

Free-ranging chickens generally have a good sense of what's poisonous to eat as far as berries and vegetation go. Penned-up chickens are more likely to eat things that aren't safe because they're hungrier for variety and fresh food.

Catching Chickens

To care for your chickens, you sometimes have to catch and hold them. This seemingly simple task isn't always as easy as it sounds. To help, this section tells you how to do it.

Getting hold of chickens at night

If you can, wait until night, walk in and take your chicken off the perch. If you normally leave a nightlight on in the shelter, use a blue bulb – chickens can't see in blue light but you can work well enough. Alternatively, wear a head torch turned down as dimly as possible. Try not to disturb the other birds as you go, especially if you need to catch more than one.

Catching chickens during the day

If you need to catch a bird and don't have time to wait until night, lure it into the smallest area possible (usually the shelter) and block off any exits with a sound piece of weld mesh. Then, staying calm, go in slowly, smoothly and deliberately and try to grasp the legs – never the neck – supporting the body at the same time (if you don't take the weight of the body immediately you may injure your chicken). Birds expect to be grabbed from overhead, and so going beneath them to get the legs surprises them. Get it right, and the bird responds without panicking.

You can also use a net or a catching stick, but in our experience nets spook birds and don't work much better than hands in close quarters. A *catching stick* has a hook at one end, which you push beyond the bird's feet and then slide back, hopefully snagging its feet and pulling the bird toward you. The stick acts like a shepherd crook, but be aware that you can break or dislocate chickens' legs with rough treatment. You may need a little practice to work out how to use one.

To catch a chick in a brooder, try putting one hand in front of the chick and use the other hand to sweep the chick into it. Don't pick up a little chick by its legs; scoop up the whole body. Putting your hand down over the back scares chicks because the motion resembles a predator swooping down, but doing so is sometimes the only way to catch them.

Retrieving chickens on the run

When a chicken flies out of a pen, new chicken owners often panic. Although chickens may take every opportunity to escape, they usually don't go far unless they're scared and in a strange place. If a chicken gets out of a pen at your home, don't be alarmed. Unless it's being chased by something it generally hangs around close to any remaining penned birds. Even if the whole flock gets out, don't panic. First, go and look at the pen and see how they got out and fix that before trying to get them back inside.

Chasing chickens should be your last resort – they're faster than you, and so try other options first. If you have a special bucket you use to feed them or bring them treats, get it out, add some feed or treats and show it to them. If they seem interested, lure them back to their pen, throw the treats inside an open door and shut it after they're inside. If they refuse to go into the pen you may be able to trick them into a garage, shed or fenced yard that you can close up after them.

Chickens want to go back to the familiar place where they roost at night when it starts getting dark. After the other chickens have settled down on the perches, open the door or gate and an escaped chicken often pops straight in. If you can't catch a chicken by nightfall, watch from a distance to see where it goes to roost for the night. When it's completely dark, you can take a flashlight and simply go to pick it up – if you can reach it. Shine the light on the bird for short periods only to pinpoint its position. Direct light stimulates and wakens the bird, which isn't in your interest.

If one of your chickens escapes in an unfamiliar place, such as a show, let it settle down a bit if you can. Don't chase it out of the area. If you travel with chickens to shows or sales, take a long-handled net or a catching stick with you, but try throwing down some choice grain such as corn to divert the chicken's attention before trying to use these implements. Or wait until it's attracted to someone else's caged birds, is crowing or is otherwise engaged before attempting to catch it. If conditions are crowded and catching the bird seems impossible, try coming back after the crowd leaves.

If you have a really hard-to-catch chicken running loose, luring it into a box or cage, with a string attached to the cage door to shut it from a remote distance, may work.

Fortunately, chickens seem to get easier to catch the older they get, especially if their caretakers have spent time taming them. Some breeds are naturally calmer and friendlier. Whichever breed you keep, taking some time to befriend your chickens is always worthwhile, in case you ever need to catch them quickly.

Carrying and Holding Chicks and Chickens

Sometimes you have to carry your chickens, perhaps when they find their way to somewhere they aren't supposed to be, if you're making a point of taming them or when they're ill or need a routine treatment. Carrying chickens is easy when you know how, but a few dos and don'ts apply.

Carrying birds by the legs with the heads hanging down is something to avoid at all times, even if you're taking them to slaughter. You may dislocate the legs or otherwise hurt them especially if the birds are heavy meat breeds. Instead, tuck the bird under one arm with your other hand holding both feet, or cradle the bird in your arms with the wings under an arm. A firm squeeze and soothing talk soon calms most birds. The head can face forward or back, whichever works best for you.

If you need to restrain a bird for treatment and don't have anyone to help you, you may need to tie the bird's legs together and lay it on a table, while restraining the wings with one hand or loosely wrapping the bird in a towel to restrain the wings. Some people hold the chicken with their knees while seated, with the feet up and the head facing away from them. If you gently rock the chicken forward, the feathers lift up which sets the bird at just the right angle for getting preparations under the feathers (Chapter 11 explains more about why this may be necessary).

Don't try to carry too many birds at a time – take your time to do it carefully and humanely – and don't squeeze birds too hard when carrying or restraining them for treatment. A chicken needs to move its ribs to breathe well and even if its mouth and nose are uncovered, you can suffocate a bird by holding it too tightly. This is often what happens when children hold baby chicks too tightly. A chick needs to be held loosely in your closed hand with its head peeking through your fingers.

If a bird is really wild and fighting, cover its head loosely with a hood or a piece of soft cloth and it probably settles down. Be especially aware of cockerels with long spurs on their legs. When they struggle, these spurs can scratch your arms quite badly, and so wear long sleeves when you handle them.

Preventing Illness and Disease

Disease can decimate a flock and so is something to strive against taking hold among your chickens. Looking after your chickens well is the first step, and you can prevent many chicken diseases through vaccination. Other diseases are caused by parasites, which you can also take steps to control. You can prevent yet more simply by keeping birds that may have been exposed to diseases apart from the rest of your flock for a time. In this section, we discuss these preventative measures.

Diseases can also spread to home flocks from wild birds and pests such as insects and rats. Chapter 9 provides info on controlling pests and predators to keep disease at bay and Chapter 11 gives details on disease symptoms and treatments.

Maintaining a healthy environment

The very best step you can take to prevent disease in your chickens is to maintain a healthy environment for them to live in. Simple things such as avoiding overcrowding, allowing plenty of fresh air and moving them onto fresh pasture regularly keep your chickens happy and healthy. Think ahead, though – chickens soon wreck small patches of ground and so don't pour all your resources into one fantastically fenced but tiny area. A flexible form of netting enables you to give your chickens more freedom and thereby to maintain the health of your flock better (go to Chapter 6 for more info about choosing your fencing options).

Giving vaccinations

Don't automatically assume that your chickens are inevitably going to contract diseases but, just as you have your other pets vaccinated, your chickens may benefit from some vaccinations too. Vaccines aren't available to prevent every disease that can affect chickens, though, and so you still need to practise good management techniques.

With vaccinations, the key is to assess the risk of disease yourself, decide how you want to proceed and act accordingly. Many home flock owners decide to do without vaccinating their chickens and live with the risk. Indeed, small-scale chicken-keeping has been going on for hundreds of years, well before vaccines were an everyday form of protection – the vaccines were developed in the first place more for the large intensively farmed flocks where disease is always a problem. If you keep your chickens just as pets or for their eggs, never show them and rarely buy or sell chickens, you're much less likely to have a problem with disease, and so vaccination is less of a priority. If, however, you intend to sell live chickens, breed them or show them, do get the basic vaccines to protect your fellow poultry-keepers as well as your flock. Some show breed societies may require you to have certain vaccinations as an entry requirement anyway.

When administering vaccinations, you need to be aware of their suitability for your birds. Some vaccinations can be given to chicks, and some need to be given to older birds. If you don't mind giving vaccinations yourself, you can usually buy the vaccines from poultry supply places. Follow the label directions exactly for administering and storing the vaccines. If you'd rather not tackle the task yourself, contact a veterinarian.

This area is a tricky one for people who just have a few chickens. Vaccines are sold in multiple doses – more than a home flock owner would ever need – and are designed for giving to larger flocks that number hundreds and thousands. The main thing to remember is not to mix unvaccinated and vaccinated chicks, because vaccinated birds can pass on a disease even if they aren't showing symptoms themselves. Chicks are especially at risk from this problem, whereas adult birds will have developed a certain amount of immunity, especially if they're free-range. If you've had no health problems in a small flock but then choose to increase your chicken numbers drastically or begin selling to other people, rethink your approach to vaccination and take advice from a vet, because as numbers increase so does the likelihood of chicken diseases.

The following is a list of common diseases that you can vaccinate chickens against:

- **Marek's disease** causes tumours and death in young birds. Almost all hatcheries offer to vaccinate chicks for Marek's for a small fee, and this is a good idea for small flock owners. If you hatch birds, you can vaccinate them yourself. The vaccination is given by needle just under the skin. After 16 weeks of age the birds don't need the vaccine.

- **Mycoplasma gallisepticum,** commonly called Air Sac disease, causes chronic respiratory problems or a more acute condition that results in many deaths. This bacterial infection is present everywhere but hardy, healthy birds are unlikely to succumb.

- ✔ **Infectious bronchitis** is a disease of the respiratory system that can cause widespread death in flocks. Chickens that have this disease can survive but their egg-laying ability may sometimes be compromised. Hatcheries often routinely give the vaccine against infectious bronchitis as a mist spray. Nearly all chickens bought at day old or point of lay have already been treated – check when you buy them. Vaccinate any expensive birds that you have and any birds you intend to use for breeding and showing. In small flocks of home-bred chickens vaccination is less important because your chickens are less likely to come into contact with other, contagious birds.

- ✔ **Gumboros disease** is an *immunosuppressant disease* – in other words, the immune system that fights off disease normally is weakened, allowing other diseases and infections to take hold. Vaccines are a sensible idea for the parent birds of large flocks of broilers, but a small flock doesn't need to be vaccinated.

Currently no vaccines are available for avian flu, but noises are being made in the large breeding companies about breeding a genetically modifying strain of chicken that can't contract it. This idea is a long way from being the norm and will doubtless cause plenty of controversy along the way.

Migrating birds and wild birds carry all sorts of potential diseases and many different strains of avian flu that can be transmitted to your chickens, and so a basic common sense approach to keeping them at bay is best. For example, keep your chickens' food and water out of sight; chase wild birds away and they soon get the message that they're not welcome. If you notice an unusual pattern of wild-bird deaths in your area, alert your local vet. If you're asked to bring a sample in, use protection while touching or carrying any carcasses and make sure that you change clothing and footwear before visiting your own chickens.

Chicken vaccines often come in large, multidose vials. Although even a 1,000-dose vial isn't too expensive for most chicken vaccines, the excess must be stored correctly and used before the expiration date – usually just a few days. If you do decide to vaccinate you may want to consider getting together with several other chicken owners to purchase the vaccines. That way you only have to pay a fraction of the cost and you don't have to worry about long-term storage.

Some diseases are more prevalent in one area than another. A local vet or a poultry expert from the Department of the Environment Food and Rural Affairs (DEFRA) can advise you. DEFRA's telephone helpline number is 08459 33 55 77.

Erecting barriers against parasites

Chickens get several types of worms, but the three most common types are large roundworms, hairworms and caecal worms. Worms are very common and, at some point, most chicken owners have to deal with them. Chickens contract worms from the faeces of other chickens or wild birds.

Unless you can move your flock every four to six weeks to clean land that hasn't been occupied by chickens, and without returning to the same pasture for one year, putting your birds on a regular worm control programme is advisable. Your vet or the instructions on worming preparations can guide you. As part of your worm control programme, when you get new birds that you want as breeders, layers, pets or show birds, worm all these birds if they're at least 18 weeks' old, or when they reach 18 weeks of age, and then at regular intervals.

The intervals at which you worm your chickens depend on many factors, such as numbers of birds in one place, frequency of rotation and the length of time you've kept birds at one premises. We recommend worming in the spring and autumn when your chickens have to reuse land within a year, unless worm problems seem to be severe or your vet recommends more frequent worming. Broiler chicks usually don't need to be wormed because they're ready for dispatch before 18 weeks.

Wormers come as a preparation that you add to your chickens' drinking water, or as pellets that you give them to eat. Some preparations are used for other animals such as dogs, cattle and sheep, and you 'spot' the wormer onto the skin. At present, these preparations aren't licensed or fully tested for chickens, although they're offered as a treatment with the get-out clause 'not licensed for chickens'. Until these preparations are licensed for chickens and the consequences in the environment are fully understood, leave them alone.

If, after a course of medication, a change in the way you manage your chickens or a reduction in the number of chickens on your pasture – or you have very few chickens and lots of space for them to roam in – you feel your worm burden is under control, you can move on to a preventative method of worming. You can do so by practising a strict rotation on your land – leaving your pasture free of chickens for one year before returning chickens to it the next.

Herbal wormers that you can buy 'off the shelf' at poultry supply shops give peace of mind as another preventative measure. Some have a very good reputation. Instructions come with the preparation, but as they have no drug withdrawal period the eggs remain safe to eat. Leaving a few cloves of garlic to infuse in the chicken's drinking water is another gentle *vermifuge* (or worm eradicator).

Coccidia – microscopic parasites that infect chickens' intestinal tracts and can cause death in young chicks – are also a problem. Almost all chicks have an infestation although some are affected more adversely than others. *Coccidiostats* (medication to prevent and/or cure coccidiosis) used to be automatically administered in medicated feeds, whether chicks showed symptoms or not (feed companies used to make medicated feeds as standard; see Chapter 8), but common practice nowadays is to use drugs that you put into the chickens' drinking water.

If coccidiosis becomes a problem in your flock (see Chapter 11 for how to diagnose it), you can obtain the coccidiosis drugs from your vet. Alternatively, adding a homoeopathic coccidiosis remedy to the water as a treatment or preventative measure is proven to be an effective and drug-free way to take your chicks past the danger zone. Homoeopathic remedies are advertised in smallholder and chicken magazines, and are available online. As chickens get older they may still carry the parasite, but most develop an immunity to its effects.

Quarantining new birds or those that have been off your property

Every time you add birds to your flock you increase the risk of introducing disease and parasites. One of the best ways to protect your flock is to quarantine all new birds or birds that have been to shows or off your property for some reason. Small flock owners are the least likely to take this precaution, but if you frequently show birds or add new birds often, sooner or later you're going to have a problem if you don't.

Keep a large cage or other area away from your flock where you can house all new birds (or birds that have been to shows) for about two weeks. Feed and care for the birds in quarantine after you feed and care for your regular flock. Watch the birds for signs of illness, and examine them for lice and other parasites. If all's well in two weeks, you can add them to the flock.

Diffusing Stress

Stress affects all animals, lowering the immune system response to disease and causing many undesirable behaviours. In this section, we take a look at things that may cause stress to your chickens and ways to manage them to reduce that stress.

Managing the moult

Moulting is the gradual replacement of both hens' and cockerels' feathers over a four- to eight-week period and is a natural process, not a disease. It's also a time for the hen's reproductive system to rest. Moulting usually happens in the autumn as the days get shorter, but lack of food or normal lighting can also cause it. Even if you keep your chickens under artificial lights, moulting eventually happens, usually after about a year of laying.

When moulting, chickens begin to lose their large primary wing feathers and head feathers first. From there, the process spreads backwards gradually. Chickens may look a little scruffy during this time, but they shouldn't look bald. You may notice lots of tightly rolled *pin-feathers,* which look like quills sticking out of the chicken, and lots of feathers on the floor. The pin-feathers gradually open into new, shiny, clean feathers.

Some breeds, particularly high-producing egg strains, have a quick, barely noticeable moult because they're selected to get through moult quickly. Individuals vary in the length of their moults, too. Hens that complete the moult quickly are generally the most productive and healthiest hens in the flock.

The process of moulting and growing new feathers is energy intensive and puts your bird under pressure, which can be considered stressful even though it's a natural phenomenon. Chickens that have a proper diet and few parasites should have little trouble with moulting. However, moulting is a time when you need to ensure that your birds' diet includes good-quality protein. Some people switch from layer ration to meat bird ration for a few weeks during the moult because it contains a higher protein ratio. If you're feeding a good commercial diet in this way, you don't need to add medications, vitamins or anything else.

Because the immune system may be less active during moult, avoid bringing in new birds or taking birds to shows; they won't be looking their best anyway. Also try to reduce stress as much as possible by leaving them alone as much as possible and minimising activity around them during this period. Chickens may want to hide away when they're moulting; they're at their most vulnerable during this time and instinct tells them to keep out of harm's way. Other than this, they usually act normally. If they seem sick or don't eat or drink well, something else is wrong. (See Chapter 11 for more about dealing with health problems in chickens.)

Hens generally stop laying during their moult, although some lay sporadically. Both of these situations are normal. The hen's system needs to produce new feathers, not eggs, and so moulting gives hens a holiday from egg laying. The first eggs she lays when returning to production may be smaller than normal, but she quickly returns to laying normal-size eggs.

Introducing new birds carefully

Bringing in new birds to your chicken flock can be problematic for two good reasons. New introductions not only risk introducing disease to your existing flock (see the earlier section, 'Quarantining new birds or those that have been off your property' for advice on how to avoid that calamity), but you also introduce stress because all chicken flocks have a ranking system, or *pecking order*. (We discuss basic chicken behaviour in more detail in Chapter 2.) Every time you add or remove a bird, the order changes, causing fighting and disorder in the flock.

Unless a chicken has been alone and pining for friends, it generally doesn't like new birds and may viciously attack newcomers. Before a new ranking order is established and the new birds become part of the flock, you may witness bloodshed or even death. You can, however, help to keep order.

First of all, never introduce a new cockerel into a flock with an established cockerel; doing so is asking for trouble and one or both of the birds may die. If you want to breed a new cockerel with your hens, divide the housing, the runs and the ladies, or remove the old cockerel. A very young male that grows up with the flock is sometimes tolerated, but don't add a cockerel that's over six weeks' old.

A cockerel introduced to a flock of all hens typically has few problems. Sometimes a hen or two challenges him, but he quickly becomes lord and master. Exceptions do exist, of course, which is where the expression 'henpecked husband' comes from!

Adding new hens to old hens is the most frequent type of introduction in small flocks but, here too, you need to be careful. Don't just toss the new birds in and hope for the best. If you've ever watched females of any species fight, you know how vicious they can get. Try to introduce more than one bird at a time, so that the bullying is divided up a bit.

Usually, young hens allow older birds to dominate them if the older ones are active and healthy. But sometimes the tables turn and an older bird gets the worst of it. Breeds vary in their assertiveness or aggression. Silkies, Polish and other breeds with topknots or crests, Cochins and some of the smaller bantam breeds, may be bullied by younger birds of large active breeds.

If you can move all the birds, old and new, to new quarters, less fighting ensues. Another way to introduce new birds is to put them in a cage or enclosure next to the old flock members for a few days (having a spare ark around for these occasions can be handy). The old birds can then let them know who's boss without actually harming the newcomers. If you can't pen newcomers nearby for a few days, put them in the house at night, after the flock goes to bed. Then keep a close eye on the flock in the morning. Or release new birds into the shelter area of a chicken house after you let the old birds out to range.

Expect some fighting, and don't interfere unless a bird is injured and bleeding, in which case remove the bird – bleeding birds may be quickly pecked to death. The flock is establishing a new order, and after they all know their places, the fighting ceases. Often a cockerel interferes in the fighting. Cockerels are attracted to the new girl or girls and sometimes protect them. Also, cockerels don't like too much strife in their households and may punish the offending birds.

Keep an eye on newcomers for a week or so and make sure that they're getting to the food and water. Adding some extra food and water dishes for a day or two is a good idea in case the older birds are guarding the food supply. Giving some interesting cover helps, too, and enables the new birds to hide out of sight of the old birds. This behaviour gives the right signals about rank and gives the new girls a breather if they've been getting hassled. If the newcomers stay huddled in a corner, you may have to remove them.

Never spray the birds with water or concoctions of scented products to try and confuse them about who's new. This practice just doesn't work, and simply causes even more stress to the flock.

Discouraging bullying behaviours

Chickens, by nature, pick on weak members of the flock. We talk about the ranking system or *pecking order* that all chicken flocks establish in the preceding section and in more detail in Chapter 2. Chickens at the bottom of the pecking order are more likely to be picked on, as are birds with different colouring, colour patterns or feathering than the majority breed, especially in brooder housing – chickens are conscious of colours and patterns. Chickens with *topknots* (fancy 'hairdos') are also frequently picked on by other types of chickens, and so watch these birds carefully and remove them if they're bullied.

If blood is drawn, chickens keep picking at the wound, often until they kill the injured bird. A dead bird may be pecked at and eaten if it's left in the pen. This behaviour doesn't mean chickens are vicious; nature just designed them that way – to be opportunistic feeders. Immediately separate any dead, injured or ill chickens from the others. Maintaining crowded conditions and frequently moving birds in and out of the flock causes yet more bullying and fighting, which increases the chances of injury.

Whenever you add or remove birds from a flock, the ranking system has to be re-established, and chickens may be injured as this happens.

Both cannibalism and feather-picking indicate poor management. As a home flock owner, you shouldn't have many problems with these issues as long as you feed a proper diet and maintain housing that isn't crowded. In nature, chickens spread out and spend most of the day searching for food and keeping busy. Plus, they have enough space to avoid higher-ranked flock members. If you give your home flocks at least some free-range time each day, you should rarely experience problems.

If your birds have to live in a small space, invent some way to use the vertical space by enriching their environment. Being able to climb up out of harm's way or to show off how high ranking they are by lording it over the topmost branch may be the right occupation for your chickens. A few carefully placed branches can give them an amazing amount of fun.

If your chickens live in a confined space, make sure that they have adequate food available at all times. Research shows that chickens on diets without animal protein and/or low roughage are more prone to feather-picking, which often leads to wounds and cannibalism. Properly balanced commercial diets supplement vegetable protein with the necessary amino acids found in meat-based diets, and a good free-range system allows the diet to be adequately supplemented with insect and worm protein.

Chickens also benefit from some edible entertainment or diversion therapy, such as pecking at a head of cabbage, a squash or a pumpkin, or looking for scratch grain scattered in the bedding. Raw vegetables, fruit and whole grains provide roughage. Suspending the larger items just above head height mimics the browsing behaviour that their jungle bird ancestors performed and keeps them off the ground to avoid them becoming soiled and unappetising.

Bullying and feather-picking can sometimes also be caused by very bright lights in brooders, stress caused by predators, too much handling, excessive noise and other conditions. Changing brooder bulbs to infrared bulbs or a red light source may help. Changing conditions to reduce stress is always desirable. Sometimes only one or two chickens seem overly aggressive, and so you may need to sell or even destroy them.

Employing Grooming Procedures

From time to time a problem occurs and you need to have a few trimming and plucking skills up your sleeve to remedy it. To that end, this section introduces trimming – both nails (typically an ex-battery problem from long-term confinement in a cage) and troublesome feathers. Yes, some chickens do need to have their fringes cut. Understanding when to employ these methods helps your hens to lead happier and healthier lives.

Cutting long, curled nails

Chickens kept in cages or on soft flooring may develop long, curled nails that can get caught in flooring or make walking difficult. Birds carrying out normal scratching and living in a free-range situation aren't troubled by this issue, but ex-battery hens sometimes have this problem when they're re-homed. You need to trim these long nails. Any pet nail-trimming device does the job, but you also need someone to hold the bird while you trim the nails.

Don't trim off too much at one time because a vein that runs about three quarters of the way up the nail bleeds profusely if cut. It grows as the nail grows, and so ensure that you take off only about a quarter of the nail at one time. If you look at the nail in good light, you can probably see the small red vein running through it. Cut just before this vein.

If you nick the vein, put pressure on the cut end with a cloth or paper towel to stop the bleeding. You can buy styptic powders, designed to stop bleeding, in pet shops and chemists. Just wipe off the blood and quickly apply the powder. At a pinch, a bit of flour may stop the bleeding. Be sure that the bleeding has stopped before you return the bird to its pen to avoid other birds pecking at the bloody area.

Trimming wings and other feathers

You may want to trim your birds' wings to keep them from flying over pen walls, though be aware that some lightweight birds may get enough lift to escape even with their wings trimmed. Also keep in mind that chickens can't be shown with trimmed wings. Only trim the large flight feathers; trimming just one wing is enough to unbalance the bird and make flying difficult. You need to do it again when the feathers grow back after the next moult.

To trim the wing feathers:

1. **Have someone restrain the chicken for you.**

2. **Pull the wing out away from the body.**

3. **Use sharp, strong scissors to cut across the middle of all the long flight feathers (see Figure 10-1).**

If you hit an immature feather and it begins to bleed, grasp the feather close to where it joins the wing and pull it out. Doing so stops the bleeding.

Cutting feathers is always better than pulling them out, which is painful for the chicken. Another advantage to cutting versus pulling is that pulled-out feathers often re-grow before the moult, whereas cut feathers don't re-grow until they're shed during moult.

With breeds that have topknots, trimming the topknot feathers over the eyes helps the chickens to see. When they can see better, they're more active, and other breeds are less likely to pick on them. However, don't trim the topknot feathers with show birds, or they'll be disqualified.

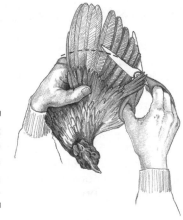

Figure 10-1:
How to clip
the wing
feathers of a
chicken.

In some heavily feathered breeds, feathers around the rear end may become matted, and even if not matted, they can interfere with mating. You can carefully trim these feathers.

Cutting feathers makes the trimming job last longer than plucking out the feathers. If a cut feather bleeds, however, you must pluck it out or it attracts pecking by other birds.

Finding a Good Vet, Just in Case

Sometimes, despite your best efforts, chickens get sick and you need help diagnosing or treating the bird. Finding help can be tricky, so knowing where to go before you have an emergency avoids you from spending valuable time looking for help when one arises. For this reason, before your birds ever become ill, ask around for a vet in your area that treats poultry.

Some small animal vets consider chickens to be livestock and don't treat them, but others may if they treat exotic birds in their practice because they're familiar with birds. Even some large-animal vets may not know much about chickens and prefer not to work with them. Livestock vets, however, often have close ties with university animal hospitals or specialist poultry vets and can refer you to a source of help.

For serious events, such as unexplained deaths in your chicken run, DEFRA can refer you to poultry researchers and labs that diagnose problems with chickens, if they don't consider your problem something the Official Veterinary Surgeons (OVS) need to look into. Depending on the problem, diagnostic work is done for free, but in most cases you have to pay a fee.

Marking Birds for Easy Identification

Marking your chickens so that you can identify them is something to consider doing – especially if you have more than just a few birds. If you're planning to show your birds, for example, identify them in a permanent manner. If you have a flock of similarly coloured birds, you may want to identify them for medication, planned mating or other reasons.

Leg or wing bands provide the best way of identifying individual birds. You can purchase bands from most poultry equipment suppliers, poultry shows and some poultry clubs and organisations that sell bands for their members. Bands may be made of metal or plastic and be flat or rounded in shape. They come in different sizes to accommodate different breeds of chickens.

You can choose between buying temporary or permanent bands:

- **Temporary bands** usually consist of a numbered or coloured coil that you spread apart and clasp on the leg for temporary identification purposes. You can also get plastic write-on bands, similar to hospital bands, which you slip over the top of a chicken's wings at the shoulder. Temporary bands are useful for keeping track of certain birds for mating, for compiling egg production records or for medicating.

- **Permanent bands** are numbered bands that you slip on the legs of chicks and remain on them for life. You can also clamp metal or plastic numbered bands through the wing web for permanent marking. Some supply houses let you choose a combination of letters and/or numbers for bands, especially if you order a lot of them, but most supply houses sell pre-numbered bands. Permanent bands are usually required at shows.

You can microchip valuable chickens in the same way as other pets – ask your vet or other poultry-keepers where this service is performed in your area. A chip is about the size of a grain of rice and is read with a machine. The biggest drawback to using them in chickens is that few people think to use a scanning machine to find an owner or prove ownership.

Chapter 11

Handling Health Problems

• •

• •

*M*any chicken-keepers worry about what to do in the event of illness or disease in their flock. We encourage you not to let your imagination run wild or allow yourself to become discouraged thinking about what terrible things may happen to your flock. The chances are pretty good that you never see any of the diseases we mention in this chapter. Injuries and parasites are much more common than diseases in smaller flocks.

Chickens are fairly hardy animals – after all, being Nature's Number One scavenger is a tough job – but when they do get sick, they often die before you can do anything about it. Usually the first sign a chicken owner finds during flock illness is a dead bird; they're adept at hiding illness. This behaviour helps the species survive, because weak birds are vulnerable. And when birds are vulnerable, they're more susceptible prey to predators and members of their own flock, which often pick on weak birds. So unless you're a close observer of your flock, you may not even notice when a bird's sick. And sometimes, if illness didn't cause death, a combination of illness and violence did.

Sometimes chickens die naturally without being diseased or attacked by predators. Old birds may have just lived out their life span. Because they're so large and heavy, broiler-type birds often die of heart failure. Genetic faults sometimes cause death as well.

In this chapter, we give you information on what to do if you encounter injuries, parasites, disease or even death in your flock. We also explain how to administer medication to a sick bird, and what to do when a bird dies.

Treating Injuries

The most common chicken injuries result from being bitten by predators or pecked by other chickens, but just like children, chickens can get themselves into some interesting predicaments. For example, if the grass is greener on the other side of the fence, they may get their heads stuck in the fence trying to reach it, injuring themselves in the process. Unfortunately, chickens, like many birds, can die from shock and stress when wounded rather than from the wound itself.

In the following sections, we tell you how you can identify and tend to the most common types of chicken injuries that home flock owners face as well as how to keep your injured birds from going into shock.

Giving your bird a health check

If a chicken has been wounded or you suspect it was in a situation where wounds may have been inflicted, check it over carefully. Feathers can hide some wounds that don't bleed very much, including deep puncture wounds exactly like the ones inflicted by the long canine teeth of dogs and foxes.

If the chicken is still up and walking and you want to examine or treat it, try to catch it without too much more stress; especially, try not to chase it because doing so can hasten the bleeding of wounds. The chicken wants to seek out a dark quiet place, and so encourage it back into its house or a corner where you can close it in. Use a feed sack or other solid barrier to manipulate the chicken rather than chase it, let the bird settle down and then catch it as calmly as possible.

When you have the bird in hand, examine it carefully, handling it gently. Part the feathers, remove loose clumps of feathers and sponge off bloody areas so you can assess the depth and extent of the wounds. Use gloves when handling injured birds in case a disease is also present.

If the injuries are extensive and dirty, as from a predator attack, the kindest thing to do is to euthanise the chicken rather than try to cure it. The easiest and most humane method of dispatching a chicken is to dislocate the neck, which we tell you how to do in Chapter 17. If you do euthanise the bird, make sure that you know the correct and legal way to dispose of the carcass and waste – also covered in Chapter 17.

Keeping an injured bird safe

If you decide that you're going to treat the bird, begin treatment quickly or take it to a vet.

Although you understandably want to do the best for your chicken, think practically about the costs involved if you decide to take your bird to a vet, because vets don't come cheap. After you've already invested in a hen house and other equipment, you may decide that the sensible approach is to cull one injured bird and be able to afford to care for the rest of your flock, and to buy a replacement, instead of being landed with an enormous vet bill. We know of someone who took her hen to the vet and ended up with a bill over £80. For this reason, finding out how to dispatch a bird humanely yourself is a good idea. You can find some good courses available at www.providence farm.co.uk.

Whether you decide to transport your chicken to the vet or treat the bird at home, do everything you can to minimise shock. To soothe an injured bird and keep it from going into shock at home, create a comfortable isolation environment:

- ✔ **Put the bird in a safe, darkened place to calm it after treatment.**

- ✔ **Unless the weather's hot, provide an overhead source of warmth, such as a heat lamp.** Hang the heat source about 45 centimetres (18 inches) above the bird. Put your hand under it at the level of the bird's back and leave it there for a few minutes. Your hand should feel warm, but not uncomfortable or burning. Chickens need room to move away from the heat source if they get too warm. If they can't move well, check them frequently to see that they aren't too warm.

  If the temperature is more than 30 degrees Celsius (86 degrees Fahrenheit), humid and the chicken is panting, it's already too hot, and you don't need additional heat. Move the chicken somewhere a little cooler but out of a draft.

- ✔ **Offer water immediately.** You can dip the beak of a chicken into shallow water but don't so any more than that to get the bird to drink. If you have homoeopathic arnica 6, add some to the drinking water. Homoeopathic remedies have healing properties but aren't medications, don't have side effects and are good for both human and animal use. You can buy them from normal chemists.

- ✔ **Don't feed the bird for a few hours after injury.** Don't try to force the bird to eat.

Injured chickens are subject to abuse by their flock mates. Bleeding wounds are very attractive to other chickens, and so always isolate a bleeding bird to prevent further injury.

Dealing with skin injuries, cuts and puncture wounds

Chicken skin is thin and tears easily. Scratches and small cuts from being caught on sharp objects, fighting or being attacked by a predator can occur. Broiler-type birds may develop blisters on their breast area.

If the wound is shallow, wash the area with warm water and soap and gently pat dry or clean the wound with iodine; alternatively, a few drops of tea tree oil in water makes another very good antiseptic wound wash. Keep some *styptic,* or blood-stopping, powder in your First Aid cupboard to use in case the wound continues to bleed, or use pressure to stop it. This inexpensive powder, often just called 'wound powder', is readily available in pet shops and other animal feed merchants. Bandages rarely work on chickens because of their feathers and the fact that the birds pull them off. If the cut is on the bottom of the foot, keep the bird on a clean surface for a few days.

Animal bites are very dangerous to chickens – even if they don't look deep or dangerous and particularly when the puncture goes through the flesh and into the body or brain cavity – and are difficult to spot. If one of your chickens sustains an animal bite but the puncture hasn't gone too deep, follow these steps:

1. **Wash the wound thoroughly with warm water and soap.**

2. **Use hydrogen peroxide, iodine or salt water (1 teaspoon in 0.5 litres/ 1 pint of tepid water makes a strong enough solution) to flush the wounds.**

3. **Place the chicken in a clean area.**

4. **Check the wound for infection several times a day.**

   Clean the wound two to three times a day if infection sets in.

5. **If the chicken can't reach the wounded area with its beak and the weather's warm, apply a wound dressing to prevent flies from laying eggs on the wound.**

   You can find wound dressing in most pet shops or chemists, or you can obtain one from your vet.

The outlook isn't very good where deep punctures have been caused by animals. Keep the chicken warm and quiet to prevent shock. If the chicken is very valuable to you, take it to a vet's surgery as soon as possible. A trip to the vet can be stressful, and so leave your bird in the car until you find out who else is in the waiting room. If your neighbour's dog attacked the bird, the last thing it needs is to be forced to sit next to a dog in the waiting room.

If your wounded chicken doesn't rally after initial treatment it has most likely sustained damage to a vital organ, causing internal bleeding that may result in an infection setting in, followed by a slow death. The kindest thing to do in the circumstances is to destroy the bird.

Tending to foot sores

Bumblefoot is a bacterial infection or abscess of the foot, which generally occurs in heavy cockerels but can affect other birds as well. It's caused by a cut or even a small scrape to the bird's foot that gets contaminated by bacteria, usually staphylococcus. Thistle spines, rough perches and wire cage floors are common causes of these cuts and scratches, and large, heavy birds can also injure a foot by jumping down from high perches.

Bumblefoot causes a large swelling on the bottom of the foot or on a toe. The swelling looks red and inflamed, feels soft in early stages and hard later, and may be hot to the touch. A black scab usually forms over the sore. The bird may limp and refuse to do much walking. In time, if you don't treat it, bumblefoot can cause blood poisoning and death.

Use gloves to examine or treat birds you suspect of having bumblefoot because the bacteria that cause the abscess can infect humans. Isolate the bird from other chickens because the bacteria can infect them too, and put the bird in a cage with clean, soft litter such as pine shavings.

If you catch the bumblefoot early on, antibiotics may be all that the chicken needs. Several registered antibiotics are available for chickens: lincomycin and amoxicillin are two common ones. You can get medication from your vet that you simply add to the bird's water – just be sure to give the antibiotics for the full period given on the label. If you prefer to take your chicken to a vet, antibiotics can be injected directly into the affected site.

Watch birds being treated with antibiotics for diarrhoea, which is caused by good bacteria that's also being destroyed, and add some probiotic yoghurt to the chicken's diet to help restore it.

In an otherwise healthy bird, you can do much yourself to help the bird before resorting to professional veterinary help. Soaking the affected foot helps, especially if the injury has progressed to the 'feeling hard' stage. To do this:

1. **Put a cup of Epsom salts in a dish of hot water – water that feels hot but doesn't burn your hand.**

2. **Soften the abscess by holding the chicken's foot in the dish for about 10 to 15 minutes until the water cools. Make sure that the bird doesn't drink any of the water.**

3. **Gently remove the scab, and open the wound by pulling it apart at the wound edges rather than squeezing it.**

4. **Rinse the wound with hydrogen peroxide, iodine wash or salt water, and gently clean out any pus.**

   The pus may be quite hard and cheesy, and may be followed by bleeding.

5. **Apply a bird-friendly antibiotic ointment (ask a vet for a recommendation) or wound powder if the wound isn't too large.**

6. **Pad the wound with a clean gauze pad, and wrap it with First-Aid tape or *vet wrap*, a bandage that sticks to itself.**

7. **Clean, flush and rewrap the wound once daily until it looks as if it's healing. Carefully dispose of all dressings and soaking fluids because they'll be contaminated with bacteria.**

We recommend leaving the cutting open of a bumblefoot abscess to a vet if soaking and pulling apart the wound edges doesn't open the wound to allow it to drain and be cleaned.

Treating head injuries

The most common areas for head injuries in chicken are the combs, wattles, eyes and beak. The following sections explain each in detail.

Torn or infected combs and wattles

A bird may tear its comb or *wattle* (the red flaps of skin hanging under your chickens' beaks) by catching them on something or damaging them in a fight. Generally, torn combs and wattles can't be repaired and may become infected, and so you may need to trim them off. Use your common sense, though, to determine whether trimming is necessary or whether just keeping the wound clean and free from flies will enable it to heal naturally. (Tea tree oil, available at any chemist or health food shop, is an excellent antiseptic and fly repellent – but keep it out of the chicken's eyes.)

If you decide that you need to trim, you can do so yourself or take the bird to a vet who can do the job for you. To do the trimming yourself, follow these steps:

1. **Wash the area with hot water and soap.**

2. **Clean the area with rubbing alcohol.**

3. **Trim off the torn area of the comb or wattle with sterilised scissors or a sterilised, heavy knife.**

4. **Apply an antibiotic cream to the edges.**

   Ask a vet or poultry expert to suggest a good antibiotic ointment or look for a bird-safe pet ointment. In warm weather you may also need to apply an anti-fly ointment to the wound area to keep maggots out of the wound. All feed and farm merchants sell this ointment.

5. **Isolate the chicken and keep it separated from other birds until the wound has healed.**

Because the comb and wattle have blood vessels, trimming them causes some bleeding, but it's minor. The procedure involves some pain; however, anaesthetics usually aren't given because the pain is quickly over and birds are tricky to anaesthetise. A vet carrying out the trimming may administer local painkillers. Most birds recover quickly and completely.

Eye injuries

Chickens sometimes damage their eyes during fights or get injured in predator attacks. A little pirate patch on a chicken would look silly if rather cute, and so the best you can do is to clean the eye and keep the chicken separated from the flock. Clean the eyes with a non-medicated eyewash for pets or humans. A chicken is fine if it's blind in one eye, but if it's blind in both eyes, the kindest thing to do is to cull it. Chickens have very beady eyes and are always on the look out for tasty morsels. At the same time, they keep their eyes open for predators, and so keeping a totally blind chicken isn't fair on the bird.

Beak injuries

Chickens' beaks sometimes get broken or cut off in freak accidents. Missing beak portions don't grow back. Depending on how much beak is left, the bird may or may not be able to eat normally. If a chicken has a small amount missing from the end of one or both beak halves, the bird will be fine. In fact, millions of birds have had their beaks trimmed because they're kept in stressful conditions where they're prone to feather pecking, and yet they're able to eat normally. However, if large portions of the top or bottom beak are gone, the bird has great difficulty eating. It needs its beak to pick up feed (not to chew or crush it), and if the bird isn't able to pick up enough to maintain a decent weight on its own, putting it down is probably more humane (see Chapter 17).

Tending to broken legs or wings

Chicken can live fairly easily with broken wings, but not with broken legs. If a bone is protruding through the skin, infection is extremely likely, and the chances of the bird making it through are poor. Wings can be amputated, but even though chickens can live with one leg, their quality of life is poor. Amputations should always be performed by a vet.

A broken wing may drag on the ground or appear twisted. You can help it to heal by folding the wing into a natural position against the bird and then wrapping the bird with gauze strips or vet wrap to hold the wing in place. Keeping the wing wrapped for two weeks is usually enough. You must separate the bird from other birds during this time, but it will feel better if it can see and hear the flock. Unless it's a show bird, the wing healing crooked or drooping isn't a problem.

A broken leg may look crooked and swollen, and the bird can't walk on it. You can splint broken legs, but preferably let a vet or someone experienced in bird rehabilitation do this. Again, you need to separate the bird from your flock until it heals – in a young chicken, the bones heal quickly. Don't, however, put the bird through unnecessary trauma. If a vet or experienced poultry-keeper thinks that the bird has a poor prognosis, having it put down may be difficult but is ultimately the kindest thing to do.

Fighting frostbite

Frostbite is very unlikely in the UK but in times of sustained cold weather a cockerel with a large comb may succumb, and so be prepared. Frostbite causes blackened areas on the ends of combs, wattles and sometimes toes. In most cases these areas dry up and fall off eventually, but do keep an eye on frostbitten areas. If they become infected you may need to trim them off and treat with an antibiotic cream. (See the earlier section 'Torn or infected combs and wattles' for instructions.) Don't trim off the blackened area unless it becomes infected – the blackened area gives some protection to the area below it. When you remove that, the area beneath it may be frostbitten next.

Prevention is better than cure, but rubbing chicken combs and wattles with oil, petroleum jelly and other things you may have heard of doesn't prevent frostbite. A better move is to look critically at your chicken housing. Check that the ventilation is removing hot air without causing a draft, think of using extra insulation or moving the house into a more sheltered position out of the worst of the weather or even find some temporary indoor accommodation for the duration of the cold snap. Chickens hate snow anyway; it's boring for them, as well as cold.

If you're tempted to heat your chickens' house, don't heat it much above freezing because doing so causes moisture problems that may be worse for your birds than the cold. Chickens fare better in dry and cold than in warm and wet conditions – their warm breath can cause excessive condensation that in turn encourages harmful moulds to grow.

Cockerels with frostbitten combs may be temporarily infertile. But the frostbite isn't what causes the infertility; it's the amount of cold to which the chicken has been exposed. Usually fertility restores itself after conditions improve and the cockerel's body recovers from the stress.

Tackling egg binding

Egg binding occurs when a hen has an egg that she can't pass from the oviduct for some reason. Frequently this diagnosis is mistakenly made when a hen looks droopy before she dies and an egg is found in the oviduct upon post-mortem examination. Hens often continue to produce eggs after getting sick, but the birds may die before the eggs are laid.

Egg binding shouldn't be a common occurrence in your domestic poultry flock if you feed and handle your birds properly, though it occasionally happens when birds are injured or have a hereditary defect. Hens weakened by poor feed, heavy worm infestations, diseases or severe cold are more susceptible. Old hens are more likely to become egg bound. If a hen has been egg bound once, she's likely to have problems again in the future.

A hen who's egg bound sits on the floor or ground. Her feathers are fluffed, and she's drowsy and lethargic. Sometimes you can see her strain as if trying to produce the egg. More often, you notice her tail pumping up and down and you can feel the bump where the egg is lodged.

If you handle a hen roughly just before she lays an egg, the egg may break inside her, and so be sure to handle hens carefully, especially early in the day. Egg binding is a dangerous situation for the hen and seldom resolves safely. In fact, you can do little for an egg-bound hen.

A hen that's truly egg bound dies if she doesn't pass the egg. Don't stick things such as syringes full of oil up her vent; instead of helping, you're likely to hurt her and cause infection. Trying to break the egg inside her and extract the pieces isn't usually effective either; it's likely to result in infection and death.

Moist heat is the safest remedy for egg binding:

1. **Put the hen in a wire-floored cage.**

2. **Place a large, flat dish of steaming water beneath the cage. Keep the water warm under her, but don't keep it so hot that the steam burns her.**

3. **Provide some overhead heat from a heat lamp, and enclose the whole cage with a blanket or plastic to keep the moist heat in. Make sure that it doesn't get too hot – use a thermometer to keep the heat between 32 and 38 degrees Celsius (90 and 100 degrees Fahrenheit), and ensure that water is available at all times for the hen to drink.**

4. **Watch for these outcomes:**

   - **If you see an egg** (the hen should pass the egg in a couple of hours with this treatment), she's likely to perk up and is ready for you to remove her from the cage.

   - **If she hasn't passed an egg and she seems more active and is eating,** you probably misdiagnosed her – something else is wrong.

   - **If she continues to act droopy and ill,** give her a few more hours of treatment. She dies unless she passes the egg, usually within 48 hours. You can't safely do much else to help her. However, a vet can give a hen an injection of calcium gluconate, which often causes her to pass the egg. After passing the egg, she may *prolapse* (when the internal organs protrude from the vent). If this happens once, it usually happens again, and the kindest thing is to cull the chicken.

Getting Rid of Parasites

Parasites are creatures that feed on a chicken's blood, other body secretions or its feathers and are common in all kinds of chickens. A few hens in an urban setting may be less likely to pick up parasites, but chickens anywhere can get them. Chickens that range freely have less trouble with lice and other external parasites but, on the other hand, are more likely to pick up some kinds of internal parasites than penned chickens.

Completely preventing parasites is difficult. Even baby chicks that come to your new housing parasite-free may eventually get parasites. Wild birds carry many parasites common to chickens, and they can be picked up from the ground, from shared carriers or equipment, from rodents and insects and even be carried on your shoes and clothing when you go to infected places.

If parasites become an issue, you have a choice of whether to treat or not. In earlier times, people expected chickens to have parasites, both externally and internally, and unless the problem became really severe, most people just accepted the fact that parasites were par for the course. Many people today are still willing to let nature take its course, and they don't worry about treating their chickens for parasites. If your own chickens seem healthy and happy and produce as you want them to, you may decide not to treat them for parasites.

However, a big difference exists between a handful of lice – which may be a mild irritant that the chickens can control through frequent dustbathing – and a debilitating infestation, which can drag down the health of the whole flock. Additionally, for people with small flocks that they handle frequently and confine for at least part of the time, parasites may be unacceptable. You don't want parasites on you or your children, and you probably want your chickens to be as comfortable and healthy as possible. You also may want optimum egg production. These considerations give you good reason to treat your birds for parasites.

The sections that follow describe all you need to know about the internal and external parasites that commonly affect chickens – and what to do about them.

Internal parasites

Worms are common internal parasites, and are bad news for chickens. Chickens with worms are more prone to cannibalism, retained eggs and loss of egg colour and size, and they may even die from blood loss and poor nutrition. Worms and other internal parasites become an even greater problem when a chicken's system is stressed by disease or a poor environment. Heavy infestations may then contribute poor growth, poor egg production or death.

Another internal parasite, Coccidia, isn't a worm; it's a microscopic *protozoa* (a one-celled animal). Coccidia infections are extremely common in chickens and cause the disease, *coccidiosis.*

Worms

Chickens that have worms can look unhealthy and thin. They may gain weight slowly even though they eat more feed than chickens without worms, and they may lay fewer eggs. Sometimes, however, you don't see any symptoms; many species of worms can live in chickens and not cause any problems. If you aren't into the habit of checking chicken droppings, you may never know that your birds have worms. Even if you do examine chicken droppings (some people actually take pictures of them), you still may not know whether your chickens have worms because often they aren't visible in droppings.

In many cases, a vet can examine a chicken's droppings in a lab to look for worm eggs or actual worms, and tell you whether your birds have them. If you're worried about worms but aren't sure whether to treat your chickens, ask your vet to conduct a 'worm count' and provide you with advice. Sometimes, however, these lab tests aren't successful, because worm eggs weren't being produced when the sample was collected.

The following types of worms are most common:

- **Gapeworms.** You generally find gapeworms only in birds that are free-ranging or in pasture pens. Gapeworms are more serious than other worms because they attach themselves to the inside of the *trachea,* or windpipe, and interfere with the bird's breathing if many are present. Infected chickens breathe with their mouths open and may produce a strange grunting sound when trying to breathe.

  Gapeworms require an intermediate host, which can be earthworms, snails or slugs. When chickens eat these creatures, they become infested. You can buy some good treatments for gapeworm, but they're not registered for chickens, and so you need to consult with a vet who can guide you in what products to use and help you with quantities.

- **Roundworms.** Roundworms are a common parasite of chickens. When foraging for food, chickens pick up roundworm eggs that then go through the digestive system to the intestines, where they hatch. Roundworm eggshells are very tough, and most disinfectants don't kill them. They can remain alive in the environment for up to two years.

  Many times roundworms don't cause any symptoms in chickens, and are more likely to harm young chicks than older birds. Heavy infestations, however, cause weakness and sometimes diarrhoea. Chickens often pass the worms in their faeces. They're easy to see – white, and up to 8 centimetres (3 inches) long, with a diameter about the size of a piece of dry spaghetti. They may be coiled up or moving when you spot them.

  Sometimes a worm comes out of the intestines and then works its way into the oviduct where it can then be encased in the chicken's eggshell and passed out in the egg. This is reason enough for occasionally treating your chickens for worms. Do so whenever you see worms in the faeces; many people treat their chickens for worms twice a year as a precaution.

- **Tapeworms.** Although fairly common in chickens, tapeworms seldom cause serious symptoms – just an increase in eating because the birds are sharing food with the worms. Chickens pick up tapeworms when they eat earthworms, snails, slugs, grasshoppers and some beetles and flies, which serve as *intermediate hosts.* The immature tapeworms, or *larvae,* that reside inside an intermediate host get into the intestines of the chicken when the chicken eats the host. They mature there and produce eggs, which the chickens pass in their faeces to then be consumed by another intermediate host. And so the cycle continues.

  Sometimes segments of tapeworms pass out in chicken faeces and are visible as small, white, flat rectangular pieces that move. However, most of the time people are unaware that their chickens have tapeworms. If you see tapeworms and want to treat your chickens, consult a vet. No registered treatment exists for tapeworm in chickens, but you can purchase some effective tapeworm killers for other livestock that you can use for chickens if you know the proper dose. A vet can help you determine this amount.

As a chicken-keeper, you can do much to prevent a worm infestation in the first place. Chickens like to have fresh pasture to graze on, and so design your housing and free-range area so that you save some space for moving your chickens around. Your first chickens probably arrive onto clean ground and have no problems, but by the second or third year a worm burden may have built up, so using the space in rotation with something other than chickens is the best way to clean up the land. If you have a vegetable plot, give the chickens access to it out of the growing season and they help to clear up any straggly vegetables and weeds while de-bugging and manuring as they go. If you have a moveable pen or enclose an area for them with flexible netting, moving it little and often enables the ground to recover each time.

If you find you do need to treat your chickens, treatment for worms generally consists of *worming* your entire flock with a medicated preparation advised by your vet. Alternatively, some well-tested herbal wormers are available and a clove of garlic left to infuse in the drinking water acts as a mild vermifuge. Both are gentle measures that can help the overall well-being of your flock when you have any infestations under control. Herbal wormers come in pellet form and replace your normal food while the treatment is ongoing, which need only be a day if your birds behave and take their medicine! The garlic treatment involves putting a whole garlic clove into the water supply and leaving it there until it disintegrates. This method is a drug-free way to deter worms and biting insects but isn't a replacement for medication if you suspect a heavy infestation.

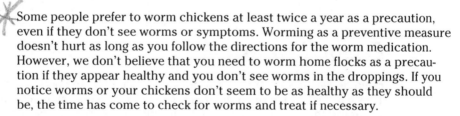

Some people prefer to worm chickens at least twice a year as a precaution, even if they don't see worms or symptoms. Worming as a preventive measure doesn't hurt as long as you follow the directions for the worm medication. However, we don't believe that you need to worm home flocks as a precaution if they appear healthy and you don't see worms in the droppings. If you notice worms or your chickens don't seem to be as healthy as they should be, the time has come to check for worms and treat if necessary.

Because they're so short lived you're unlikely to want to worm meat chicks, but if for any reason you do, be sure to follow label directions about how long to keep the birds before you can dispatch them for eating; you don't want pesticide residues to remain in the meat.

Coccidiosis

Most types of animals are infected by species of Coccidia, but each species seldom infects more than one type of animal. Chickens, on the other hand, can be infected by nine kinds of Coccidia, but they all cause basically the same problems.

Chickens get Coccidia by ingesting *oocysts,* which are immature Coccidia passed in faecal matter. The oocysts contaminate feed, litter and soil and can last for a year in the environment. They can be spread by shoes, clothing, equipment, wild birds, pests such as rats and infected chickens. The Coccidia line and damage the intestinal walls, where they interfere with the absorption of food, causing bleeding. The severity of the disease is determined by how many Coccidia live in the intestinal tract.

Coccidia are most often a problem in young, growing birds but occasionally can cause problems with older birds, especially if they get bacterial diseases such as ulcerative colitis. The symptoms are the same for older and younger birds. When birds become adults, they develop some resistance to coccidiosis but they may still carry some Coccidia. Birds under 3 weeks seldom show symptoms. Slightly older chicks from 3 weeks to 30 weeks may have bloody diarrhoea, anaemia, pale skin colour, listlessness, poor appetite or dehydration. Young birds with heavy infestations of Coccidia often die.

Good treatments for coccidiosis are available. Feeding baby chicks a starter feed medicated with *coccidiostats* such as Amprolium and Decoquinate (which kill Coccidia) used to be common practice but few poultry-keepers do this routinely nowadays. Instead, at the first signs of symptoms your vet can prescribe medication for the chickens' drinking water. If older birds seem to be infected, you can treat them with these medications as well. Chicken-keepers have also had great success with a homoeopathic coccidiosis remedy.

External parasites

External parasites are the creepy crawlies you find on the outside of chicken. They can suck blood or eat feathers. The bloodsuckers can make chickens anaemic if they occur in large numbers, and the feather-eaters irritate the birds, making them pick at their own feathers and skin, which damages their looks for showing. Birds with heavy external parasite loads often don't lay well or gain weight well. Young birds can even die from external parasites. Most external parasites that affect birds don't live on humans, but a few take a bite out of you if they get a chance.

Wild birds are the source of many kinds of external parasites, which is a good reason to keep them from nesting and roosting in chicken shelters. Signs of external parasites include seeing them crawling on the chickens or in the coop, being bitten by them yourself, noticing chickens with broken, chewed-looking feathers and reddened skin patches and seeing chickens doing a lot of scratching and picking at themselves. Hen's eggs with tiny spots of blood on the shell can be another indication. Other signs are more subtle, such as a drop in egg production, anaemia with pale combs and wattles and looking fluffed up and sick.

Chickens in the wild help to control external parasites by taking dustbaths. Dust smothers and dislodges the parasite and cleans the body of oils, dust and debris that some parasites feed on. Help your confined chickens to do the same and keep parasites away by giving them a large, deep box of sand to wallow in. Add a few drops of tea tree or eucalyptus essential oil to the dustbath as an extra deterrent. If your chickens free-range, they make their own dustbaths (generally in the middle of a flower bed) but you may have to help them out in wet weather when their outdoor dustbaths turn to puddles. With their dustbaths, your chickens are only too pleased to keep on top of the parasite problem and provide you with an amusing spectacle at the same time. Who needs TV when you have chickens to observe?

Don't try to eliminate parasites by spraying your housing with old-fashioned remedies such as Jeyes fluid or old engine oil. These products are environmental pollutants that cause more harm than good, and using them this way can be illegal – especially if any storm drains are located nearby. They also can have toxic effects on your birds through being absorbed into your bird's skin or adversely affecting their lungs.

Lice

Lice are long, narrow, tiny insects that move quickly when you part a chicken's feathers, the most common type being _Menopon Gallinae_. When your chicken is suffering from a heavy infestation, you can see the lice scurrying around on the bird. Their eggs are small dots called _nits_ that you find glued to the birds' feathers.

Several types of lice get on chickens, some of which specialise in certain areas of the body – you find head lice, body lice and lice that live on feather shafts. They spend their entire lives on the birds and die quickly when they're off them. Unlike human lice, chicken lice don't feed on blood; they eat feathers or shedding skin cells.

Check for lice regularly by upending your chicken (see Chapter 10) and parting the feathers near the _vent_ (the opening under the tail). If the skin looks smooth and the feathers dry and fluffy, all is well. If you see reddened and crusty skin, or a white crystalline deposit attached to the shaft of the feathers that looks a bit like grains of sugar and lice scurrying away, you need to act. Checking your birds regularly isn't too arduous if you've tamed them.

To control the lice you have to apply directly to the birds; treating the environment doesn't work. Treat them by plucking the worst of the feathers out and burning them to destroy the lice and their eggs. Then use whichever louse powder suits your needs (you can buy this powder from smallholder supply or farm feed stores). Diatomaceous earth is an effective non-toxic powder as are ashes from a wood fire (make sure that they've completely cooled though) – keeping your chickens louse-free without resorting to chemicals is possible. Hold your chicken as for an examination and puff the dry powder in amongst the feathers paying particular attention to the vent and tail area. Put some louse powder in the dustbathing spots and the nest boxes, too. Some vets administer drugs not licensed for poultry that are effective, but have other repercussions in the environment.

Mites

Most *mites* are bloodsuckers that burrow into the chicken's skin and feed on their blood. A few mites eat feathers and they sometimes feed on people, too, even though they prefer chickens. Some types feed at night on the birds and then hide in cracks in the chicken house during the day; others stay on the birds. They can cause anaemia, decreased egg-laying and damage to skin and feathers. Some types even invade the lungs and other organs. Heavy mite populations can cause death to chickens.

The following are three types of mites that commonly infest chickens:

- ✔ **Northern Fowl mite *(Ornythornyssus sylviarum).*** Fortunately not very common, but very difficult to shift after it has become established, the Northern Fowl mite remains on the chicken throughout the mite's life and causes scabby skin, anaemia and much discomfort to the chickens. It attacks the feathers of the birds, which become lacklustre and shabby. They look as if they've been gnawed at. Seek advice from a vet who may be able to prescribe treatments not licensed for poultry. You also need to cleanse the chicken house thoroughly; if the house is an old one, a better option may be to incinerate it and start again.

- ✔ **Red mite *(Dermanyssus gallinae).*** Sucking chickens' blood by night and hiding in cracks and crevices by day, the red mite is a particularly pernicious creature that can quickly multiply to infestation levels, especially in warm weather, causing anaemia and even death. The red colouring that gives this mite its name is the blood that used to belong to your chicken. If you see it in any quantity you can easily imagine just how a red mite infestation can make a chicken anaemic.

  If you notice any mites on your hand when you collect eggs or clean out the chicken house have a really thorough search with a torch. The red mite is visible to the naked eye but has a legendary ability to hide. Check the underneath of perches and the joints where they attach to walls. Red mites don't live on the birds, but they happily lodge themselves in any mucky encrusted feathers, so watch out there, too. The larvae are tiny and grey and are attracted by warmth, and crawl onto the hand of the chicken-keeper. These mites don't live long on humans but have a nibble and definitely take away the enjoyment of the daily egg hunt.

  Avoiding red mite altogether is impossible if you keep chickens for any length of time, but good house hygiene and vigilance ensure that you can nip it in the bud quickly. After a thorough clean, run a lighted blow torch over all the corners and crevices of the house structure to frazzle any lurking eggs – these eggs are tough and can survive for years until the perfect conditions fire them up again. If you can suspend the perches from the framework of your chicken house you can confine red mite to one area and deal with it effectively, breaking the life cycle of the mite and minimising the problem. Any of the various louse and mite powders available help in the battle to control red mite but can't beat it by themselves.

✔ **Scaly leg mite** *(Cnemidocoptes mutans)*. The microscopic scaly leg mite lives, as you may expect, on chickens' legs between the skin scales. The debris and irritation they cause force the leg scales to stand out from the leg, making the skin look thick and scabby. In severe cases it may cause the chicken to have problems walking.

This mite is usually noticed in older birds, but if one chicken has it, the whole flock is in danger and you need to treat them all – even if scaly leg mites aren't always evident, which they often aren't in the early stages. For advanced cases, soak the leg in warm water with a softening agent such as Epsom salts or soft soap to soften the crusts enough for you to pick off. Be careful, though, because if you cause bleeding you need to isolate the bird. When you've removed the worst of the old dead skin and debris, smear the legs in petroleum jelly or some oil-based product to smother the mites and kill them. Check out the ventilation in your house, too, because scaly leg mites thrive in damp conditions.

Recognising and Dealing with Disease

Studies done on commercial chickens show that 90 per cent of diseases chickens suffer are carried to them by their human caretakers. Disease organisms can survive on shoes, clothing, unwashed hands, car tyres and equipment. Your first measure to take is one of prevention – heed the advice on handling chickens that we provide in Chapter 10 at all times, but be especially mindful of it when you visit a poultry show, an auction or your friend's flock and have less control over your environment. Diseases can also spread to home flocks from wild birds and pests such as insects, mice and rats, so you need to prevent their intrusion as well (see Chapter 9).

The good news is that you're unlikely to see most of these diseases in your flock, and lots of birds have been vaccinated already against some of the most common diseases when you buy them. To be aware of these diseases is good, but don't dwell on them – hundreds of people have been successfully keeping small flocks of hens on smallholdings and in back gardens for years without major problems. The very large, intensively housed flocks are more prone to diseases getting out of hand than small domestic flocks.

Read the following sections for tips on recognising and dealing with diseases in the knowledge that your flock is very unlikely to succumb to any of them, provided that your general standard of husbandry and hygiene are good.

Checking for signs of disease

A good habit to get into is to look over your flock on a daily basis as you feed or care for your birds, checking for signs of disease. Catching a problem early can stop its spread and may give you a chance to treat chickens before they're too ill. Because some chicken diseases, such as avian flu, can infect humans too, noticing illness in your flock can protect you and your family. The more time you spend observing your chickens when they're acting normally, the quicker you get at noticing when something unusual is happening.

Some general signs of disease in chickens to be aware of are:

- ✔ Inactivity; drowsy or weak appearance; unusual tameness; inability to walk; reluctance to move around; sitting on the floor at night when the bird normally roosts.

- ✔ Sitting or standing with feathers fluffed up; head tucked under wing or hanging down; tail down.

- ✔ Swollen eyes; discharge from eyes; cloudy eyes; cloudy spots on pupils of eyes.

- ✔ Discharge from nose; scabby or crusted nostrils.

- ✔ Laboured breathing; breathing with beak open; fluid running out of mouth.

- ✔ Sores; blackened or red areas on skin.

- ✔ Green, white or bloody diarrhoea; pasted, clumped feathers around the *vent* (anal area); sore, swollen or distended vent area.

- ✔ Not eating or drinking.

- ✔ Drop in feed consumption.

- ✔ Sudden drop in egg production.

- ✔ Lots of poorly shaped or coloured eggs.

- ✔ Less active than usual.

To give you a starting point for diagnosing what may be wrong with your chickens when they appear ill or are dying, check the following list. It describes some common diseases, specific symptoms, how they're contracted and whether a treatment is available. After pinpointing the problem, you can then consult a vet or other poultry expert for confirmation and advice about treatment and what to do with the rest of the birds in your flock.

Use this list as a guideline for recognising and dealing with disease (and check out Chapter 10 for information about vaccinations):

✔ **Avian Pox/Fowl Pox:**

- **Symptoms.** White spots on skin; combs turn into scabby sores; white membrane and ulcers in mouth, on trachea; laying stops; all ages affected.

- **How contracted.** Viral disease; mosquitoes; other chickens with pox and contaminated surfaces.

- **Treatment.** Supportive care; warm, dry quarters; soft food. Many birds with good care survive.

- **Vaccine available.** Yes: recovered birds are immune and don't carry the disease.

✔ **Botulism:**

- **Symptoms.** Tremors quickly progressing to paralysis of body, including breathing; feathers pull out easily; death in a few hours.

- **How contracted.** Caused by a bacterial by-product and eating or drinking botulism-infected food or water.

- **Treatment.** Expensive antitoxin available from vet; if found early try 1 teaspoon Epsom salts dissolved in 29.5 millilitres (1 ounce) warm water dripped into crop several times a day.

- **Vaccine available.** None: locate and remove source, usually a decaying carcass, meat near water or insects that fed on the meat or the water the carcass is in.

✔ **Infectious Bronchitis:**

- **Symptoms.** Coughing; sneezing; watery discharge from nose and eyes; hens stop laying.

- **How contracted.** Viral disease; highly contagious; spreads through air, contact and contaminated surfaces.

- **Treatment.** Supportive care; 50 per cent mortality in chicks under 6 weeks.

- **Vaccine available.** Yes: give to hens before 15 weeks of age because vaccination causes laying to stop temporarily and results in eggs of poor quality when laying resumes.

✔ **Infectious Coryza:**

- **Symptoms.** Swollen heads, combs, and wattles; eyes swollen shut; sticky discharge from nose and eyes; moist area under wings; laying stops.

- **How contracted.** Bacterial disease; transmitted through carrier birds, contaminated surfaces and drinking water.

- **Treatment.** Birds need to be put down because they remain carriers for life.

- **Vaccine available.** None.

✔ **Marek's Disease:**

- **Symptoms.** Affects birds under 20 weeks primarily: causes tumours externally and internally; paralysis; iris of eye turns grey, doesn't react to light.

- **How contracted.** Viral disease; very contagious; contracted by inhaling shed skin cells and/or feather dust (also called *dust and dander*) from other infected birds.

- **Treatment.** None; high death rate; any survivors are carriers and need to be put down.

- **Vaccine available.** Yes: given to day-old chicks.

✔ **Moniliasis (Thrush):**

- **Symptoms.** White cheesy substance in crop; ruffled feathers; droopy looking; poor laying; white crusty vent area; inflamed vent area; increased appetite.

- **How contracted.** Fungal disease; contracted through mouldy feed and water and surfaces contaminated by infected birds; often occurs after antibiotic treatment for other reasons.

- **Treatment.** Yes; ask vet for antifungal medication; remove mouldy feed and disinfect water containers; to prevent further outbreaks, ensure replacement of good bacteria after treatment with antibiotic.

- **Vaccine available.** None.

✔ **Mycoplasmosis/CRD/Air Sac Disease:**

- **Symptoms.** Mild: weakness and poor laying. Acute: breathing problems; coughing; sneezing; swollen, infected joints; death.

- **How contracted.** Mycoplasma disease; contracted through other birds (wild birds carry it); can transmit through egg to chick from infected hen.

- **Treatment.** Antibiotics may save birds – see a vet.

- **Vaccine available.** Yes.

✔ **Newcastle Disease:**

- **Symptoms.** Wheezing; breathing difficulty; nasal discharge; cloudy eyes; laying stops; paralysis of legs/wings; twisted heads/necks.

- **How contracted.** Viral disease; highly contagious; contracted through infected chickens and wild birds and also carried on shoes, clothes and surfaces.

- **Treatment.** None; birds under 6 months usually die; older birds can recover; recovered birds aren't carriers.

- **Vaccine available.** Not at this stage. Newcastle Disease is a *notifiable disease* (by law you must notify the government) – contact Department of Food and Rural Affairs (DEFRA) immediately via their website (www.defra.gov.uk) or call their helpline on 08459 33 55 77.

✔ **Omphalitis (Mushy Chick):**

- **Symptoms.** Newly hatched chicks; enlarged, bluish inflamed navel area (where a tummy button would be if chicks had one); bad smell; drowsy, weak chicks.

- **How contracted.** Bacterial infection of navel from unclean surfaces or chicks with weak immune systems; can spread from chick to chick on contaminated surfaces.

- **Treatment.** Antibiotics and clean housing sometimes help, but most chicks die; remove healthy chicks immediately to clean quarters.

- **Vaccine available.** None: use caution handling – this disease can infect humans.

Considering bird flu

When people get in an uproar about bird flu (also known as 'avian flu'), few realise that it's already around us. In fact, bird flu is anywhere you find wild birds. Dozens of strains of this virus exist – as do dozens of strains of the human flu virus. Most strains of bird flu are harmless to humans and other animals. In birds, however, the outcome is varied. The flu can cause sudden death, or birds can recover, depending on the strain and many other factors.

A particular strain of avian flu called H5N1, which is of Asian origin, is of concern to humans because it can infect people who have close contact with poultry. It can cause severe illness and even death. Outbreaks of the disease have been found in many parts of the world. So far, the disease hasn't spread easily from person to person as human strains of flu do, but medical experts are worried that the virus may mutate and cause epidemics.

Avian flu is carried by wild birds, particularly waterfowl, that can migrate long distances. The UK is monitoring wild bird populations as well as domestic poultry and cage birds, and asks that chicken-keepers be vigilant for signs of the disease. Keeping your home flock of chickens from contact with wild birds whenever possible is the first step in prevention. (See Chapter 9 for ways to exclude wild birds from your flock.)

Some signs of the H5N1 strain of avian flu include difficulty breathing, swollen heads and necks, swollen bluish combs and wattles, poor coordination and paralysis, no appetite but excessive thirst and other common signs of illness in birds such as ruffled feathers. If most birds in your flock die suddenly in a short space of time (48 hours) with no evidence of a predator and with the above-mentioned signs of disease, you need to consult a veterinarian or the

Department of the Environment Food and Rural Affairs (DEFRA). You can visit its website (www.defra.gov.uk) or telephone its helpline on 08459 33 55 77.

Making decisions about treatment

When faced with diseased chickens, you have in most cases to decide whether treating them is worth the time and money. In some situations, you may not have a choice. If, for example, you suspect that your birds have Newcastle Disease or the H5N1 strain of avian flu – both notifiable diseases that we describe in the preceding section – you're obliged to contact DEFRA immediately (by telephone at 08459 33 55 77 or via its website, www.defra.gov.uk).

The problem with many chicken diseases is that they're *viral* – so they're contagious – and no effective treatment exists (or no product is registered to treat chickens). Also, if your birds are diagnosed with a serious disease, the Animal Health Agency (a department in DEFRA) can order them to be destroyed. The diagnosis may come because you wisely sought help with a suspected disease. Or, if disease is sweeping the area, your official veterinary surgeon (OVS) may send people to inspect all chicken flocks. (If you keep more than 50 birds you're required to register your flock with DEFRA to make its task easier when the need for eradication schemes arises.)

When your chickens have a disease that the government considers to be a threat to other poultry and poultry businesses, the Animal Health Act 1981 gives DEFRA the right to cull your animals. In this situation, DEFRA informs you of what to do, how to clean your premises and how to dispose of the chickens. You're unlikely to be compensated for the loss of your birds if you're ordered to destroy them.

If the choice is yours, consider what's best for you as well as for the rest of your flock. Many small flock owners, particularly those who keep birds as pets, develop a sentimental attachment to their chickens and want to try and save them. Sometimes, though, the better option is to destroy one or two ill birds rather than risk the health of your remaining ones. If the disease is a treatable one, you can choose to keep and treat the ill bird, knowing that you can treat the others if they catch the disease, but do isolate the sick bird from the rest of the flock and stay away yourself from other people's birds. Helpful websites (such as the Compendium of Animal Health & Welfare in Organic Farming at www.organicvet.co.uk) can help you diagnose whether your chicken's condition is serious and plenty of online chicken-keepers' forums provide a place where you can discuss issues with other chicken owners.

In some cases you may feel that your best option is to dispose of all the birds, and to thoroughly clean the premises or wait the required time until replacing the flock is safe. If you have sick birds in your flock but none have died, you may need to dispatch a living but ill bird to find out what's ailing the flock. Then you can submit it for a *necropsy* (the animal equivalent of an autopsy). Your vet or a university poultry specialist can guide you with all decisions on cleaning, disposing of dead birds and submitting birds for testing.

If your birds have been diagnosed with a highly contagious disease and you're trying to treat them, tend to them *after* you care for the rest of the flock and wash your hands thoroughly with soap and water between locations. In some diseases, such as Marek's, the 'dust and dander' – that is, the feather particles and skin cells – in the air carry the disease, and so you may also want to change your clothes and shoes after caring for them. Don't use the same tools or containers to carry food and water between locations without disinfecting them, and don't let children or anyone else handle a sick bird without gloves. If many birds appear ill, you may want to remove those that don't seem ill to another location instead of moving the sick birds. Handle them with gloves also.

Using simple natural remedies

Positive health management – 'preventing problems from occurring in the first place' in plain English – is the first rule of good health. The cornerstones of positive health management are maintaining hygiene in the chicken house, keeping a sensible number of birds, supplying fresh feed without mould and sensibly rotating your available pasture. By doing this, you give your flock the best possible chance of staying in good condition. However, keeping a few simple First Aid and preventative remedies up your sleeve will help your flock to fight mild bacterial infections that may come along and keep their immune systems in tip-top condition.

In the event of a mild tummy upset, adding cider vinegar to the water (100 millilitres in 2 litres/3.5 fluid ounces in 3.5 pints of water is a rough guide) for a few days changes the gut flora, flushing out negative bacteria. You can then replace it with good bacteria by administering a one-off feed of probiotic yoghurt, especially after you've given the chicken an antibiotic treatment. (Be sure to use plastic containers only when using cider vinegar because the acid within it dissolves zinc in galvanised metal.)

Keeping a clove of garlic in the water container is a general health tonic and blood cleanser as well as a deterrent to both internal and biting parasites. Just leave it there until it disintegrates and then add another. As a bonus, you never see vampires around chickens when you keep garlic in the water!

Give your birds access to a constant supply of both soluble and non-soluble grit at all times – your chickens know just how much they need. If you don't, they can bung themselves up with fibrous, stringy grass which then gets lodged in the crop and becomes smelly – a condition called *sour crop* – and may become infected.

Whatever the weather, chickens love to dustbathe, which is entertainment for them (and you!) as well as a natural behaviour and an all-round visit to a beauty parlour. Chickens love to look after their feathers, not just out of vanity, but necessity too. Their evolution has depended on them being able to move quickly and leap out of harm's way onto a high perch, and their feathers are crucial for that. Your birds turn fine sand, wood ash, clean shavings and earth into dustbaths; adding drops of tea tree oil and eucalyptus to their bathing material helps both if you suspect you have a louse problem and in helping to keep them away in the first place.

All these steps can be useful to you as a chicken keeper, but they're no substitute for expert advice when things get serious.

Administering Medications

You can treat parasites and some diseases with medication. When treating your chickens with medication, always follow the label directions exactly. Don't guess at amounts; measure. If you have trouble converting measurements from millimetres and centimetres to common household measures such as teaspoons, look for conversion charts (many are available online) or ask someone to help you. You also can determine the correct conversion by substituting water for a liquid medication or sugar for a solid one. To do so, measure the dosage into a syringe or cup marked with millilitres or centimetres, and then pour that into common measuring items such as teaspoons or measuring cups.

Keep medications and medicated feeds in their original containers with the directions for use, and discard them when they expire. Store them exactly as the label directs (some need to be kept refrigerated) where children, pets and other livestock can't get to them. Using expired or poorly stored medications can harm or kill your chickens.

Check label directions for the amount of time you need to wait before eating eggs or meat from chickens treated with medication. Followers of organic poultry-keeping practices have a policy of doubling or trebling the withdrawal period before using a medicated animal for food.

If you keep chickens for meat, avoid treating them with medications unless you absolutely have to. Use antibiotics only to treat genuine diseases – not as a preventive – and then only if a vet prescribes them. If you need to medicate your meat birds because of illness, whether by injection, adding medication to the water or feed or any other way, be sure to read and follow exactly the label directions on the product as to how long you must wait afterwards before dispatching the birds for meat. Then review your health management system to make sure that you avoid the problem in the future.

Don't experiment on your chickens with medications meant for humans or other animals. Birds' bodies metabolise medicines differently from other animals, and you may kill your chickens with your kindness. However, be aware that sometimes vets prescribe medications intended for other animals for use on your chickens simply because some drugs that are effective on them haven't been registered as a poultry product. Vets use their experience to work out an appropriate dose for the size and type of bird.

Encountering Death

When you keep chickens as pets or as layers, most birds live to a good age and eventually die of natural causes. Nonetheless, if you find yourself with a dead bird on your hands, remove it from the flock immediately. Take a close look at that bird and your remaining birds for signs of disease. Move the dead bird to an area with good light so you can inspect it closely and, unless you like smelly things, do this as soon as you find the bird.

When you're inspecting the bird, make the following observations:

- ✔ **Can you see any signs of a predator attack?**
- ✔ **Does the bird have any wounds, large clumps of missing feathers or an unusually twisted neck or legs?**
- ✔ **Do you see signs of entry into or damage to your housing or fencing?**
- ✔ **Does the bird's neck appear broken?** Sometimes predators don't actually touch a bird to kill it. Their presence may have caused a bird to fall off a roost at night or panic and fly into a wall or fence. In these instances, a broken neck is the cause of death.
- ✔ **If the dead bird shows no signs of injury, is it a very old bird?**
- ✔ **Is it a hybrid broiler chicken?** Heart failure in broiler birds often leaves them on their backs with their legs extended.

Some chicken diseases can be transmitted to humans. Anytime you handle a chicken or find dead birds you suspect are ill, wear disposable gloves, or use plastic shopping bags or something else that forms a barrier between you and the bird. Even when you've worn gloves to handle dead or ill birds, still wash your hands thoroughly with soap and hot water afterwards.

Sometimes you need to send a dead bird to a vet or a poultry specialist for a *necropsy* (animal autopsy). For example, when many birds are dying with no signs of injury and you want a definite answer to what's happening, a necropsy may be necessary. Ask your vet where you can send the bird (private vets experienced with poultry may do necropsies themselves) and find out the cost. You can then decide whether knowing why the chicken died is worth it to you. In cases where the dead bird may affect public health, such as suspected avian flu, vets generally do the necropsy or lab tests for free.

Even though technically you can do a necropsy yourself, we don't advise it. You're likely to be unsuccessful; some diseases don't leave many internal clues that a lay person can find, and so they're best diagnosed with lab tests.

If a necropsy is needed, the bird should be as 'fresh' as possible, which means it needs to be newly dead or stored properly. If the bird isn't going to be looked at straight away, place it in a sealed plastic bag. Double or triple bagging is a good idea. Place it in a cool place – refrigerated between 0–4 degrees Celsius (32–40 degrees Fahrenheit) is ideal, but don't freeze it – away from the flock and other animals.

After looking at the birds' internal organs and noting any signs of illness, the person performing the necropsy forms a diagnosis based on the observations. Because the signs of illness may be the same in many different diseases, the expert may only give you an educated opinion, not a definitive cause of death. In other cases, lab tests along with the necropsy results pin down the exact cause of death. If everything inside appears normal, the bird may have died from something other than illness.

When a professional does a necropsy and gives you a definitive answer or a high probability of what's wrong, the person can then advise you on what to do with the rest of the flock as well as suggest treatments if any are available.

If your chicken had a notifiable disease, whether because it has human health implications or because it's a notifiable disease that DEFRA requires you to report, the professional will report it. An official veterinary surgeon (or OVS) may then come to your home and inspect your flock and tell you what you must do with the birds.

After a bird is killed or dies, you must follow regulations about what to do with the dead body. You're not allowed to bury or dispose of a dead bird except through official licensed premises. The 'Animal Health' department within DEFRA (check its website at www.animalhealth.defra.gov.uk) can point you to the right places in your locality; for example, a registered hunt kennel may be allowed to take the dead bird off your hands. Don't be surprised if they refer to your dead bird as a 'by-product'; they do so because it goes to the same place as waste from the food industry.

If you're in an urban area and have no good way to dispose of a dead bird, contact a veterinarian. Vets usually have a way of disposing of dead pets and may let you bring the bird in, for a fee.

Part IV
Breeding: The Chicken and the Egg

'And to think we were attracted to you by your elaborate courtship display!'

In this part . . .

Some people love chickens so much that they want ever more, and want to know how to get their chickens to produce these extras. The chapters in this part cover breeding. Chapter 12 discusses mating chickens. Chapter 13 takes a look at the two methods of incubation: using incubators to do it yourself and using broody hens to do it the natural way. And Chapter 14 covers all you need to know about neonatal care . . . raising baby chicks.

Chapter 12

Breeding Your Chickens

· ·

In This Chapter

▶ Discovering how to sex chickens

▶ Checking out the reproductive system

▶ Looking at methods of mating and hybridising chickens

▶ Understanding fertilisation

▶ Selecting and preparing chickens for breeding

· ·

*W*hen you experience the joy of chicken ownership, you may decide you want more chickens, and what better way to get them than to produce your own? However, successfully mating chickens and producing chicks isn't as easy as you may think. Keep in mind that many modern breeds of chickens don't even sit on their own eggs – the desire to raise a family has been bred out of them. Therefore, you need an incubator to produce new chicks or a hen from a breed that still has the brooding instinct. (We describe incubation, both artificial and natural, in Chapter 13.)

In this chapter, we discuss mating chickens, how to choose mates, how to prepare chickens for breeding and how to recognise mating and reproductive behaviour.

Sexing Chickens: Cockerel or Hen?

You probably aren't surprised to hear that to mate chickens you need birds of both sexes. If you can't keep a cockerel for some reason, you can't raise any chicks from your hen's eggs, no matter how hard you try.

For some people, determining whether they have both sexes isn't easy. We've seen many people waiting to get eggs from a cockerel, and an equal number waiting for their 'cockerel' to crow. The first order of business in mating chickens, therefore, is determining the sex of your birds – called *sexing*.

Sexing young chickens

By the law of averages, a batch of chickens sold as *as hatched* or that you hatch yourself should be 50 per cent male and 50 per cent female, but real life doesn't happen quite that way. Imagine, for example, that you toss a coin ten times. You may get heads five times, or seven times or two times. The same is true with a batch of chicks. So don't assume that every batch of chicks contains equal numbers of male and female chicks.

Sexing newly hatched chicks is extremely difficult, unless the chicks are *sex/colour-linked.* Some breeds of chickens can be bred so that one sex is one colour and the other sex is another. This sex/colour linking obviously makes sexing quite easy as soon as the chicks hatch, and mistakes are rare. Large poultry breeding centres use it extensively. Unfortunately, however, many breeds can't be sex/colour-linked. (We talk about this subject in more detail in 'Producing sex/colour-linked colours', later in the chapter.)

Hatcheries often employ workers trained in how to sex baby chicks by examining their hidden reproductive parts. This work is delicate and exacting, but these hatcheries are usually pretty accurate when delivering sexed chicks. The problem is that if you order, say, 22 pullets (young hens) and 3 cockerels (young males), you probably won't be able to tell them apart when they arrive. Hatcheries don't mark chicks after sexing them.

Vent sexing

To vent sex a very young chick, you must open the *vent* or *cloaca,* the common opening for faeces and eggs to pass from the body, and look for a tiny organ similar to a penis called the *intermittent organ* on the inside. If you see one, you have a male. The problem is that so many variations in the intermittent organ exist that even sexing experts may take up to a year of practice to achieve a reasonable degree of accuracy. For someone hatching just a few clutches each year, mastering vent sexing is a tall order.

As a home flock owner, you're not likely to need to vent sex a tiny chick in many situations. This intrusive task is prone to disasters if badly or clumsily done, and so is a task better left to the experts. You can easily displace the chick's delicate internal organs and the chick may die or, worse, be irreparably damaged. The smaller the chick, the less likely you are to succeed in a correct diagnosis, and never attempt vent sexing with bantam chicks. The larger breeds, such as Rhode Island Red and Light Sussex types (Chapter 3 has more about breed types) and meat chicks, are easier to vent sex, but you can also sex them by the feather growth on their wing tips (as explained in the next section, with an equal degree of accuracy).

Sexing by appearance

Unless you have sex/colour-linked chicks whereby each sex is a different colour, sexing all chicks accurately before they get their feathers is impossible. When chicks hatch they're covered in a fluffy down. The first proper feathers that start to grow are the wingtips. Males and females show slightly different growth patterns here, and so looking at the appearance of the wing tip feathers gives you another way of identifying the sexes (see Figure 12-1). This method is a safer way to identify your chicks than vent sexing and can be just as reliable when you master the technique. Just follow these steps:

1. **Find a well-lit place in which to inspect the chick, under good light and above something light and of a contrasting colour from your bird, such as a sheet of card.**

2. **Hold it gently between your fingertips, and pull out the chick's wing just enough to inspect the feathers on the outer edge of the wing.**

   The first adult feathers will be appearing still wrapped in their protective cuticle.

3. **Check the *primary feathers* (the hard, spiny feathers on the outer edge) against the *covert feathers* (the next layer in) by inserting your card between them.**

   If the primary feathers are the same length or shorter than the covert feathers, you have a *cockerel* (a young male). If the primary feathers are distinctly longer, you have a *pullet* (a young female). (See Chapter 2 to understand better which types of feathers are which.)

This method works only in the first 24 hours after hatching; any later and the wing tips have changed again. A hatchery uses this method to sex day-old meat birds, for example.

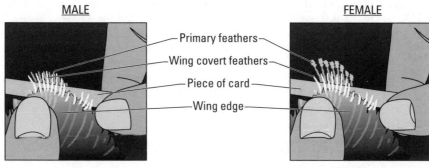

MALE

FEMALE

— Primary feathers —
— Wing covert feathers —
— Piece of card —
— Wing edge —

Figure 12-1: Using wing tip feathers to sex a chick.

Primary and covert feathers are the same length.

Primary feathers are distinctly larger than wing covert feather.

As the chicks mature and get more feathered up, (but you have to wait four to six weeks before the signs are really obvious), telling the sexes apart becomes ever easier. Here are some additional ways to do it:

- ✔ **Watch the hackle feathers.** Keep an eye on these feathers around the neck as they grow in. Males have pointed hackle feathers; females have rounded tips on hackle feathers. These feathers tend to be shinier in the male as they near maturity.

- ✔ **Examine the tail feathers.** At 3–4 weeks' old the female birds often have more tail feathers than the male. Growth patterns change over time. The male bird's tail feathers become more obvious and flamboyant the nearer to maturity he is.

- ✔ **Pay attention to patterns.** In many breeds of chickens, the females may have a different colour pattern to the males and as chicks feather up, you can determine the sex by the pattern. You can look up the colour patterns of the breed you're trying to sex to see what to look for.

- ✔ **Check out the combs and wattles.** Combs and wattles of males tend to grow faster and larger than those of females, although some females can have quite large combs. The combs of hens often flop over in large-combed breeds, but that can happen in cockerels too.

- ✔ **Consider size.** The legs and feet of males may grow faster, and overall, males tend to be a little larger than females at any age.

Behaviour can also be a clue. Males begin squaring off with each other from quite an early age. However, all fighting chicks are not males, and those that are never observed fighting aren't necessarily females. All chicks go through a ranking or pecking-order process early in life. In a group with a lot of males, some males low on the ranking scale act more like females. Finally, many people believe that males are generally more active as chicks than females.

Pick out birds that you think are one sex or the other, and mark them with a reusable leg ring so you can observe the individuals more closely. You can get these rings from poultry supply shops. In time, the correct sex reveals itself, and then you can judge how accurate your observations were. As you gain experience with chickens, particularly when you raise a certain breed for a long time, you get a feel for identifying the sexes at a young age, although you may still get a surprise from time to time.

Sexing mature chickens

Although sexing adult chickens may be hard for the novice, when you've been around chickens for a time you can begin to sex mature birds more easily. First, start with the obvious:

✔ **If the bird crows, it's a cockerel.** Adult males rarely go long without crowing in daylight.

✔ **If the bird lays an egg, it's a hen.** In rare situations where a hen has had some trauma to her ability to produce female hormones, she takes on the characteristics of the male, including crowing. The bird isn't fertile though and still looks mostly like a hen.

Chapter 2 describes the differences between cockerels and hens. Here are some more clues to look for:

✔ **Examine the feathers.** In cockerels, the neck or *hackle* feathers and the feathers along the back, called *saddle* feathers, have pointed tips. In hens these feathers have rounded tips. The hackle and saddle feathers may be a contrasting colour from the body colour in some breeds. In cockerels, the feathers have a shiny iridescence. Several long, thin, curved feathers, called *sickle* feathers, arch over cockerels' tails and are also iridescent.

✔ **Compare the combs and wattles.** The combs and wattles of cock birds are generally larger and more erect than those of hens.

✔ **Check for spurs.** Cockerels have a long, hard, toe-like extension sticking out of the inner leg, called a *spur.* Spurs are 4 to 8 centimetres (1.5 to 3 inches) above the toes and may curve slightly upwards. These start as a small bump in young males, become apparent as the cockerel becomes sexually active and grow quite large in older birds. Cockerels use their spurs for protection and fighting. But beware: old hens occasionally develop small spurs, too.

Reviewing the Reproductive System

Familiarity with the reproductive system helps you when you're trying to mate chickens and produce a new generation, and so in this section we discuss the basic biology of the reproductive systems of both hens and cockerels. The reproductive system of chickens differs from that of mammals and even from some other types of birds.

Remember that when we use the word 'egg' in this chapter we're referring to the familiar egg that humans eat, unless otherwise noted. These eggs are a package of genetic material from the hen, and a food supply for the embryo for when the genetic material combines with sperm from a cockerel and begins an embryo. In chickens, the true egg, or the female's genetic contribution to new life, is called the *blastodisk,* which we discuss in the later section 'How an egg is fertilised'.

Cockerels

Cockerels vary in age of sexual maturity. Fast-growing breeds may begin crowing, a sign of sexual maturity, at 20 to 25 weeks, but slower-growing breeds may take much longer. When a cockerel crows loudly and frequently, he's mature enough to mate. Examine his legs too – some cockerels develop a distinct red tinge on their scales, which is another sign of a sexually mature male.

A male chicken doesn't have testicles on the outside of the body as mammals do. Instead, two large testes are located near the kidneys along the cockerel's back. Each testis is attached by a tube or *sperm duct* to a small bump on the inside of the cloaca, called the *intermittent organ*. In the sperm duct just above the intermittent organ is an area called the *vas deferens,* which stores sperm produced by the testes until the cockerel mates.

Unlike that of most animals, chicken sperm are viable and active at the body temperature of the bird – about 41 degrees Celsius (106 degrees Fahrenheit). A young, mature cockerel of good fertility produces more than 30,000 sperm each second, but as he ages or if his health declines, sperm production also declines. As with all species, some cockerels are more fertile than others, and some are infertile.

Cockerels have plenty of sperm to mate numerous times a day with different hens. However, when a cockerel has more than 15 to 20 hens to 'service', he doesn't get to each of them every day. Cockerels sometimes pick favourite hens or ignore certain hens.

Light stimulates cockerels' *pituitary gland* (a hormone-producing organ), and so fertility increases in the spring and early summer when daylight is long. You can, however, manipulate cockerels by artificial light to keep them at top production throughout the year. When you intend to produce fertile eggs outside of the normal spring/early-summer period, give your cockerels the same lighting treatments you give your hens. (See Chapters 6 and 15 for information on lighting.)

Hens

Hens begin early embryonic life with two ovaries but the right one becomes dormant even before she hatches, leaving just the left ovary to produce eggs. Before a hen even hatches, all the eggs she ever has exist as tiny spots of genetic material in that ovary. She may have hundreds of them.

The ovary sits up along the hen's backbone, on the left side of the abdomen, just above an organ called the *oviduct.* The upper end of the oviduct, called the *infundibulum,* is roughly funnel-shaped, with protruding 'fingers' along the edge that catch the eggs released from the ovary. Fertilication of the egg occurs in this area. The rest of the oviduct is 25 to 35 centimetres (or 10 to 14 inches) long and runs along the hen's back, where it ends at an opening inside the cloaca. Pouches along the lower part of the oviduct collect and hold sperm for up to two weeks.

When the hen reaches sexual maturity – which varies with the breed and time of the year or lighting – the ovary starts forming egg yolk around various egg-germ cells. About 19 days are necessary to produce an egg yolk. Yolk is composed of fat, protein, carbohydrates and water taken from the hen's body. Egg-yolk production goes through several different stages in the ovary for as long as the hen is laying. When a hen is killed, you can see these yellow yolks in clusters surrounded by the ovary membrane.

When the egg yolk is completed, the follicle holding it splits and releases the yolk into the oviduct. The now ball-shaped yolk is covered with a film called the *perivitelline membrane*. The genetic material to begin a new chick, called the blastodisk, is collected in a white spot just under that membrane on one side of the yolk. (Check out Chapter 15 to see an illustration of the parts of an egg.)

Fortunately for omelette-lovers, even if a hen never mates and no eggs are ever fertilised, she still produces eggs. A hen is sexually mature and ready to mate when she produces an egg that's full-sized for her breed. If she's from a line of production layers, she may lay regularly for about a year.

Hens don't lay an egg every day, because about 25 to 26 hours is needed for each egg to make its way through the oviduct. The egg-laying, therefore, happens later each day until it reaches about 2 or 3 p.m. At that point, the hen 'resets' and skips a day or more of laying. She then begins laying early in the morning again. Each hen has her own individual cycles, even among egg-production breeds.

Hens of some breeds may lay in sporadic bursts, or only in the longest daylight periods of the year. If a hen gets the hormonal urge to sit on her eggs and raise a family – known as *broodiness* – she lays around 10–12 eggs, starts sitting on them and stops laying. Some breeds are more prone to broodiness, with Cochins and Silkies probably being the top contenders for mother-of-the-year.

Hens stop laying when they *moult,* that is, when they shed old feathers and grow new ones. The moulting period can last between four and ten weeks, depending on the breed. Poor nutrition, stress and illness can also cause hens to stop laying.

Hens continue to lay for several years, but egg production usually drops each year by an average of 25 per cent. Some very old hens (8 years of age is old) may still lay a few eggs in spring; others may stop laying for good when they reach around 5 years' old.

How an egg is formed

It takes about 25 hours for the egg yolk to proceed through the oviduct and be laid. All eggs, whether fertilised or not, follow the same procedure. As the egg goes along its route it first comes to the *magnum,* a large area where about half of the 'white', composed of water and protein, covers the yolk in a process that takes about three hours. The white part of the egg cushions and insulates the yolk.

The egg then travels to the isthmus and lingers there about an hour while it gets more 'white' added and is moulded into the familiar egg shape. It then goes to the uterus, called the *shell gland* in birds, where it stays for about 21 hours. The egg package gets some final 'white' material, a membrane around that package, and then a second tough membrane forms. Finally, the shell covers the whole package. Pigment glands may colour the egg at this point. The egg then passes into the vaginal area for a short period before being laid. In this area a clear, waxy coating covers the shell to help protect it from moisture loss and lubricate its passage.

About 25 hours after it begins its journey, the egg passes from the hen's body through the cloaca. It leaves the hen's body at a temperature of about 41 degrees Celsius (106 degrees Fahrenheit), the temperature of her body.

How an egg is fertilised

When a hen mates with a fertile cockerel, sperm enter her oviduct and are stored in several glands near the end of the shell gland or uterus area of the oviduct. Some may also begin to swim up the oviduct looking for an egg. Sperm can be stored and remain capable of fertilising an egg for more than two weeks, although fertilisation isn't as likely to occur after ten days, and the best fertilisation comes by day four. For you, that means a hen doesn't need to mate every day in order to lay fertile eggs. Sperm from multiple cockerels can be stored, and each day a different cockerel's sperm can fertilise an egg.

When a formed egg on its way down the oviduct passes over a gland storing sperm, some sperm are 'squeezed' into the oviduct and begin swimming towards the *infundibulum,* where the meeting of sperm and egg takes place. By the time the passing egg hits the sperm pouches a new yolk may be ready to be released from the ovary.

If a new yolk has dropped into the infundibulum when the sperm reach it, they have about 15 minutes to fertilise an egg. The blastodisk, or genetic material, is a tiny white spot just under the membrane surrounding the yolk. Only one sperm unites with the blastodisk, but through a quirk of nature, many sperm need to penetrate the membrane – about 30 or so – for a good chance of fertilisation to occur. Usually hundreds of sperm penetrate the membrane. Eggs are always fertilised before the whites and the shell are added to the yolk.

When the sperm unites with the blastodisk it becomes a *blastoderm,* or embryo. The cells begin dividing and making a baby chick immediately and continue to divide until the egg is laid and cools down below the hen's body temperature. Unfertilised eggs have a yolk with a small, solid white spot, which you can see if you crack open an egg and examine the yolk soon after it's laid. A fertilised egg has a white ring with a clear spot in the centre on the yolk.

Sometimes the ovary releases two egg yolks at the same time. They can both be fertilised in the infundibulum and proceed down the oviduct, where both yolks are covered by one shell, forming a double-yolk egg. These eggs can be hatched, but the chances of both chicks surviving are very slim.

Watching Reproductive Behaviour

In chickens, as with most animals, behaviour is initiated and influenced by hormonal and chemical signals from the body that get their clues from the environment.

Courtship and mating

Unlike many birds, the male chicken doesn't develop special plumage in the mating season. Healthy, sexually active cockerels may have a bit more iridescence to their feathers, their combs and wattles may be firmer, darker and glossier, but external plumage changes are slight. You may, however, notice a cockerel's legs take on a red tinge when he's mature enough to mate. This feature isn't so easy to spot on cockerels with dark legs.

Chickens don't have an elaborate courtship ritual. A male may circle a female in a strange, strutting manner, almost like walking on its tiptoes, occasionally dragging a wing. He may pick up pieces of food and drop them in front of her. She then usually crouches down, spreads her wings out a little and pushes her tail to one side. The cockerel jumps on her back, holds on to the feathers at the back of her neck with his beak and pumps his cloaca against hers for a few seconds. This move is referred to as *treading*.

Both birds' cloacas open, allowing the intermittent organ of the cockerel to get close to the vaginal opening of the hen, and sperm is deposited. Deed over, the cockerel jumps off, fluffs his feathers and struts off. The hen also fluffs her feathers, and then continues what she was doing before she was interrupted.

As part of this ritual, hens are supposed to submit when the cockerel tells them to, but occasionally they try to avoid the cockerel or fight him. Depending on how many hens he has available and his mood, the cockerel may accept this rejection, or he may chase her and forcibly mate her. Some hens may seduce the cockerel, crouching in front of him as an invitation. Some hens also crouch before human caretakers or other animals, especially if a cockerel isn't present.

Although a cockerel rarely avoids any hen altogether, he may pick favourites in a flock and mate with them more often than others. Hens also may pick favourites if they've a choice. Cockerels vary widely in the quality of 'husbands' they make to their flock. Some are very solicitous and gentle with their hens, giving them the best foods, leading them to nest sites, being fairly gentle during mating and even sitting on the eggs if they have to. Other cockerels are rough with hens and look after their own interests first. Where several cockerels are present in a flock, harems form with small groups of hens surrounding one cockerel. If a dominant cockerel sees another cockerel treading he may well run over and knock the cockerel off that hen and mate her himself. Each cockerel guards his harem and at the same time steals other hens from other groups.

Nesting and brooding behaviour

All hens, even those bred for maximum egg production, have a normal hormonal surge an hour or so before laying an egg that causes them to seek out a nest and do a little homemaking. They sit in a nest, turning around to shape it, arranging bedding with their beak and making crooning sounds. After hens lay their eggs, they may sit for a few more minutes and then get up and go about their lives. This behaviour is normal for 95 per cent of the time with laying breeds and a good deal of the time for dual-purpose and other breeds, whether the egg is fertile or not.

At some times of the year, however, usually when the days are at their longest, the hormonal surge in some hens doesn't drop much after a hen lays an egg – meaning that she's becoming broody. For the first one to five days, she may not spend all her time on the nest, but spends a lot of time there or nearby, and briskly defends it from those – for example, you – who try to remove eggs. After she's accumulated five to ten eggs in the nest, she sits tightly on it, leaving it only to eat and drink, and normally only when nothing she perceives as a threat to the nest is around.

Some hens are happy to share a nest with other hens. Silky hens often choose one nest to lay in and crowd into it together to raise babies.

Most egg-laying breeds don't get broody. Other breeds vary; some have a greater tendency than others to produce broody individuals. (We discuss natural incubation in more detail in Chapter 13.)

Looking at Mating Methods

You can breed chickens in two main ways:

- ✔ Group, or flock, mating
- ✔ Pairs or trios (one or two hens and a cockerel)

Unless you're just breeding chickens for fun – as pets or to let the kids see the miracle of life – the most important things to have in a mating or breeding programme are the following:

- ✔ **A clear purpose or goal:** Do you want to produce meat birds, layers or show birds?
- ✔ **Proper identification:** Individual birds need to be permanently identified.
- ✔ **Good record-keeping:** Written records are necessary; relying on memory just doesn't work. You need to keep track of *pedigrees* – the ancestors of your chickens – and the results you get when you cross the birds.
- ✔ **Knowledge:** A basic understanding of chicken reproduction, genetics and biology.
- ✔ **Healthy chickens:** Well-nourished, sexually mature birds.

If you have these things, you're ready to produce new generations of chickens, whether for profit or personal satisfaction.

Flock mating

In flock mating, you have a group of birds that you allow to breed freely. You can use a purebred cockerel with a flock of purebred hens of the same breed and produce fairly uniform purebred chicks, or produce hybrid chickens by using a different breed of cockerel from the hens.

Single cockerel

For best results, limit your flock to about 15 birds, made up of 14 hens and 1 cockerel. When you put more than 14 hens with 1 cockerel, overall fertility may be reduced because he mates with each hen less often. Smaller groups are even more desirable when breeding chickens of the heavy, loose-feathered breeds, because they tend to have lower fertility to begin with. Young cockerels can handle more hens than older ones without a drop in fertility.

Multiple cockerels

If you have more than one cockerel in a group, unless you balance the male–female ratio the cockerels spend more time fighting and protecting territory and less time mating, and so fertility may suffer. Even in a flock where the goal is just to produce as many meat birds as possible, having more than one cockerel may complicate things. You don't know whether one cockerel is doing much of the fertilisation and the other is just eating feed. And if a genetic problem crops up, you don't know which cockerel caused it. This system works best if you have a larger group and more cockerels, keeping the numbers to a maximum of ten hens to one cockerel. More than three cockerels generally even out the pecking order and a working truce is reached. The section on 'Courtship and mating' earlier in the chapter has more info.

A disadvantage of flock breeding is that it may keep you from identifying hens that aren't laying or hens that are producing less than ideal chicks for your goals. If your hens are producing a large percentage of infertile eggs, you may not even be able to tell whether some hens are being mated or which hens are producing fertile eggs.

Flock breeding is best used when you want quantity rather than quality and when space doesn't permit you to separate chickens into small breeding groups.

Pair and trio mating

Serious breeders of show birds and people trying to improve a breed put pairs or trios of birds – one or two hens and a cockerel – into small pens, large cages or arks. Keeping close records on this type of mating programme helps you identify which hen produced which results with the cockerel.

Remember that hens can store semen for two weeks or longer. Depending on her prior living arrangements, to make sure that you're getting the offspring from a certain hen and cockerel, you must isolate the hen for three weeks before introducing the cockerel – or at least not save her eggs for hatching for that length of time.

If a hen hasn't been with a cockerel for some time, the first one or two eggs she lays after being put with a cockerel probably aren't fertile. If you're using an incubator for hatching (as we discuss in Chapter 13), start collecting eggs for hatching with the third egg. If you intend to leave mother hen with her eggs, remove the first couple of unfertile ones.

With pair or trio breeding, you can rotate a good cockerel between cages every 3 to 4 days, and chances for fertilisation remain very high for the hens involved. If the cockerel is with hens for at least 24 hours every 3 to 4 days, you should be able to use him with several cages of hens with good results.

 When you keep hens in cages, small pens or arks, make sure that you supply them with a good nest box with clean bedding and move outdoor arks frequently to avoid poaching the ground. Producing lots of fertile eggs is no good if they get broken or become extremely dirty.

Artificial insemination

Artificial insemination isn't for most home flock owners. You can do it with chickens but you have to keep the cockerel close by because, unlike sperm from other species, chicken semen doesn't store or travel well.

Neither collecting semen nor inserting it in hens is all that difficult, but artificial insemination is a specialised craft – you need to observe and practise with an experienced person. If you're considering the idea, ask at an agricultural college that does poultry research if someone can demonstrate the procedure for you.

Artificial insemination is sometimes done with large, heavy-feathered breeds that have difficulty breeding or in the cases of a very special cockerel who's lost a leg – chickens need both legs to mate successfully. This method is mainly used commercially on very heavy turkey breeds whose enormous breasts render natural breeding impossible.

Selecting Birds for Breeding

Unless you just want a science experiment for the children, take some time to choose the right combinations of hens and cockerels from which to breed. You also need to prepare your birds carefully for breeding so that optimum fertility results. In this section we discuss the best ways to choose birds to breed.

Choosing the right combinations

No matter what kind of birds you intend to produce – show, meat or layers – the parents of those chicks always need to be the healthiest and soundest chickens that you have. Never breed birds with serious genetic faults such as extra toes, deformed beaks, drooping wings and so on. Even pass over for breeding purposes those birds with problems that you aren't sure are genetic. For example, cockerels need two normal legs to breed successfully (maybe that twisted leg was caused by slippery floors in the brooder and not by bad genes). Whatever the reason, that bird still isn't a good candidate for breeding.

Both the hen and the cockerel contribute genetic material equally to the new generation of chicks, and should be healthy and of normal size for the breed. Always breed the best cockerel you have to the best hens you have for the best results. The best cockerels to breed from – the most fertile, productive cockerels – are usually active birds that are a bit aggressive in protecting the hens, but a beautiful cockerel with hens that aren't anywhere near his breed standards isn't going to produce good-quality show birds unless you're very lucky. Equally, a beautiful bantam cockerel doesn't produce good meat birds, even if you mate him with huge hens.

Remember that like begets like: if you want to produce good laying hens, try to get eggs from your best-laying birds. That means putting *those* hens with that good cockerel – not the hens you don't mind separating from the flock because they don't lay well anyway. The cockerel you choose to produce those new layers needs to be from an egg-production breed as well.

Producing purebred chickens

If you're trying to produce new birds for showing, you need to study the breed standards for that breed of chicken. Identify which colours you're allowed to show and breed only those colours. Often, crossing two colours of a breed may get you an odd-coloured offspring that doesn't show well, and so you also need to study colour inheritance in your chosen breed. Know what type of comb, what size, what number of toes and other features your breed needs, and breed good representatives of the breed together.

The Poultry Club of Great Britain – a charity dedicated to preserving all breeds of poultry – has the information you need about specific breed standards for showing. (You can find the address and website in Chapter 3.) You may also want to attend poultry shows and look at good representatives of the breed you want to reproduce.

If you want to breed purebred chickens, but one of your birds is lacking in a particular area, you can sometimes compensate. If one bird has a fault, the other parent should be better than average in that area. For example, if one parent has feet that aren't feathered quite right, the other requires feet that are perfectly feathered for its sex. Don't forget about sex differences. Good tail carriage in a hen, even though her feathers lack the elaborate sickle feathers of the cockerel, can mean better tail carriage in both the cockerels and hens in her offspring. Genetics can be funny, though, and despite your best efforts, even two great-looking chickens that match the breed standards may not produce good-quality chicks.

So that you can track which birds produce which offspring, you need to keep good breeding records. When that ugly chick turns into a best-of-show winner, you need to know who its mum and dad are so you can produce more winners like it. The normal way of identifying chickens is to use numbered leg rings. (See Chapter 10 for info on identifying individual birds.) Good breeders also keep pedigrees – a list of the ancestors of each breeding bird.

A breeding record can be an elaborate computer program or as simple as a file card you write on. You need to include the bird's identification number, the cockerel's identification number and the date when you put the hen with the cockerel. You may also want to include columns for recording the number of eggs produced and set, when they were incubated, how many chicks hatched and any other information that's important to your personal goals for breeding, such as how many chicks turned out to be 'blue' coloured. No one perfect record style exists; feel free to personalise records to fit your needs. Table 12-1 shows an example of a breeding record.

Table 12-1		Sample Breeding Record Hen B4402: Spunky/Silky/ Blue Splash/D.O.B. 15-03-10/Pen 2				
Mated To	*Date Mated*	*Number of Eggs Produced*	*Number of Eggs Set*	*Date Set*	*Number of Chicks Hatched*	*Comments*
C304	30-03-11	12	11	25-04-11	10	All blue
C310	15-05-11	16	10	30-06-11	4	1 blue, 3 white; incubator problem

Inbreeding – the crossing of close family members – occurs often in the animal kingdom, but if you do it too often, the fertility and health of the offspring start to suffer. Related – but not too closely related – individuals, such as 'second cousins', are often the best choices for mating. Occasional *outcrosses* – breeding to totally unrelated birds that are carefully chosen to complement your own chickens – is also a good move.

When you're new to the world of showing chickens, you may want to ask a long-time breeder of your breed to look at your flock and help you choose which birds to mate and which offspring show the most promise. Knowing how to select the best birds for breeding show birds is as much an art as a science and taking the time to master this is part of the fun of the hobby.

When producing show chickens, you need space to keep young birds until you can see what type of adults they develop into. You can cull some chicks (Chapter 13 gives advice on how to do so) at hatching or in the brooder, such as those that have more toes or the wrong kind of comb, but usually you need to wait until the adult feathers are in and the body frame is filled out to judge accurately how good the birds are. Some people dispatch the rejected ones; others sell them or give them away as pets. If the birds come from large breeds, you may want to use them for meat when they grow large enough.

Producing hybrids

If you're interested in producing your own meat birds, or even some good laying hens, you may want to consider hybridising. *Hybridising* generally means mating two different purebred birds, although it sometimes involves mating a hybrid chicken to a purebred chicken or mating two different hybrids.

Crossing two purebred chickens together often produces what's known as *hybrid vigour* in the offspring. That means they're healthier, grow faster and produce more eggs or meat. Of course, you have to cross two breeds that both have desirable traits for the kind of bird you're trying to produce. For example, crossing a little Polish cockerel with a Cornish Game hen doesn't produce good meat birds just because the offspring are hybrids.

To get good meat birds, you need to cross two good meat breeds. The heavier utility breeds such as Light Sussex, Welsummer and Indian or Cornish Game give a good meat bird. Crossing heavy-bodied breeds that grow reasonably fast is the goal for meat production.

Be aware that your hybrid meat chicks don't grow as fast or with as much breast meat as the commercial meat chicks that you purchase from a hatchery. That's simply because the parent birds of those commercial meat chicks have been carefully selected over many generations for high-speed meat production. The particular strains of the parent birds are owned by large companies that carefully cross the male and female lines and provide eggs for other hatcheries to hatch and sell to you as broiler chicks. The huge amounts of time and money that have gone into this industry isn't something you can replicate in your back garden.

High-producing laying hens are often bred, not surprisingly, by crossing two high-producing egg-laying breeds. Such crossing occurs commercially most often with brown-egg-laying breeds, which are usually crosses of Rhode Island Reds with other brown-egg layers. Commercial white-egg layers are predominately White Leghorns. Be aware that when you cross a brown-egg layer and a white-egg layer, you generally get offspring that lay brown eggs of some shade.

Of course, some people want it all. Getting good meat birds that also lay well is the goal of many home flock owners, which means crossing a good laying breed with a decent meat breed. You probably never get a hybrid chicken that has fast-growing, deep-breasted, meaty males and hens that produce large quantities of large eggs, but you may get reasonably good meat and egg production.

Don't try to save commercial hybrids such as Hubbard or Sasso broilers or Isa Browns and breed them to try to produce more like them. Breeding two hybrids together produces unpredictable results: you get birds like both grandparents and lots of things in between. The meat birds don't become good breeders if you keep and allow them to mature – they're meant to be harvested when young and are prone to problems such as prolapsing in hens and heart problems in the males if you allow them to live beyond about 18 weeks.

Producing sex/colour-linked colours

The Isa Browns and many other high-production brown-egg layers are examples of sex/colour-linked crosses. Breeders found that when they crossed certain colours of chickens, the baby chicks that resulted were one colour if male and another colour if female. This fact enables breeders to sex chicks as soon as they hatch. The ability to do so is most important in egg-laying breeds, where you allow the hens to grow to maturity and dispose of the males early on. Commercial hatcheries often dispatch male chicks of sex/colour-linked laying breeds shortly after hatching.

As a home flock owner, you can use the sex/colour link to your advantage. Although distinguishing the down colour of chicks can be a little difficult in some crosses, most sex/colour-link crosses are easy to sex when hatched. As soon as you sex the chicks, you can keep only the sex you want or separate the sexes early for different styles of rearing.

The following are two basic sex/colour-linking methods that work in most (but not all) breeds that have the colours mentioned:

✔ **Black sex links** are produced by crossing a 'black' cockerel, such as a Black Orpington or Black Silkie, with a barred coloured hen, such as a Barred Plymouth Rock or Cuckoo Maran. This cross produces black chicks but the males have a white spot on the head. As they get feathers, the females are all black, sometimes with a few red feathers on the neck. The males are barred black and white with a few red feathers on the neck.

✔ **Red sex links** are produced by mating red or 'gold' cockerels, such as Buff Orpington, Red Sussex or Rhode Island Red, with hens that have what's called the *silver gene,* or *Columbian,* pattern, such as Light Sussex or the named 'Silver' types. Some white birds carry the silver gene, but determining this fact by sight is hard for most chicken owners. Instead, look for hens with white feathers tipped in black on the neck and tail (the Columbian pattern of, for example, the Sussex breed) or hens with the *silver lace pattern* (neck feathers that are black with white edges and body feathers that are white with black edges – the Silver Laced Wyandotte, for example).

In the red sex link, male chicks are white (with very light yellow down), and females are buff or red (with dark yellow, reddish or brown down). When mature, males are white with a few black tail feathers; females are red or golden red, some with a few white feathers or white down feathers under a red outer colour, depending on the breeds that you crossed.

Don't try to keep sex/colour-linked chickens and breed them to each other, expecting the sex linking to continue. In the second generation, the chicks are all different colours in both sexes.

Preparing Birds for Breeding

After you choose the birds you want to mate, you need to optimise their health and condition to get top-quality eggs that hatch into healthy chicks. Examine them for parasites, frostbitten combs and missing toes, all of which can affect breeding. For example, birds with frostbitten combs – combs that have black, shrivelled areas – are often temporarily infertile.

Breeding preparation is especially important as the birds get older. You need to do everything you can to optimise fertility. Nevertheless, some very nice-looking, lustily crowing cockerels may prove to be infertile, and some hens that lay regularly may not produce eggs that can be fertilised. You only discover these outcomes through mating the birds under ideal conditions and getting no chicks. It's just nature's roulette.

Feeding future parents

If you're not feeding a properly balanced commercial ration, you may want to start doing so before attempting to breed from your chickens. Getting a good layer ration so that eggshell strength and quality are good and the egg yolk is rich in vital nutrients to sustain chicks until they hatch is very important for hens. Even if your birds are on a layer's ration, offering oyster shell grit, with added calcium, enables them to top up. The birds themselves know best what their needs are.

Your hens and cockerels may also have problems breeding if they're fat. Overly fat hens may develop liver problems and stop laying; cockerels that are too heavy may have physical trouble mating. Feel the breast area of the bird in question. If the breast is very plump and the breastbone is difficult to feel, the bird is probably too fat. This often occurs when chickens in small pens get fed a lot of high-calorie snacks, such as corn. The answer is to cut back on feed and increase the opportunities for exercise. Hanging up a cabbage in the pen just above head height provides them with a good incentive for healthy exercise.

If your hens and cockerels are thin, increase their rations and check for internal or external parasites. Thin birds feel light, and their breastbone is very prominent, without much padding. You may need to worm the birds before breeding. Healthy birds that are thin may show an increase in fertility when rations are increased slightly. (For more information on internal and external parasites and how to check for them, see Chapter 11.)

As long as you provide your chickens with a well-balanced diet, you don't need to add vitamins or supplemental treats to encourage chickens to mate or to ensure fertility.

Providing optimum lighting and temperature

Lighting and temperature are crucial to maintaining fertility in chickens. Although chickens mate all year round, cockerels may not be fertile when the days are short and the temperature is cold. Likewise, hens may not lay well when the days are short or the temperature reaches extremes.

If you want fertile eggs outside the normal spring/summer period, you need to set up artificial lighting in your shelters so that the chickens get 14 to 16 hours of light per day. Likewise, if the temperature falls below 4 degrees Celsius (40 degrees Fahrenheit) or rises above 30 degrees Celsius (86 degrees Fahrenheit) you need to provide supplemental heat or cooling to get high-quality eggs for hatching.

The sexual maturity of chickens may be delayed if they reach 20 weeks of age during the period when the days are getting shorter. Becoming fertile may take until spring, or until you add artificial lighting. This is particularly true of rare showing breeds. If chicks hatch early, in late February or early March, they may become sexually mature with natural lighting in late summer. Egg-production breeds are more likely to become fertile in autumn and winter than other breeds. If chicks hatch later than March, you may want to add artificial light to your housing to extend day length to 14 to 16 hours. (For more information about lighting, see Chapter 15.)

Trimming feathers

In some cases, you need to trim feathers around the vent to allow the chickens to mate successfully. This problem is especially true of loose-feathered breeds, such as Cochins or very well-feathered Orpingtons. Examine both the cockerel and hen of heavily feathered breeds and trim or pluck out feathers around the vent area so that they don't inhibit contact between the two birds' vent areas.

Chapter 13

Incubating Eggs and Hatching Chicks

Almost as soon as people get chickens, they want more. Some people start thinking about how much fun they'd have letting a mother hen raise some cute little chicks. Other people want to add interest to their chicken hobby by buying an incubator and hatching eggs for fun or profit. And we agree – seeing little chicks emerge into the world and helping to nurture them through to maturity can be a wonderful and rewarding part of keeping chickens.

In this chapter, we describe the two methods of incubation: natural incubation by hens and the use of an incubator. We also explain how chicks develop in the egg and fill you in on what to expect during the hatching process. So if you want to bring new chicks into the world, this chapter is for you!

Plan for your baby chicks before you start developing those eggs. You need a space for a hen and her brood to be separated from the rest of the flock and protected, or a brooder to keep them warm and safe (see Chapter 14 for more on brooders). Before you get too far into the incubation process, read Chapter 14 for the full scoop on raising baby chicks, so that you know what to expect.

Making More Chicks: Incubation Basics

Incubation is the 21-day period from when a hen lays a fertilised egg to when a chick is big enough to survive outside the egg. Incubation can be achieved by the hen sitting on eggs and *brooding* them (keeping them warm) or by another source of heat that keeps eggs reliably warm, such as an incubator.

A long time ago, people noticed that some chickens were better mothers than others. This discovery led to people using the best mothers to hatch eggs from the not-so-good mothers, who were less inclined to sit on their own eggs. Likewise, it enabled farmers to develop the breeds of chicken that may not have been great mothers but which continually laid eggs.

When electric incubators arrived, chicken breeding got an even bigger boost. These incubators enabled farmers to multiply their numbers of chickens rapidly by combining eggs from several hens in one place. Plus, they were able to incubate eggs from any hen – even one who didn't want to sit on her eggs. Over the years, incubators became quite reliable and easy to use; today, even home flock owners can buy affordable incubators and use them.

The vast majority of home flock owners who are thinking of hatching chicks do so because they want to, not because they have to. You can obtain baby chicks from a number of sources, and so if you want baby chicks without the 'hassle' of hatching, you can skip the incubating and hatching process and buy them at a day old. But if you're up for something a bit more challenging (and fun!), hatching your own eggs may be for you.

Some people hatch eggs because they can't buy chicks of the breeds they want. Some rare breeds may only be available this way. Other people want to make their own hybrids – for example, to make a good meat bird that does well free-range, or to produce a hen that lays extra-large, deep brown eggs. Still other people want to replace old layers with hens just like them. If truth be told, most home flock owners just think, 'We have this beautiful cockerel and some lovely hens. Wouldn't they make some cute baby chicks?' You may want to do your own hatching once, just to see how it's done. Or you may want to show your children something of the wonder of nature. But whatever your motive, watch out – it can be an addictive hobby!

Choosing Your Hatching Method

You can hatch eggs by letting hens do it or by using an incubator. Here, we explain each method to help you choose the right one for you.

Don't believe those old wives' tales of putting eggs in your bra (or your wife's bra) to hatch them. Bras don't make good incubators. Humans aren't built for hatching eggs – human body temperature (37 degrees Celsius/98.6 degrees Fahrenheit) is much lower than the body temperature of a hen (41 degrees Celsius/105.8 degrees Fahrenheit). And whatever you do, don't use the microwave, even on low!

Looking at the two methods: Hens versus incubators

Before you can choose the hatching method that's right for you, you need to know a little bit about them. In this section, we tell you about each method.

Relying on mother hens

Mother Nature has perfected hens to hatch eggs; even the best incubator can't match the hatching ability of a good hen. But an incubator can still be a great help, especially when you don't want your hens to take a break from laying so they can sit on eggs, or when you have hens that don't want to sit on eggs.

A hen normally lays about 10–12 eggs before she starts incubating them. The first few days after she lays her eggs the hen doesn't sit on the nest all the time so that the development of the first eggs and the last aren't very far apart. After laying a certain number of eggs (it varies from one hen to the next), she then begins to sit tightly, only leaving the nest for a few minutes each day to eat and drink. She defends the nest fiercely.

The only things you need to do during the hen's incubation period are to protect her from the elements and predators and give her food and water. She does the rest. If you want to give the hen every advantage, though, isolate her from the rest of the flock in her own safe little area so that the other birds don't disturb her, break eggs or add their own eggs to her clutch.

 If this natural way of incubating eggs appeals, pick hens from breeds that are more inclined to brood than others (see the 'Understanding why some hens brood, and others don't' section later in this chapter for more on which breeds brood best). Keep a few of these hens around just to hatch eggs if you want.

For more information on the hen method, read the section 'Letting Mother Nature Do It: The Broody Hen Method' later in this chapter.

Employing the incubator method

Even the most advanced incubators require more attention than letting a hen hatch eggs, but incubators still offer some advantages over hens:

- You can use an incubator at any time of year; some hens lay eggs all year but refuse to sit on them in any season but late spring or summer.
- You can use an incubator when your hens don't want to sit on their eggs.
- Most incubators hold many more eggs than one hen can sit on.
- Incubators hatch chicks in a clean, protected environment.
- You can see the hatching process if you use an incubator with windows.

When you use incubators, you store fertile eggs until you have the amount you want to hatch (see 'Taking care of fertile eggs before incubation', later in this chapter). Then you place them in the incubator. You need to monitor the heat and humidity in the incubator closely, and if you don't have an automatic egg turner in your model, you need to turn the eggs at least twice a day.

We go into more detail about the incubator process in the section 'Going Artificial: The Incubator Method' later in the chapter.

Determining which method is best for you

The decision of whether to let a hen incubate your eggs or use an incubator depends on several factors:

- ✔ **Whether you have room for an incubator.** For best results, you need to place your incubator in an insulated room. If you don't have room for an incubator in a suitable place, going the hen route is your best bet.

- ✔ **What breed your laying hen is.** If you have hens of a high-production-layer breed, they're not going to sit on their own eggs (check out Chapter 3 for breed information). The urge to do so has been bred out of them. If you have room for more hens, you can get a hen or two of a breed that's known for sitting and put fertile eggs from the other hens under them (see the 'Understanding why some hens brood, and others don't' section later in this chapter for more info). If you can't have – or don't want – more hens, you need an incubator.

  If you have breeds of chickens that sit on their own eggs, you can wait until they feel broody and let them raise a family. We discuss broody behaviour in more detail in the later section "Understanding why some hens brood, and others don't'.

  Hens stop laying while they're sitting on their eggs and while they're caring for their chicks. So, if you want to keep egg production high, or if you want to raise chicks at a time when the hens don't feel broody, you need to use an incubator.

- ✔ **Whether you want to see the hatching process.** If you or your children want to observe eggs hatching, use an incubator. You can't see much as a mother hen hatches chicks because they hatch beneath her, and too much disturbance causes her great stress and may harm the chicks. You can, however, observe a mother hen caring for chicks and watch their antics in a natural environment after the hatching is complete, when she takes her babies out into the world.

- ✔ **What time of year you want the eggs to hatch.** Left to nature, most chicks hatch in late spring and through the summer months, and the hen may not begin to lay again until the following spring. If early laying is important, use an incubator to hatch eggs early in the spring. If you want chickens for meat, the time the hen picks to hatch the eggs may not be the best growing time for your needs. That doesn't mean you can't eat the excess chickens from any type of hatching and rearing, but if you want consistent meat production at convenient times for you, use an incubator to hatch meat chicks.

   If you want show birds to be at a certain age for a show, you may need an incubator to plan hatching at the best times. If you want to sell chicks, particularly those that grow to be point of lay pullets, before winter sets in plan for them to hatch in spring and very early summer, which can mean using an incubator to catch the high-demand period for chick sales.

- ✔ **What your priorities are.** If you're a home flock owner who isn't good at keeping track of things such as turning eggs and who isn't in a rush to have chicks at a certain time, using sitting hens is great. Sitting hens are the ecological way to increase your flock, especially if you only need a few chicks, because they don't require electricity to do their job.

Letting Mother Nature Do It: The Broody Hen Method

Some hens can lay and incubate eggs without any help from you. In fact, some may sneak off and surprise you with a cute little brood. But with a little help, you give a hen a better chance of success. After the initial organisation, though, step back and try not to interfere. Too much fussing by the human caretaker may cause a hen to abandon her nest.

Understanding why some hens brood, and others don't

Hormones affected by the lengthening daylight hours trigger the nesting instinct in hens. Hens have to be laying actively to get broody, and so they need to be in good health and receiving good nutrition.

For about an hour before an egg is laid, and for 30 minutes afterward, a hen of any breed gets a bit broody, going to a nest and moving nesting material, sitting in it and crooning. In some breeds at certain times, the hormones that influence motherly behaviour don't go down after the egg is laid; instead, they intensify over a week or so until the hen is sitting firmly on a nest.

Humans have selected certain breeds of chickens to produce fewer of the hormones that influence broody behaviour. This selection process was done to keep them producing eggs because, when a hen starts sitting on a *clutch* (group) of eggs, she stops laying. Table 13-1 shows some breeds of chickens, listed according to how often they get broody.

Table 13-1   Chicken Breeds and How Often They Sit

Breeds Unlikely to Sit	*Breeds that May Sit*	*Breeds that Often Sit*
Any colour Leghorn	Australorp	Any of the Sussex types
Cornish Game	Japanese	Araucana
Hamburg	Maran	Brahma
Isa Brown; any sex/colour-linked production layer	New Hampshire	Cochin
Minorca	Old English Game	Jersey Giant
Polish	Plymouth Rocks	Orpingtons
Rhode Island Red or White	Wyandottes	Silkies

Note: This table doesn't cover all breeds – just some of the more common ones. Breed descriptions generally note whether a breed sits or not. (See Chapter 3 for more breed descriptions.)

In breeds that sit on their eggs, laying a certain number of eggs and having the right environmental conditions may trigger broody behaviour. Still, even in these breeds, each hen is an individual, and some hens are better mothers than others. Some people think that hens that were naturally hatched and raised by their own mothers take better care of their nests. This has been proven true in other species, and so it's probably partly true in hens, too. Surprises happen, though, and even 2-year-old hybrids that have never seen their own mothers sometimes get in touch with their broody side and become excellent mothers.

Hens of naturally sitting breeds seem to be influenced by hens around them, and several hens may start sitting at one time. Several hens may sit on the same nest, though you can discourage this behaviour by having plenty of nests and enclosing each hen as she settles down to sit firmly. More than one hen in a nest often results in broken eggs and confusion about ownership of chicks.

Encouraging your hens to brood

If you have a hen from a breed of chicken that's known to sit on eggs, natural environmental conditions such as long days and warm weather encourage her to get broody. If you want a hen to sit in other conditions, provide a source of artificial light to increase gradually the length of the hen's day to 16 hours.

Don't encourage sitting in the coldest months of the year unless you can heat the coop to about 4 or 5 degrees Celsius (40 or 41 degrees Fahrenheit). Hens don't sit tightly for the first several eggs, and so those eggs may get chilled enough to die. Plus, extreme cold may keep even the later eggs from hatching.

Hens like dark, comfortable nests with plenty of nesting material. Just installing one of these nests makes some hens ready for a family. You may want to set up some of these nests away from regular egg collection nests, so that you can enclose the hen to protect her as she sits.

Some people keep hens for sitting in their own separate area, especially when the hens are a different breed from the hens producing most of the eggs. The more active breeds sometimes pick on sitting hens, and hens and chicks need protection from predators that a normal hen house may not be able to offer.

Hens don't need a cockerel around to get the broody feeling. They sit on infertile eggs – or even on fake eggs or stones. But some home flock owners notice that when a cockerel is around, he may encourage a hen to sit on a nest, even occasionally standing guard over her. On the other hand, some cockerels may harass sitting hens, especially if few other hens are around. In rare circumstances, cockerels have even been known to take over the sitting!

Adding eggs to the nest

If you have a hen that isn't broody, you may want to give her eggs to another hen that is broody. Most hens readily accept eggs from other hens. You can remove the eggs that the non-broody hen laid and place them with the ones you want to hatch, or just give the broody hen a few new eggs to sit on and remove a few of hers.

Don't give a large hen more than 10–12 eggs and give a bantam breed no more than 6–8 large eggs or 10–12 of her size if you want optimal hatching (see Chapter 3 for what constitutes a bantam breed).

If you want to save eggs to hatch from another hen, the eggs you add need to be very close to the same stage of development as the ones the hen laid. To achieve this, start collecting the desired eggs about the time you notice the broody hen starting to lay. Keep the eggs you collect in a place between 7 and 18 degrees Celsius (45 and 64 degrees Fahrenheit) and place them with the small ends down. Don't shake them or subject them to rough handling.

Turning them by tilting them one way and then the other at least once a day is a good idea. Store the eggs for a maximum of ten days; after that, they may not hatch as well. (See the 'Taking care of fertile eggs before incubation' section later in this chapter for more on how to store fertile eggs.)

If you're buying eggs to set under a hen, order them only when you have a sitting hen. Even if this approach means a few extra days for her to hatch the new eggs, you're better off with fresh fertile eggs than you are with storing fertile eggs for a long time.

You can delay a hen's sitting by leaving just one or two eggs in the nest. Remove those eggs before adding the new ones. Hens remain sitting a few days beyond the original hatch date (that's why they can also hatch duck and goose eggs, which take longer to develop), but if one or two chicks hatch more than two days before the rest, the hen is likely to abandon the late eggs so that she can lead the chicks to food and water.

You can replace each egg the sitting hen lays with a fake egg, or leave the eggs in the nest until you're ready to remove them all. You can buy plastic, wood and ceramic eggs from poultry supply shops or farm and smallholder suppliers; you can often find fake eggs in craft supply stores as well. Try to use natural-coloured fake eggs. Don't use polystyrene eggs. If you just remove the eggs without replacing them with fakes, the hen may not settle down to brood the eggs right away and may even abandon the nest. You want a hen to be sitting firmly on the nest before you give her the eggs you bought or saved to hatch. She sits firmly when she thinks the clutch is big enough – an empty nest gives her the wrong signals.

Mark your eggs with a pencil to avoid any confusion – mistakes happen and a different hen may decide to join in the fun and lay beside your sitting hen. It's not supposed to happen, but chickens haven't usually read the rule book. When you're ready to set the eggs, wait until dusk or night and then reach carefully under the hen and remove the eggs she has. She may peck at you, grumble and slap you with her wings. Then, protecting the new eggs in your closed hand, slide them under her carefully. That's all it takes.

Occasionally, a hen throws out a new egg; this can happen when the egg is a different colour or size from hers. If it isn't cracked, put it back in the nest.

Giving a sitting hen what she needs

A sitting hen needs isolation from the rest of the flock, or at least from non-sitting flock members. Chickens that aren't protected may be picked on by other hens, disturbed by the cockerel or picked off by predators because they refuse to move off the nest. Other hens may crowd into an unprotected nest to lay with the sitting hen, adding more eggs than she can sit on, or breaking eggs in a shoving match. Your best bet, therefore, is to get the hen sitting in a place where you can protect her from the start of incubation.

If you find a free-ranging hen sitting somewhere, do your best to protect her from the environment and surround her with some kind of small mesh fencing with a board or tarpaulin on top to protect her from predators. If you have a hen sitting inside, protect her from other flock members and predators.

A hen doesn't need a very big enclosure while sitting – a spare small coop or ark does the job (see Chapter 5 for more on types of chicken housing). If you don't have a spare coop, keep the hen separate from the other birds inside the original henhouse using a crate, cage or even a cat travelling basket to confine her, along with her food and water. This situation may be a bit cramped, but is the best solution. You must, however, provide her with her own place when her chicks hatch – even if you don't have this available from the start – because other chickens can kill them. Fortunately you have 21 days to organise this place.

While sitting, a hen needs enough room to stand and stretch and flap her wings when she gets off the nest, and she needs room for small feed and water dishes. About 0.3 square metres (3.2 square feet) for full-size hens and 0.2 square metres (2.1 square feet) for bantams does the trick. You can make the enclosure with a circle of fencing with small openings to keep the chicks in, or you can use a cage with small wire openings.

If a hen starts to incubate eggs before you've been able to provide her with protection, you may need to move her. Some hens don't object to a careful move of the whole nest to a protected place. When you move a sitting hen, do it at dusk or at night. Moving slowly and quietly, try to slide something (such as a big cardboard box) under to confine her with the eggs, and move the whole thing to the new spot. Be careful as you do this: wilder hens may fly off the nest as you move it and be hard to catch. The hen doesn't know where you've moved the eggs to and can't find them.

By morning, a confined hen should've decided to stay put, but make sure that she can't get to the old area. Some hens may abandon the eggs and go back to sit where the nest used to be. If you must move a hen quickly during the day, get a good nest area ready, move the eggs and the hen into it, try to darken the spot and then leave – don't fuss about her. Many hens settle down straight away, but if your hen doesn't, you can't do much about it.

Hens don't carry eggs to a nest, though they may roll eggs a short distance back to a nest by reaching out with their beaks. If you see eggs out of the nest, mark them with a pencil and put them back in the nest. If the hen rolls the marked egg outside again, she may be trying to discard it because she knows that it's no good, especially if no other eggs are out of the nest. Sometimes the hen is just clumsy. Other times, other chickens or predators may be disturbing the nest. If an egg comes out of the nest a third time, discard it.

Avoid washing eggs that you want a hen to hatch. You can try to clean the egg with a soft, dry cloth, but washing it can remove its protective coating and cause bacteria to invade the shell. Also refrain from handling the hen's eggs if you can avoid it because they get chilled every time you remove them. Special egg-washing disinfectants are available, but these are designed for use with incubators; a hen detects and discards an egg washed in disinfectant.

Clean the pen area if it gets dirty, and ensure that the hen always has food and water available. Place it a little way away from her, so that she has to get up to eat – getting up once in a while is good for her, and she may need to have a good empty out. A sitting hen doesn't eat or drink very much, because she isn't active and eating too much food creates the need to get off the nest more often. Food, water and a clean pen provide all the care she needs.

Caring for a hen and chicks

Twenty-one days have passed, and the big day is at hand. If a hen has remained sitting firmly on the nest until now, you have a very good chance that at least some of the eggs are going to hatch.

Hatching day

Chicks hatch under a hen. You know it's happening when you:

- ✔ See pieces of eggshell outside the nest.
- ✔ Hear peeping sounds.
- ✔ See a tiny head sticking out from under the hen.

The chicks hatch out over the course of 36 hours or so. Try to avoid disturbing the hen frequently during this time – you want her to stay on the nest until all the healthy eggs have hatched. So listen carefully and watch the hen for tiny heads popping out from under her, but don't shoo her off the nest to look. The hen stays put until all the chicks hatch and dry or until about two days after the first chick hatches.

The hen knows that she must move then so that the chicks can get food and water. If she's in a small area, she may go back to the nest with the chicks and some eggs may still continue to hatch if it's warm and they didn't get too chilled in the hen's absence. Ideally, however, you want the eggs to hatch as close to each other as possible, and so to achieve this aim, set the eggs all at the same time.

If you intend to collect the chicks and care for them in a brooder, collect them on the second or third day so that you don't disturb the hen and any eggs that are continuing to hatch (see Chapter 14 for more on brooders).

For the first day or two, chicks spend a lot of time hiding under their mother. Keep children or other visitors quiet when they visit, and don't allow them to pick up the hen or try to scare her away from the babies in order to see them. Not only are the chicks just as cute in a few days, but also the hen's more relaxed and the chicks more active.

The early weeks

Baby chicks have a 36-hour store of yolk to sustain them after they hatch, which allows time for the later chicks to hatch. However, they need food and water they can reach the moment the hen moves from the nest with them. Place chick starter feed in shallow pans and have a water container with a narrow opening nearby. You can leave a dish of the hen's food with her, but make sure that you leave no large open water dishes in the area because the chicks can drown in them. The hen shows them how to eat and drink.

If the chicks feel scared or cold at any time, they hide under the hen, who fluffs up her feathers to help cover them all. They don't normally need supplemental heat other than the hen, but if the temperature is below freezing (0 degrees Celsius/32 degrees Fahrenheit), suspending a heat lamp over one corner of the pen can be a real help. Ensure that you place plenty of bedding between the cold ground and the chicks.

The best thing you can do for the chicks is to keep them in a dry area. If they get wet and chilled, even the hen may not be able to warm them enough, and they don't do well. If the pen is outside on the ground, heavy dew in the morning can be a problem; mow short any vegetation that's inside the pen.

We don't believe in letting hens roam the yard with their babies straight away. Far too many dangers for the little ones lurk there – cats, dogs, children, snakes, hawks, even crows and seagulls pick off chicks. Cockerels rarely hurt chicks, but other hens often do. The mother hen does her utmost to protect them, but she can't always succeed.

Because catching baby chicks that are even a few days' old in a large area is difficult, you're better off keeping them in a smaller, more manageable area. A hen with chicks needs at least 0.6 square metres (6.4 square feet of space) in the beginning; after the chicks are feathered, they need more space. Several hens with their own chicks may share a large area, but some hens try to steal other hens' chicks.

Make sure that the chicks can't get out of the fencing or shelter on their own. The mother hen can't protect them if she can't get to them. Don't think that you can keep chicks confined with a low barrier – they quickly discover how to hop up and over. Chicks are also good at finding any small opening they can squeeze through.

Transitioning teens

You can separate chicks from their mother at any time, though they prefer to stay with her until they're properly grown up. If you do want to separate them, your best bet is to wait until the chicks have a decent covering of feathers at about 4 to 6 weeks of age – then they don't need any extra heat. If it's really cold – say, below freezing (0 degrees Celsius/32 degrees Fahrenheit) – and you really want to separate them from the hen, you may need to provide some heat until they're 5 or 6 weeks' old. At this age, you can take young chickens to their own pen, or take the mother hen out of the rearing pen and put her back with the flock.

When the chicks have a good covering of feathers, you can let them out with their mum to free-range, if that's your way of keeping chickens. Chicks are still a little vulnerable at this stage, and so be prepared to lose some. If you have penned chickens and want to introduce the mother hen and her fully feathered family to the flock, put them in the adult housing with the hen around the time the other birds are going to roost, or place their ark within the existing pen for a few days before letting them in together at dusk.

Keep an eye out the next day to see whether any young birds are getting picked on. If they're wounded and bleeding, separate them from the flock immediately. As long as they're not wounded, let them get to know their place in the flock. Make provision for the newcomers by providing extra cover, drinkers and feeders until they establish themselves. Sometimes older males don't tolerate young *cockerels* (male chicks), and so you may need to pen them separately.

For more information on raising chicks, see Chapter 14.

Going Artificial: The Incubator Method

You may opt for artificial incubation when a hen refuses to sit on her own eggs, or you may just find an incubator easier. Incubators have come a long way since they were invented in 1843, and chicken producers have discovered a lot about incubating eggs in the last 50 years or so.

With artificial incubation, you collect the eggs and put them into a container (the incubator) that keeps them warm and at the proper humidity until they hatch. If the incubator has windows, you can watch the hatching process!

Choosing an incubator

When buying an incubator, you can choose between models for almost all budgets and even build your own (you can purchase the 'technical bits'). You can buy deluxe models that you almost 'set and forget', as well as cheaper models that require more careful attention from you. Smallholder supply catalogues, online shops, farm shops and pet stores all carry incubators, and you can usually find second-hand incubators for sale in newspapers, magazines and online.

Considering your needs and wants

Before you buy an incubator, you first need to consider the following factors:

- How often you expect to be using an incubator.
- How many eggs you want to hatch at a time.
- What type of poultry, if any, other than chickens you want to rear.
- What your budget can afford.
- How much you want to be involved in the project.

If you or your children want to see the eggs hatching, you need to get an incubator with an observation window.

If you think that your egg incubation project may turn out to be a one-time thing or something you do maybe once a year, buy a cheaper incubator or a good second-hand one. You can always upgrade later. If you're already deep into the chicken hobby and plan to be hatching eggs frequently, buy a better incubator. If you want to hatch the eggs of turkeys, ducks or geese in addition to chicken eggs, make sure that the model you buy can accommodate the bigger eggs.

Identifying the right incubator for you

Incubators come in three main types. The following list describes them starting with the cheapest version (Figure 13-1 shows a couple of examples):

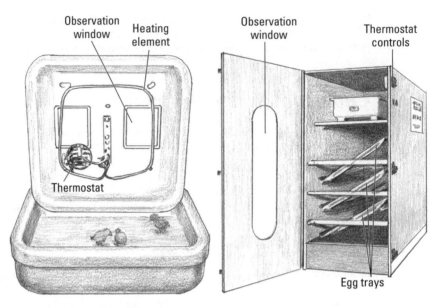

Figure 13-1:
Types of
incubators.

An example of a still-air,
tabletop incubator

Cabinet-style incubator

✔ **Still-air incubators.** Still-air incubators are the cheapest models, and are usually made of polystyrene or plastic. They range from models that can hatch just 4 eggs to models that hatch 36 or more eggs, depending on egg size. Many of these incubators have clear tops or observation windows.

Despite the name, air in still-air incubators is moved through the incubator by cooler air being drawn in from the bottom and rising as it heats up, escaping through some ventilation holes on the top. Most models have a reservoir in the bottom where you add water in order to provide humidity. Some models also incorporate automatic egg turners; otherwise, you need to turn the eggs by hand.

Still-air incubators require careful attention daily, and you need to follow the instructions exactly. They can provide a good hatch if you pay attention to the details.

✔ **Forced-air incubators.** Forced-air incubators move heated air through the incubator with a fan system. Although more expensive than still-air incubators, forced-air incubators generally provide for a better hatch because they eliminate the hot and cold spots that still-air incubators sometimes have. They range in capacity from 12 to 50 or so eggs, and some have observation windows.

Many forced-air models also come with automatic humidifiers and egg turners, and some deluxe models have warning buzzers and other bells and whistles. You still need to give some attention, but less than you would with the still-air models. These models are probably the best type for most home chicken flock owners.

✔ **Cabinet-style incubators.** Cabinet-style incubator units are for the serious chicken hobbyist. These incubators are large units with forced air, automatic humidifiers and egg turners and they have several shelves to accommodate eggs at different stages of hatching or of different sizes. They occasionally have observation windows, and they cost a considerable amount unless you're fortunate enough to find a good deal on a used model.

Automatic egg turners are standard in some incubator models and vary from model to model, but you can add them to others. They increase the number of eggs that hatch by turning the eggs consistently and gently, tilting the eggs in a different direction every few hours. They don't actually roll the eggs over. Turn off automatic egg turners three days before the chicks are due to hatch.

Most incubators require electricity (although gas-powered ones do exist) and are meant to be used in a well-insulated room without vast temperature fluctuations.

Accessorising your incubator

You need to buy a few incubator accessories if they aren't supplied with the incubator. Some necessary accessories include the following:

✔ **Hygrometer.** You can buy *hygrometers,* which measure humidity, that are made specifically for incubators or for household use. Many of the remote-style thermometers also measure humidity.

Humidity is crucial to a good hatch, and so make sure that you buy a good instrument. Many of the better incubators have built-in hygrometers, some of which add water automatically from a reserve tank to adjust humidity.

✔ **Thermometer.** Place a thermometer at a level near the top of the eggs. You can place your thermometer on a stand or hang it to achieve this level. Position it in the centre of the incubation space so that it doesn't touch the eggs or the walls of the incubator.

Typically, when an incubator has a window, you read the thermometer remotely from outside the incubator because the minute you open the incubator, the temperature drops; therefore use a thermometer that shows measurements large enough to read from the observation window, if you have one. Many remotely read thermometers are on the market, too – you place a probe or sensor in the incubator, and it transmits a reading to a device located outside the incubator. You can find such thermometers made specifically for incubators or for general household use.

The correct temperature is critical to the success of incubation, and the thermometer must show each degree. Even a difference of 1 degree can affect hatching. Buy a good thermometer and handle and store it carefully.

You may also want to get an *egg candler,* a light that helps you look inside an egg to see if it's developing. You can buy or make these instruments. (See the section 'Looking Inside the Egg' later in this chapter for more on candling.)

Setting up and caring for your incubator

After you buy an incubator, carefully read the instructions that come with it and keep them safe so that if you decide to use the incubator again in a year or so, you have the instructions to refresh your memory.

Your next step is to find a place to set your incubator near an electrical outlet. The incubator will be there for a month and so you don't want to put it in the middle of a high-traffic area where it can get knocked off or jostled. Ideally place it in a room that's about 15 degrees Celsius (59 degrees Fahrenheit) and away from windows, heating vents and doors that may cause wide fluctuations in room temperature.

If you have children or pets, you may want to put the incubator where it's out of sight and out of mind. Children are prone to opening the incubator too often, handling eggs or tampering with the controls. Pets can hear the chicks peeping in the shells in the last days of incubation and may damage the incubator trying to get at the eggs. They may also jostle controls while inspecting the incubator or, in the case of cats, sleep on top of it, blocking the vent holes.

Have the incubator set up and running for at least 24 hours before adding eggs and monitor it carefully to make sure that it's working correctly.

Cleaning your incubator

Every time you use the incubator – even if it's brand new – you need to clean it. Remove any debris and then wash the incubator with hot soapy water and rinse. Be sure to wash and rinse the screens the eggs lie on, and egg-turning racks, too. Consult your manual to see which parts you can safely submerge in water before doing so.

If the weather is nice and sunny, leaving the incubator open for a few hours in the sun is a good way to kill germs and dry the incubator at the same time. Otherwise, dry the inside of the incubator with clean paper towels.

Don't use strong disinfectants or cleaning solutions on incubators. If any chemical residue remains it can harm the embryonic chicks by passing through the porous eggshell. Instead, we recommend that you use disinfectants specially formulated for the job of sterilising eggs and incubators. You can find these disinfectants at the places you buy incubators from and in most poultry feed and equipment merchants.

Adjusting the temperature

Having the right temperature – for the entire time of incubation – is one of the crucial steps to good incubation. As a guide, for still-air incubators, the temperature needs to be 39 degrees Celsius (102.2 degrees Fahrenheit). In most other incubators, the temperature needs to be 37.5 degrees Celsius (99.5 degrees Fahrenheit), but read your incubator's instructions for information on the correct temperature for that model. Make sure that you set the temperature of the incubator correctly before adding eggs to your incubator.

Outside temperatures, how often the incubator is opened and the stage of development the chicks are in can all affect the temperature inside the incubator. If the incubator has a thermostat, as most do, the correct inside temperature should be maintained automatically, but do still check the thermometer often and adjust the thermostat right away if the temperature is incorrect. A short period of cooling, as when turning eggs manually, doesn't harm the chicks unless it happens too often, but even a short period of time at temperatures above 40 degrees Celsius (104 degrees Fahrenheit) can kill most of or all the chicks.

When chicks get near hatching, their bodies produce some heat, and this change may cause the temperature to rise in the incubator. Pay close attention and adjust the temperature if necessary close to hatching time.

At hatching time, people often open the incubator frequently to check on the chicks. Doing so too often can chill them and may even cause them to die during hatching.

Paying attention to humidity

Chicks require the right humidity to develop correctly in the egg and to be able to hatch. The humidity in the incubator needs to be about 55 per cent until the last two days, when you need to bump it up to 65 to 70 per cent. A hygrometer can help you to keep these levels of humidity right.

You can usually increase humidity by adding water to some sort of reservoir. In still-air incubators, the reservoir is generally a pan under the screen on which the eggs sit. Other types of incubators have different reservoir systems. When you add water to these reservoirs, make sure that you heat the water to the incubator temperature first. Some people like to add some specialised incubator disinfectant to the water to reduce the likelihood of bacterial contamination.

If water is dripping off the inside of the incubator or the eggs seem wet, the humidity is probably too high, which can kill your chicks. You need to increase the ventilation by opening additional vent holes or even cracking the lid open a little for a short time – just keep track of the temperature to make sure that it doesn't drop while you're doing so.

Making sure that the incubator has adequate ventilation

Chicks require oxygen to breathe. They pull oxygen in through the pores of the egg before they hatch. As they grow they require more oxygen, and so you may need to increase ventilation. As they hatch and breathe air, ventilation is critical. For this reason, all incubators must have some way for fresh air to enter the incubator and stale air to leave. Still-air incubators usually get their fresh air through small openings you can open or close; other incubators pull in air with a fan through a vent.

Determining whether an incubator is ventilated correctly can be difficult. The first step is to carefully read your incubator instructions. Unless the instructions direct you to do so, don't cover any ventilation holes in the incubator. Most incubators require that you open additional vents near the hatching date. If instructions are missing, look for capped vents and open them on day 18 of incubation. Adjust the temperature and humidity if opening vents causes them to fall too low. If the incubator has no capped vents, it probably supplies enough ventilation.

Finding and storing fertile eggs

After you have your incubator ready, the next step is to source fertile eggs. Finding fertile eggs may take a little time if you don't have your own hens and a cockerel, and so start looking well before you want to hatch eggs. Even if you have your own hens and a cockerel, you may not be getting fertile eggs. (In Chapter 12, we talk about how eggs are fertilised and how to make sure that your own chickens are laying fertile eggs.)

Finding fertile eggs

Fertile eggs are usually available in the spring and summer, but they may be very hard to find at other times if you're not producing your own.

Try to find a local source of fertile eggs if you can. Sending eggs through the post and getting a high percentage of them to hatch is a bit of a gamble – the average rate of hatch under ideal conditions for fertile eggs sent by post is only 50 to 60 per cent. Normally, you don't get any guarantees when you have fertile eggs mailed to you because the seller can't control the shipping temperatures or the way the eggs are handled in transit.

The following list names some good places to obtain fertile eggs:

- ✔ **Friends.** If you have friends with both hens and cockerels, they may be able to save fertile eggs for you.

- ✔ **Local farmers.** Some farmers sell free-range eggs at farmers' markets. If one such market takes place near you, ask the owner to save you some eggs for hatching. Eggs that have been handled correctly – for the intended purpose of hatching – are more likely to hatch than those that someone has washed and cooled for consumers to eat.

- ✔ **Online chicken forums and smallholders' websites.** Online sites are good sources for breeders looking to sell fertile eggs.

- ✔ **People who sell show stock and rare breeds.** These people may also sell fertile eggs.

- ✔ **Poultry auctions.** As well as live poultry, auctions generally have facilities for buying and selling fertile eggs in the right season.

- ✔ **Poultry or smallholding magazines.** These publications usually have a comprehensive list of breeders and enthusiasts from all over the country that can supply fertile eggs.

If you're saving your own eggs or you have the chance to pick which eggs you want, always choose the cleanest ones. You can brush off dirt with a dry cloth; some people even use fine sandpaper. If eggs are heavily soiled, best not to use them. Discard also any cracked eggs and any eggs that have very thin shells or are oddly shaped; the latter seldom produce good chicks.

Eggs you buy from the supermarket don't hatch because those hens aren't kept with cockerels. Even the eggs labelled organic are most likely from female-only flocks. We've heard of eggs hatching from free-range hens that were bought at farmers' markets, but don't count on hatching eggs that were meant for eating.

Taking care of fertile eggs before incubation

The most important thing to remember about fertile eggs meant for hatching is that they're living things – you need to handle them gently and keep them at the right temperature or they die. Don't shake them or toss them around – if you do, you may kill the embryo.

Ideally, avoid washing fertile eggs that you intend for hatching. Washing eggs removes the protective coating that the egg gets as it leaves the hen, and the egg's pores can then draw in bacteria. The warmth of incubation turns the contaminated egg into a factory for bacteria reproduction, often killing the embryo and spreading to nearby eggs. Washed eggs are therefore far less likely to hatch than unwashed eggs, and the unhatched, washed eggs may also affect the hatching of eggs around them. Starting with a perfectly clean egg is better than having soiled eggs, however. If you're determined to set eggs but the eggs available are soiled, using a specialist egg wash disinfectant can help, but is very much a compromise measure.

When a hen lays an egg and it cools below her body temperature, the egg goes into a state of suspended growth until conditions are right again. During that time, store eggs that you intend for hatching at temperatures between 7 and 18 degrees Celsius (45 and 64 degrees Fahrenheit). If the egg temperature drops below 7 degrees Celsius (45 degrees Fahrenheit) for very long, the embryos become weak chicks that probably don't survive the whole incubation period, or that have deformities if they do; if the egg temperature goes above 18 degrees Celsius (64 degrees Fahrenheit), the embryos may start to grow.

In cold or very warm weather, collect eggs for hatching from your hens as soon after they're laid as possible and then move them to the right conditions. If you're collecting your own eggs, you may need to put them in a cool basement to suspend growth. Always keep them out of the glare of sunlight.

When storing eggs for hatching, keep them with the pointed end down (an egg box is good for this purpose) and rotate them from side to side twice a day to keep the early embryo from sticking to the shell in a bad position. You can store eggs like this for about a week without much drop in vitality, but after a week of storage, the percentage of eggs that hatch drops sharply. After two weeks of storage, few eggs hatch.

If you're getting eggs from anywhere but your own hens, have your incubator set up and ready to receive the eggs as soon as you get them. If you're collecting eggs from your hens, store eggs until you have about a week's worth; only then set up the incubator and put them into the incubator all at once. Using a pencil or non-toxic marker, mark the eggs with the date you set them. If you're going to be hand-turning the eggs, put an *x* on one side so you know which eggs have been turned.

Try and avoid having several different hatch dates in the same incubator unless you have a cabinet incubator with multiple shelves or drawers. Eggs at different stages of incubation require slightly different care.

If you do have eggs at different stages of development in the incubator (more than two days apart), having a second incubator is handy so that you can move the eggs that are scheduled to hatch. You can get incubator boxes called 'hatchers' for this purpose. Hatching is naturally a messy process and having a separate space for this process is a good idea if your incubator is in constant use.

Caring for eggs in the incubator

A mother hen seems to know instinctively what her eggs need. In very hot weather, for example, she gets off the eggs to let them cool a little; if it's cold, she sits tightly. Her body provides the perfect humidity for the eggs, and she fills it with water herself. When you take over the job of incubation, you can never be as good as a hen, but with careful attention to details you can have a successful hatch from an incubator.

Turning eggs

Hens don't actually turn their eggs with their beaks on a regular basis as many people think. (They do occasionally rearrange them with their beaks, but it's usually for their own comfort.) Instead, their coming and going from the nest and shifting positions to get comfortable alter the position of the egg several times a day.

Turning eggs is a matter of debate, but most experts believe that you should change the position of eggs two or three times a day for the first 18 days of incubation. Automatic egg turners can do this task for you, or you can do it yourself by rolling the eggs to a new position. The turning keeps the embryo from becoming attached to the outer membranes and the eggshell. If you're turning the eggs yourself, do it quickly so that you don't chill the eggs too much.

If you have egg racks for turning eggs in your incubator, place the eggs in the racks with the small end down. If you're using an incubator without racks, lay the eggs on their sides. Cluster them in the centre of the incubator if lots of room is available.

Wash your hands before handling eggs. Oil or bacteria from your hands can cause hatching problems. Warm hands are much friendlier to eggs than cold ones. (How do *you* feel when someone touches you suddenly with cold hands?) And be sure to wash your hands again *after* touching the eggs. Eggs can have harmful bacteria on them.

On day 18 day of incubation, stop turning the eggs, or if you're using an automatic egg turner, be sure to turn it off. If you have multiple clutches of eggs in one incubator, now is the time to remove the 18-day-old ones to your hatcher incubator, which should be warmed up and have the correct humidity for the eggs. At this stage, the chicks are getting into position to hatch, and they don't have much room to move around anyway. If you change the position of the eggs at this point, the chicks have to reposition themselves again for hatching, and doing so wastes valuable energy and may even mean that hatching is impossible for them. Chicks can hatch from the wrong position in the egg, but doing so is much harder, and many of them die trying.

On day 18 day of incubation, you also need to increase the humidity in the incubator from about 55 per cent to 65–70 per cent. You may want to increase the ventilation – refer to your incubator's instructions to see whether they advise doing so. Also set up and warm your brooder on day 20 so that you can transfer the hatched chicks to it (see Chapter 14 for more on brooders).

Hatching eggs

Eggs that you put in the incubator at the same time should hatch within 18 hours of each other (see the section 'Knowing what to look for: Stages of embryonic growth' later in this chapter for more on embryo growth). Chicks struggle to get out of the egg, and some time may be necessary for a chick to hatch fully. If a chick requires help in hatching, it usually isn't a strong, healthy one.

When chicks start hatching, people get excited and they want to open the incubator and handle the chicks. Stop right there! Every time you open the incubator, you lower the temperature and humidity and make things harder for those still hatching. Chicks are fine in the incubator for quite a few hours while the others hatch, and so leave them alone until they're dry and fluffy. Remove the dry fluffy ones every six hours and put them in the brooder. If any eggs are left after 18 hours from the time the first chick hatched, you can leave them for another 24 hours, but don't expect much.

Looking Inside the Egg

When you're incubating eggs, you may want to know what's going on inside them. If you're new to incubation or you have children, you may want to open an egg every couple of days and look inside. Opening an egg kills the embryo inside, of course, but does give you a fascinating look at the miracle of a chick forming in just 21 short days – from a glob of cells to a baby chick that can run around and feed itself. If you decide to open eggs and look inside, we suggest opening them on days 3, 7, 12 and 16 of incubation.

You may even want to set extra eggs so that you can sacrifice some for this little biology lesson. Perhaps you come upon some eggs that didn't develop embryos, and so set a few more. For example, if you intend to open four eggs, you may want to start eight more eggs than you want to end up with.

If you don't want to open eggs and sacrifice the chicks inside just for a biology lesson, you'll be relieved to know that you can get a glimpse inside without opening an egg. In this section, we show you how.

Egg ultrasound: Candling an egg

Candling is a way of looking inside an egg without opening it and killing the embryo. Candling involves shining a bright light on the egg in a darkened room and seeing a shadow inside. You can see the size of the air cell, veins in an egg with an embryo and the dark mass that indicates the yolk and the embryo.

Candling works best with light-coloured eggs, but you can candle brown eggs, too. If you do so quickly in a warm room, you don't harm the developing chick. You can buy candlers from poultry supply places, but you can also make one relatively easily.

To make a candler, follow these steps and refer to Figure 13-2:

1. **Line a cardboard box with aluminium foil or use a metal can with reflective inner walls.**

   The foil makes the light stronger by reflection.

2. **Place any good, strong source of light inside the box.**

   An LED flashlight or reading lamp works fine, as does a 60- to 75-watt incandescent light bulb.

3. **On one side of the box, make a 2.5-centimetre (1 inch) hole.**

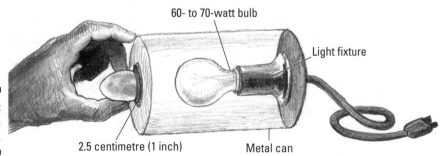

60- to 70-watt bulb

Light fixture

Figure 13-2:
Making an
egg candler.

2.5 centimetre (1 inch)

Metal can

To candle an egg, follow these steps:

1. **Wash your hands.**
2. **Turn off the lights in the room.**
3. **Remove an egg quickly from the incubator and hold it outside the box against the hole in your candler box.**

The beam of light shines into the egg, allowing you to see shadows inside the egg. You're looking for the size of the air cell, dark embryo masses and blood veins inside the egg.

Don't keep the egg out of the incubator too long – no more than three minutes. Mark the egg before you put it back in the incubator, so you know which eggs you looked at. If you intend to candle a few eggs more than once, you may want to number or otherwise identify each egg. You can candle an egg as many times as you want, but the more you handle the egg, the less likely it is to hatch.

Knowing what to look for: Stages of embryonic growth

A tiny chick develops very quickly. Almost as soon as the sperm penetrates the female reproductive cell – called the *blastodisk* or *true egg,* which is a small white spot on the yolk – the cells begin dividing to produce a chick. (For more information on fertilisation, see Chapter 12.) By the time the egg is laid 26 hours or so later, a ring of cells has already begun to form. If the hen incubates the egg or you put it in an incubator, cell growth rapidly continues. But most eggs cool and go into a state of suspended growth for at least a few days.

As a chick develops, the air space in the egg enlarges, and the egg loses weight. Looking at the egg's air cell changes during candling is one way to see whether a chick is developing (see Figure 13-3).

After 24 hours of incubation at 37.5 to 39 degrees Celsius (99.5 to 102.2 degrees Fahrenheit), the chick's head, eyes, digestive system and backbone begin to form. These features are still quite small – you need a magnifying glass to see these structures well – but rest assured that they're there (see Figure 13-4).

After 44 hours of incubation, the heart begins beating. Two circulation systems begin to function – one in the embryo and one in the *vitelline membrane,* which surrounds the embryo and acts similarly to a placenta in mammals, absorbing food and oxygen from the egg white and transferring it to the chick.

Candling eggs at five days

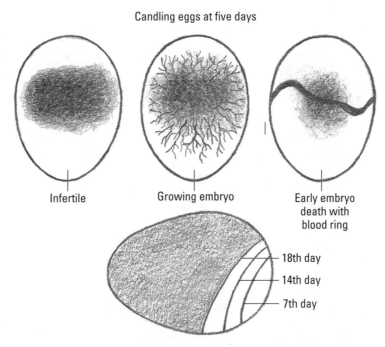

Figure 13-3:
Air space
changes
in an egg
during incu-
bation; what
a fertilised
egg looks
like during
candling.

Infertile

Growing embryo

Early embryo
death with
blood ring

18th day

14th day

7th day

By the end of day 3, the beak begins to form, limb buds grow and you can see temporary 'gills' on the embryo.

By the end of day 4, the embryo begins flexing and moving and generally rotates 90 degrees to lie on its left side. All the internal organs form by the end of the day, as well.

By the end of day 5, the reproductive system is formed and begins to differentiate into male or female, although you can't tell the sex by looking at the embryo.

By the end of day 7, you can see digits on the wings and feet that make toes and wing sections.

By the end of day 10, you can see the tracts where feathers grow, and the beak hardens.

By the end of day 14, the toenails form.

By the end of day 16, the embryo occupies much of the space inside the egg and begins to use the egg yolk as a food supply, because the egg white is used up.

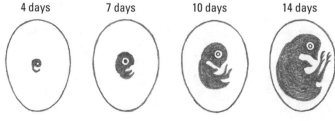

| 4 days | 7 days | 10 days | 14 days | 17 days |

Figure 13-4:
The stages
of embry-
onic growth.

On day 18 of incubation, a chick begins to prepare to hatch. Its head moves toward the large end of the egg and the chick curls its feet up close to its head. The head gets tucked under the right wing. From day 18 onwards, you may be able to hear peeps coming from the egg if you listen closely.

By the end of day 19, the yolk is drawn up into the chick's body in preparation to hatch.

On day 20, the chick pierces the membrane around it, sticks its beak into the air space in the big end of the shell, and begins to breathe. As air is used up, on day 21, the chick uses its *egg tooth* (a hard area on the end of the beak) to break a hole in the eggshell, letting in outside air. The chick chips away at the shell in a circular pattern, frequently stopping to rest. This process can take from 30 minutes to several hours. When the chick has cut the shell completely around in a circle, that top circle pops off, and the chick uses its legs to push the back part of the shell off. At this point, the chick is wet and tired, and it may lie there for an hour or more to rest and dry off before getting up and moving around.

Helping a Chick to Hatch

People sometimes feel sorry for a chick that appears to be having trouble getting out of an egg. It may have *pipped* (made a tiny hole) and have its beak out, but it may seem unable to proceed any farther.

You may feel a great temptation to help these chicks, but doing so can cause more harm than good. You can't just pull the shell off. Hatching is a slow process even in ideal conditions – you have to be patient.

Generally, when eggs pip but then fail to hatch, the temperature, oxygen level or humidity is too low. The temperature may drop because someone is constantly opening the incubator to check on the chicks' progress, or poor ventilation may cause the chicks to become weak because of a lack of oxygen. Where humidity is too low, dry chicks can get stuck to the egg and can't get out.

Other possible reasons for a chick struggling to hatch are that the humidity is too high (which can drown hatching chicks because their lungs get saturated in damp air and they can't breathe), the membrane dried out too much around the chick or the chick is in a bad position for hatching. If the pipped hole isn't in the large end of the egg, the chick is in the wrong position. If you see no hole but hear peeping inside, the chick again may be in the wrong position. If a hole is in the right area but the chick can't seem to finish hatching, the membranes may be dried out.

In these cases, helping – slowly and carefully – can save a healthy chick. If you decide to try to help the chick, here's what to do:

1. **Make sure that the chick is still alive – it moves or peeps if it is.**

2. **Make a warm operating area with a padded, clean surface.**

3. **Sterilise a small pair of nail scissors and a pair of tweezers with rubbing alcohol or by boiling them for a few minutes. Have some clean, warm water nearby, too.**

4. **With the scissors and tweezers, carefully pick little pieces of shell off the thick membrane underneath, around the hole in the egg. The chick has already pierced the membrane in one spot if it has started to hatch. Work like the chick would, circling the large end of the egg and removing half the shell.**

   No membrane may be around the air cell, allowing you to remove shell pieces easily. If the chick has started a hole in the side or small end of the egg, it's in the wrong position, which is why it's having difficulty hatching. You need to be even more careful in this case.

   Be careful not to cut the chick. The membrane underneath the shell is loaded with blood veins. If you cut a vein in the membrane, it bleeds profusely and weakens the chick or kills it. Stop at once and put the chick back in the incubator. You can't do anything to stop the bleeding. It probably stops, but does weaken the chick. You can start trying to help again in an hour or so if the chick is still alive and not yet free.

5. **When you've removed half the shell and the head and neck are exposed, moisten the membrane that's left with a little warm water and place the chick with some shell still attached back in the incubator.**

   The chick probably just lies there for a while, and then becomes active and wiggles out of the rest of the membrane and egg. If the chick isn't up and walking within an hour or so, it's probably too weak to survive.

Some chicks struggle to hatch even when conditions are right. Many experienced chicken owners believe that chicks that can't hatch on their own in good conditions are doomed to die or to live a weak, unhealthy life.

Pinpointing the Problem When Things Go Wrong

Hatching chicks should be fun, but in the natural world things don't always go to plan. If something does go wrong, you need to identify the problem so that you can get it right next time. The following list details the most common problems:

- ✔ **If the eggs hatch early,** the temperature was too high in your incubator, or you counted the days incorrectly.

- ✔ **If the eggs hatch late,** the incubator was a little cold, or you counted the days incorrectly.

- ✔ **If the eggs are hatching over a wide time spread (more than 18 hours),** your incubator may have had hot or cold spots in the incubator, or you may have improperly stored the eggs before putting them in the incubator.

- ✔ **If many eggs don't hatch by day 22,** several possibilities may be to blame. Pull some eggs out and open them. If no embryo is inside, the eggs probably weren't fertile to start with, or you stored them too long or improperly. If embryos started forming and then died, the temperature or the humidity of the incubator wasn't right, the ventilation wasn't adequate or the chicks were killed by bacterial contamination or a disease. Not turning the eggs enough or handling them roughly can also cause embryos to die.

If you open eggs and find some living embryos, put the unopened eggs back and give them some more time. If many eggs don't seem to be hatching, always check a few eggs before discarding them all.

- ✔ **If chicks hatch but they appear 'sticky' with egg goo on them,** the humidity may have been too high, the temperature may have been slightly low, the ventilation at hatching may have been poor or they may have an inherited condition that causes this symptom.

- ✔ **If your chicks are malformed or crippled,** the temperature may have been slightly too high during incubation, or disease, chemical contamination of the incubator or not turning eggs enough during incubation may be to blame. Bad legs can also be caused by hatching chicks on slippery surfaces.

- ✔ **If hatched chicks have large, bloody or mushy navel areas when hatched,** the incubator temperature may have been slightly low, the humidity may have been too high or the chicks may have a disease called *omphalitis,* which is caused by bacteria invading the navel area. For this reason, always hatch chicks on a dry, clean surface.

If your attempt at hatching with an incubator isn't successful first time, clean the incubator thoroughly and start again. Getting to know your machine takes time. You may just have been unlucky and had a power cut that was long enough to arrest development at a crucial time.

If you have chicks that are living but weak or malformed, they may be in pain. Don't leave them to die slowly. The kindest thing to do is to cull them. This job isn't pleasant but is part of the process when you take on hatching chicks.

The quickest way to cull chicks is to dislocate their necks. Place the chick's neck, just below the skull, against a hard right-angled surface – the edge of a door is fine – and push firmly to remove the head from the neck. The head may detach completely from the body or the neck may be broken inside the neck skin, in which case check that a gap is present between the skull and the hard neck bone. Culling chicks in this way isn't nice, but you know the job is done.

Chapter 14

Raising Chicks

. . . .

In This Chapter

▶ Getting familiar with brooders – commercial and home-made

▶ Helping the hen to help the chicks

▶ Providing food and water for chicks

▶ Knowing what's normal when rearing chicks

▶ Combining chicks and children safely

. . . .

*A*ll babies need food, water, warmth and protection, but some need more help than others. Many baby birds hatch out naked and helpless, relying on the parent bird to forage and bring food back to the nest for them. However, in the pheasant family – to which chickens belong – when chicks hatch they're already wearing warm suits of down (also called *fluff*), have their eyes open and are fully alert. After merely an hour or so, the chicks stand up and move around, and they're fast on their little feet after just a few hours. Chicks are also able to feed themselves. Their major needs from some-one else are warmth and protection.

Being warm and protected are the primary things with which chicks need outside help, and so we cover those subjects in this chapter. We also discuss suitable feed for chicks and the growth stages of chicks, and take a look at any adjustments in the care they may need as they grow.

You can give warmth and protection to chicks in two ways: by putting them in a brooder or by helping a hen to protect them and letting her keep them warm. A good *broody hen* (when the hen sits on the eggs) is probably a good *mother hen* (when the chicks have hatched) and does all the work of keep-ing the chicks warm. This option is the easiest and most natural one, but sometimes when you have chicks no broody hen is available, in which case you need to use an incubator. If you have a hen raising your chicks, jump to 'Helping a Hen Provide Warmth and Protection', later in the chapter.

Understanding the Basics of Brooders

Baby chicks need to be kept quite warm during their first few weeks of life. Without the proper warmth, they don't eat or drink and soon die. They snuggle under their mother's breast, where the feathers have thinned out and they can get their backs up against her warm body. They leave her warmth to find food and water and do a little playing, but as soon as they get cold, back under her they go.

Chicks raised without their mother need a source of warmth too, and that's where you come in. Unless you have the money and means to heat a whole room constantly to 35 degrees Celsius (95 degrees Fahrenheit), you need to provide a brooder for your chicks. A *brooder* is an enclosed area that enables chicks to find that perfect warm spot but also allows them a little space where the air is cooler. A brooder also protects the chicks in the absence of a mother hen.

You can find many types of brooders (check out the later 'Buying a commercial brooder' section for more details). Most use a heat lamp, but some use gas or propane to heat the unit. Alternatively, you can make your own brooder. Fashioning one from materials found around the home or from local shops is relatively easy to do (see the later 'Making your own brooder' section).

Knowing when chicks need a brooder

Chicks need to go straight from an incubator or hatcher to a brooder. (We discuss incubators in Chapter 13.) If you're getting your chicks from another source (buying day-olds from a commercial hatchery, perhaps), put them into the brooder as soon as you get them. Set up the brooder a few days in advance of the chicks arriving so you can see how it operates and adjust the temperature as necessary to keep it around 35 degrees Celsius (95 degrees Fahrenheit).

Your chicks need a brooder for at least a month, depending on the weather and where you house them. If you're looking after chicks at a very cold time of the year, you may need to keep them warm even longer, especially at night.

As well as being vulnerable to the cold, baby chicks can overheat, too. If the weather is very hot (above 32 degrees Celsius/90 degrees Fahrenheit), take special care to see that the brooder doesn't get too hot. Temperatures above 38 degrees Celsius (100 degrees Fahrenheit) are too warm. Older chicks may be uncomfortable at high temperatures as well, especially if the humidity is high. In this case, you need to provide some ventilation or other cooling to lower the temperature. They may still need the brooder to confine them and keep them safe, however.

Comparing brooder sizes and shapes

Large-scale chick-raisers use dedicated buildings where they brood chicks. As a home flock owner, however, you're likely to rear fewer than 100 chicks at a time, which means you just need some form of container in which to do it. If you have a large group of chicks you can partition off a corner of a room or building or make an adjustable circular pen (see 'Making your own brooder' later in this chapter).

Average-sized chicks need about 225 square centimetres (say 36 square inches) (that's 15 centimetres x 15 centimetres/6 inches x 6 inches) each in a brooder to begin with. This area looks like a lot of room at first, but chicks grow rapidly, and it doesn't look so big when they're a month old. Some brooders are expandable, and so you can increase the size as the chicks grow.

Brooders need to be at least 45 centimetres (18 inches) deep, especially if the source of heat is overhead, as is the case with heat lamps. If your brooder uses heat lamps, be sure to adjust them upwards as the chicks grow so their heads can't touch them and bear in mind that as you add bedding, the floor level for the chicks rises. A deeper brooder can prevent older chicks from jumping out and it needs some sort of cover to prevent the chicks from escaping.

Chicks need to be able to move from an area of optimal warmth to a cooler area within the brooder so they can find just the right temperature that makes them comfortable. Rectangular brooders allow the heat source to be at one end with progressively cooler temperatures towards the opposite end of the brooder. With square or round brooders, the heat source needs to be in the centre. Chicks usually fare better in a container where the corners have been rounded off, which stops them from huddling together in the wrong place.

Getting the temperature just right

For the first week, baby chicks need a temperature of 35 degrees Celsius (95 degrees Fahrenheit). Each week thereafter, lower the temperature by 5 degrees until you reach the surrounding room temperature outside the brooder. Gas or propane brooders generally have a thermostat, but if a heat lamp is your brooder's source of heat you need a good thermometer in the brooder to keep track of temperatures until you're confident enough to 'read' your chick's behaviour. Use a non-mercury thermometer (in case it breaks) with markings large enough for you to read from outside the brooder. You can get brooder thermometers at poultry supply shops, but any thermometer that you can read with ease is workable.

When checking the temperature, measure it at the height of a chick's back, about 5 centimetres (2 inches) or so from the floor. Check the thermometer often in the first days of using a brooder so that you know how the temperature inside fluctuates with the temperature outside.

Be aware that a thermometer may not correctly register the heat produced by infrared heat lamps because those lamps are designed to heat the chicks, not the air. With an infrared lamp, you need to watch the chicks to see whether they're warm enough. Your chicks' behaviour is the most accurate guide to whether the temperature is correct (see the next section). We discuss infrared heat lamps in more detail in the section on 'Choosing bulbs' later in this chapter.

If you need to adjust the temperature inside the brooder, raising or lowering the heat lamp or changing the number size or wattage of the heat lamp bulbs can help. Most brooders have solid sides to help hold in heat. You access the brooder from the top and ventilate it there, too. Hot air rises, and adjusting the open area at the top helps regulate the temperature. If the temperature still keeps getting too hot, a few holes drilled in one or more sides of the brooder, a few centimetres/couple of inches from the bottom, helps airflow because cool air is pulled in as hot air rises. Just make sure that the chicks can't squeeze through the holes.

Gauging temperature by your chicks' behaviour

After you gain a little experience, you don't need a thermometer to tell you whether the temperature in a brooder is right. The actions and sounds of the chicks let you know if the temperature is too hot or cold, and watching them for a few minutes tells you whether everything is all right (see Figure 14-1). Keep these tips in mind:

- ✔ If the chicks walk casually and quietly around, eating and drinking and some of them are peacefully snoozing near the heat source, the temperature is probably fine.

- ✔ If the chicks huddle in a pile near the heat source or in a corner, cheeping shrilly, they're cold. Remember that noisy chicks are unhappy chicks. Cold chicks also eat less and are less active.

- ✔ If the chicks spread themselves along the edges of the brooder, far from the heat source and have their beaks open, panting, they're too warm. Hot chicks drink more, eat less and aren't very active.

Eventually, if you don't correct the temperature, your chicks start dying.

Figure 14-1:
Watch
chicks to
see if the
tempera-
ture in the
brooder is
right.

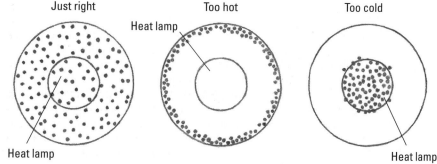

Lighting the brooder

If you use ordinary heat lamp bulbs to provide heat in a brooder, you probably don't need to install additional lighting. If you're heating the brooder by infrared or red heat bulbs, propane or gas, and if your brooder is in a place with no natural light, however, you may want to add some overhead lighting. For chicks designed to be pets or layers, keeping the extra overhead lights on for about 12 hours a day is sufficient; however, for modern broiler chicks, depending which type you choose, the lights may need to be on for longer. These chicks need to eat more to sustain their rapid growth, and chicks don't eat in the dark. You can eliminate this problem by heating your brooder with white heat lamp bulbs, giving chicks enough light to eat round the clock if they want to.

If possible allow the natural daylight hours to regulate your chicks' habits alongside their brooder lamp, whatever the type. Doing so eases the transition to the time when they no longer require extra heat and enables chicks destined to become layers to develop naturally. As the chicks become older and are ready to move to a pen, get them used to less light at night. You can dim room lighting each night by using a dimmer switch, or moving the heat bulbs farther away. If the housing you're transferring the chicks to gets dark gradually at night as with a natural dusk, the birds probably adjust quite well without any elaborate preconditioning. If, however, the housing gets very dark with the flip of a switch, you need to install a nightlight, or do some darkness preconditioning to avoid a panic situation.

Providing suitable bedding for chicks

Chicks need some bedding in the brooder to absorb moisture, dry droppings, reduce smells and provide grip while they strengthen their walking muscles.

You can use many things for bedding, but pine wood shavings are easy to find and are the best choice for many people. Other good options include clean sand, coarse sawdust and fine grit, such as chick grit.

Don't use paper (newspaper or any other kind), cardboard or plastic on the bottom of brooders unless you use it as an insulating layer with an additional bedding material on top. These materials get slippery when wet, and the chicks can develop *spraddle legs,* a permanent disfigurement. Other things not to use include cat litter (which is often treated with dyes and scents and may swell if ingested), hay, leaves and straw. Hay, leaves and straw, in particular, may contain harmful substances, get slippery and mouldy when wet and aren't absorbent. Avoid shavings made from unknown woods, too, which may be chemically treated or come from hardwoods. When moist and warm, hardwood shavings can develop a harmful mould. If you ever notice mould on bedding, remove it straight away.

How often you need to change the bedding in a brooder varies with the type of litter, the age and number of chicks, the type of brooder and many other factors. In general, if the bedding is damp and smelly, you need to change it completely. If the bedding appears dry and fluffy but dirty-looking, just add a fresh layer on top. After a while, the litter gets wet or too deep for the brooder, and you need to change it.

When changing litter, you usually need to remove the chicks. Place them in a large box where they're safe as you work. Unless the temperature is really cold, they stay warm enough for a short time without a heat source. If you wash the brooder with disinfectants or strong soaps, make sure that you rinse the brooder well and dry it before adding bedding and returning the chicks.

Always scrub a brooder between batches of chicks. Use a good disinfectant, such as common bleach (1 part bleach to 4 parts water), rinse well and let it air dry.

Buying or Building Your Brooder

You can buy brooders, or parts to make them, from poultry supply sources; you can also put together very functional brooders with common items available at hardware or do-it-yourself shops. Whichever option you choose, be sure to place your brooder in a weatherproof building or you waste heat and energy.

Buying a commercial brooder

If you want to raise many batches or large batches of chicks throughout the year, you may want to look into buying a commercial brooder.

You can buy units ranging from small brooders with plastic tops and built-in heat sources for a handful of chicks to brooders designed for hundreds and hundreds of chicks. Be realistic, and choose something that's appropriate for the number of chicks you're likely to hatch or buy. For example, no matter how enthusiastic you are about raising chicks, shelling out for an enormous industrial brooder is pointless when you plan to raise just a few chicks. Think practicality, too – chicks create a lot of dust as they dry off and fluff up, and the bigger the brooder, the more you have to clean.

If you do decide to plump for a really large brooder, you won't find them for sale in pet shops or poultry feed supply outlets, although these places may act as manufacturer's agents and be able to order one for you. If not, you need to search around online.

Making your own brooder

If you plan to brood just a few chicks once or twice a year, a home-made brooder is the most economical and practical way to go. Home-made brooders, of course, need to be inside a dry building – ideally with an electricity supply. However, you probably don't want to keep a brooder in your house. Baby chicks smell and make dust, and a brooder creates an added fire hazard. Chicks can also be a source of salmonella bacteria.

A brooder may work in the basement or garage, but a barn or shed is better if you have one. Some people like to set their brooder on the floor, but we like them elevated so that we can reach the bottom without stooping. If you do set yours on the floor, place something such as cardboard or layers of newspaper under the brooder to insulate it from the cold floor or ground.

Building the body of the brooder

Whatever kind of brooder you build, key things to bear in mind are to make it:

- ✔ Easy to move and clean.
- ✔ Large enough for the batches of chicks you intend to brood.
- ✔ Strong enough to keep out predators.
- ✔ Not easily ignited or melted by a heat source.

The most common way to build a brooder body is simply to make a box out of wood. You can put a bottom on the box or set it on a tray or the floor. Use fairly thick wood for good insulation and a piece of small wire mesh in a removable frame for the top. You also need pieces of wood, glass or Perspex to arrange on top to cover as much of the opening as necessary to keep heat in but still allow for ventilation. If you're brooding in a warm – and safe – environment such as an enclosed barn or sturdy shed you may not need a cover at all. Brooders with solid sides work best because the chicks aren't exposed to drafts. However, if you're brooding chicks in very hot weather, a wire side or two may help with ventilation in the daytime.

Here are some common brooder body ideas:

- ✔ **Farm water tanks.** You can make excellent brooders from the large tanks used to water livestock that you find in farm supply stores. These tanks come in various sizes and are made of metal or heavy-duty resin material. Pick a fairly deep tank in the size that allows about 15 x 15 centimetres (6 x 6 inches) per chick. A tank about 1.3 metres long by 1 metre wide (4 feet long by 3 feet wide), for example, is big enough for 25 chicks. In addition to the tank itself, you need small-mesh wire for the top, and wood, clear plastic or other solid material to cover part of the top. Water tanks are often oval or round in shape; oval ones are easier to fit covers on.

  Water tanks can be expensive but last many, many years. Look around for used tanks for sale at farm sales or auctions. You may even find one that a farmer has discarded. A leaky metal tank works just fine for a brooder and is often free.

- ✔ **Children's paddling pools.** You can make a quick brooder out of a child's swimming pool. Choose a hard plastic- or metal-sided one – not a blow-up pool or one made from flimsy plastic – and look for one at least 45 centimetres (18 inches) deep. Leaky pools are fine, and so keep your eyes peeled for discarded pools. You don't need the liner for metal pools, just the sides.

- ✔ **Plastic storage tubs and containers.** With a little care, you can use these containers for small batches of chicks, but they may melt if a heat bulb is too close to the sides. Don't use a top without holes for ventilation!

- ✔ **Parlour board.** If you're likely to use your brooder several times, a good idea is to invest in some *parlour board* – the strong plastic walling sheet used in farmer's milking parlours. You can cut parlour board in half lengthways and curl it into a ring to make a circular brooder. The walls are high enough to keep chicks in and you can adjust the size simply by letting the walls in or out and using strong bulldog clips or similar grips to hold them in place. Just add more strips to make a bigger brooder. The material is resilient, long lasting and easy to keep clean by washing and sterilising, and you can store it flat when you're not using it. Many professional poultry-keepers use this type of system.

If you're really into recycling, by all means, think outside the box, so to speak. For example:

- ✔ We've seen an old chest freezer being used as a brooder. It's well insulated, and so it holds heat well, and is easy to clean. You can remove the motor to make it easier to move (though it's still a bit heavy) and you can lower or raise the lid to regulate heat. You can put in a wire top under the lid and attach two blocks of wood to the edge of the top opening so that the lid doesn't ever shut completely by accident and shut off the air supply.

- ✔ We've also seen dryer drums, old bathtubs and stacks of tractor tyres used as brooders.

- ✔ You can even use large aquariums for small groups of chicks. Make sure that the glass bottom is well covered with bedding, though.

Take these ideas, use your imagination and you may find the perfect brooder.

Avoid using cardboard boxes or rings of cardboard unless you use it as a one-off and discard it after use. Boxes are impossible to clean, and they get wet, deteriorate and can catch fire. Rings of cardboard tend to get damp on the bottom and start to fall apart.

Never use containers that once held toxic materials, such as pesticides. Some containers absorb some of the toxic contents and then emit them as a gas, even after you wash them. It doesn't take much of this type of gas to kill baby chicks.

The tops you use on brooders need to be adjustable to provide ventilation, and strong enough to keep out any predators and support the weight of other farm animals, such as barn cats, which are often attracted to the heat of the brooder and may want to sleep on top of it. You need to run the electric cable for your heat source out of a corner of the body or through the top.

Heating the brooder

Before the advent of low energy light bulbs you could easily find a bulb that gave off enough heat to keep a small number of chicks warm. An old-fashioned 150-watt bulb does the trick in a warm room, but never use low energy bulbs because they don't give off heat.

Chicks seem to like heat to come from overhead, perhaps because their backs nestle against a hen's warm body when they're under a hen and so overhead heat feels natural. And because the chicks' lungs are close to the ribs and backbone and blood circulation runs through the lungs, using overhead heat is an efficient way to warm the body. For this reason, home-made brooders generally use a heat lamp as the heat source. If you locate the brooder in a heated building, heating your brooder in cold weather is cheaper. Floor heat isn't a good way to heat chicks, and so don't set a box on a heating pad.

Regarding reflector lights

Almost every hardware or farm shop carries a *reflector light:* a bulb socket with a metal 'shade' around it to reflect light and heat. For use in a brooder, choose one with a ceramic socket, a BS safety guarantee and a metal ring on the back of the socket from which you can suspend the light. The best reflector lights made for brooding or as a heat source also have guard wires in the front of the reflector to keep the bulb from touching bedding if it falls.

Choosing bulbs

To heat a brooder, you need to buy special heat lamp bulbs or reflector bulbs. Fluorescent bulbs of any kind (including low-energy bulbs) don't work because they don't give off heat.

Heat lamp bulbs come in infrared and white light wavelengths. Infrared bulbs give off less light with the heat, and new chick-rearers may have difficulty regulating their heat because although the heat is felt by the chicks, it doesn't register well on a thermometer. Using infrared bulbs means that you have to rely on watching the chicks to be sure that the temperature in the brooder Is rIght (see the earlier section 'Gauging temperature by your chicks' behaviour' for more details).

Don't worry about the light that reflector bulbs give off. Yes, it's bright directly under the light, but chicks don't seem to mind. Reflectors point the light down, and the other areas of the brooder aren't quite as bright.

The number of bulbs you need depends on the temperature of the room the brooder sits in, the size of the brooder, what the brooder is made of and how many chicks you're going to brood. A 250-watt heat light bulb is usually the highest wattage that a reflector lamp is rated to hold. This size bulb in a 3-square-metre (15-square-foot) brooder sitting in an area where the temperatures don't fall below 7 degrees Celsius (45 degrees Fahrenheit) should provide enough heat for 25 chicks. It may be enough for 50 chicks in a 5-square-metre (25-square-foot) brooder if the outside temperature is warmer.

You may need to experiment to find out how many bulbs you need in a homemade brooder, which is why you're wise to set up your brooder before you have the chicks. Keep in mind that you don't have to heat the whole brooder to the optimum temperature – only an area sufficient to let all the chicks rest under it without piling on top of each other.

A small batch of chicks probably needs just one lamp. Use more than one lamp if you need to, but hang the lamps close together instead of at opposite sides of the brooder. You want to concentrate the heat in one area and keep another area cooler.

Usually, one bulb lasts the entire brooding time – even when it's on 24 hours a day – but keep a couple of spare bulbs to hand. Bulbs can fail, and it always seems to happen when the shops are closed!

Installing heat lamps

Suspend the lights over the chicks so that they point down. Don't attach them so that they reflect light sideways. Chicks get the greatest warmth when the light reflects directly downwards. Raise and lower the height of the lights to adjust temperature. Hang the lights in the centre of a round or square brooder, or at one end of an oval or rectangular brooder.

Don't suspend the light from its cord – doing so puts unnecessary strain on the wires, which can cause a breakdown or worse still, a fire. Always use a chain or wire to suspend the light from the ring on the back of the reflector (you can buy lengths of lightweight chain from hardware or DIY shops). You may well have to run the chain or wire up through a hole in the brooder lid.

If you don't have something strong above the brooder to suspend the light from, lay a pole or rod across it, making sure that it can't roll or fall into the brooder. If you have a sturdy lid, you can suspend the light from it, but if you raise the lid every time you need to put food and water into the brooder, suspending a light from it is very awkward.

Whatever you use, start with suspending the light about 30 centimetres (1 foot) from the brooder floor. If the temperature is right for the chicks, leave it there. (Make sure that the lamp isn't so close to the chicks that they can touch it, though.) As the chicks grow and need less heat, raise the bulb, or adjust the heat by opening or closing more of the top. Raising the bulb above the top of the brooder to make the chicks comfortable is fine as long as the heat can get through your lid. Having a wire top in a frame facilitates this by allowing you to place something solid over it to conserve warmth or remove part of the solid material to let heat out. The wire top also keeps chicks from getting out and predators from getting in.

Installing gas heating elements

If you're brooding chicks without access to electricity you can use the heat from burning gas instead of a heat lamp. You need a gas heating element with a reflective hood that you attach to a portable propane gas cylinder. These units are manufactured especially for brooding chicks, and are classified according to how many chicks you need to keep warm – 50, 100, 200 and so on. These brooding systems are very adaptable, providing you place them in a well-ventilated area, but out of draughts.

The same principles apply as for electric heat lamps. Suspend them above the chicks from a sturdy point using wire or chain and raise or lower to adjust the temperature.

Putting safety first

Unfortunately, heat lamps and other methods of warming animals have caused many home and farm fires. Always use extreme care when using such sources of heat. We mention several safety tips throughout this section, but here's a rundown:

- ✔ **Always use BS kite marked or EU approved equipment, which rates the safety of electrical items.**
- ✔ **Never exceed the wattage listed for a bulb socket.**
- ✔ **Check wires for frayed and broken areas before use.** Don't use lamps or heating equipment with damaged wires.
- ✔ **Use proper extension leads if you need to, but don't run extension leads outside unless they're specifically for outdoor use.**
- ✔ **Keep wires out of wet areas and anywhere someone can trip over them.**
- ✔ **Never suspend lights by their wires.** Make sure that you suspend them from an independent wire or chain so they can never fall into the bedding. Leave the guards on the reflector.
- ✔ **Keep heat bulbs away from anything that can melt or ignite.** This includes paper, cardboard and the sides of thin plastic containers.
- ✔ **Keep water dishes as far from the bulb as possible.** If cool water splashes on a hot bulb, it breaks.
- ✔ **Keep wires out of brooders and away from where animals can chew them.** Chicks rarely peck at cords, but they sometimes try to roost on them. Damaged wires can electrocute you, other animals or chicks and start fires.
- ✔ **When using gas or propane heat, ensure that the ventilation is good, but not draughty, and use a carbon monoxide detector in the area.**
- ✔ **Keep anything combustible from touching a heat source.**

Helping a Hen Provide Protection

If mother hen is present when the chicks hatch, you can rely on her to take care of their warmth, as long as the weather isn't unseasonably hot or cold (see Chapter 12). She needs just a little help from you to protect the chicks.

By separating the mother hen from the flock during incubation and housing her and her eggs in a spare ark or coop in the same area as your other chickens, you make life easy for yourself. Just make sure that any wire mesh has holes small enough to keep the tiny chicks in with mum – she'll be frantic if they get out and she can't protect them. Although cockerels seldom harm chicks, other hens often consider them intruders and attack them.

In a free-range situation, if a hen hides and hatches eggs somewhere, the other chickens probably don't bother her, but her babies are at big risk from predators. The mother hen tries to protect them, but she's no match for sneaky or overpowering predators, and so the best thing you can do is to move her to a safe place when you first notice her sitting on eggs (Chapter 13 gives information about moving broody hens). Other mother hens may also steal chicks, even if they already have their own chicks and especially if the chicks are similar ages. A hen does have a limit as to how many chicks she can care for, and if chicks end up with surrogate mums, you may never know who's related to whom.

In pens or arks containing baby chicks with their mother, you need to protect them against the danger of drowning while drinking. Filling a dish with marbles or pebbles and topping it up with water enables the chick to drink (mum can manage shallow water too) but stops it from getting too wet and chilled – or drowned – because it can't be submerged in the shallow water. You also need to provide finely ground starter feed or chick crumb instead of adult feed. The chicks can't cope with big pellets and leave them. Both these water and feed issues are another reason why separating chicks from older birds is a good idea.

Feeding and Watering Chicks

Because a few chicks often hatch out before the rest, nature builds in a little time before the chicks have to be fed to give other eggs more time to hatch. The hen stays on the nest waiting for more eggs to hatch, and the hatched chicks usually stay under her. They don't need to eat or drink right away. Only after a day or so does the hen leave the eggs that didn't hatch and leads her young ones out to forage.

Baby chicks have the remains of the egg yolk in their abdomen that sustains them for a couple of days – which is why chick-rearing companies can safely send day-old chicks by mail. However, if you hatch them at home or get them soon after hatching from a local source, offering feed and water is a good idea as soon as the fluff dries and the chicks begin moving around.

Choosing a starter feed

Baby chicks need a chick starter feed, which may be a coarse-milled mixture of wheat and other grains, or chick crumb, which is a compound feed that's been put through a 'kibbling' machine to reduce it to the right size for tiny beaks.

You can buy chick starter feed or crumb at farm shops and pet food shops, but check before you go that they have it in stock. This feed is specialist and demand may be seasonal, and so be prepared to order in advance of your chicks hatching.

In an emergency, hard boiling an egg, letting it cool and mashing it up small provides an acceptable substitute if your chicks arrive before your feed does. This method is the old-fashioned way to provide all the nutrients your chicks need and keeps them alive until your feed arrives. Adult chicken feed, feed for other animals, bread crumbs and so on, may keep chicks alive, but they don't grow well with these types of food.

With the right feed, chicks grow quickly. Chick starter rations contain between 18 and 20 per cent protein, and are suitable for any chicks. Although you can get other higher-protein rations, they're generally for fast-growing turkeys. Avoid these products – they can cause tummy upsets.

Most baby chick starter rations don't require any added grit. *Grit* is a fine gravel that helps birds digest food. However, if you don't use a commercial feed, you need to supply chick grit – unless the chicks are free-ranging with mother hen and can find their own. You can purchase baby chick grit in feed shops or use canary or parakeet grit from a pet shop. Coarse sand or fine gravel are just as good, as long as they're free of salt.

Considering medicated feed

Another choice to be made is whether to buy medicated or unmedicated feed. Medicated feeds aim to give baby chicks a good start in life by including anti-biotics and medications to control parasites and to help with stress-related diseases such as coccidiosis. Health problems such as these are associated with intensive commercial situations, where feeding a medicated starter ration used to be common practice, and so we discuss it here only because you may be offered medicated feed as an option from a poultry feed supplier. As a home flock owner, providing medicated feed is rarely necessary unless you know that your chicks have a health problem (see Chapter 11) or a vet advises it.

Most home flock owners can deal with the more common chick problems and avoid the stresses associated with intensive situations without the need to use medicated feed. Normal chick feed is usually fine. A better way of helping chicks in their early days than relying on medicated feeds is to look at your management system and keep stress to a minimum for your chicks.

Medicated feed also has a *withdrawal period* (the time the body takes to elimi-nate the drugs). Regard the withdrawal times on the label as the minimum. If you're using this feed for a bird you intend to eat, the more time you can give in addition to the minimum, the better. Meat chicks grow fast and are ready to eat in as little as three months which doesn't leave much time for drug residues to subside.

Medicated feed must be labelled as such, but feed bags often look similar, and so double-check the label when buying. It may be printed on the bag or attached to a paper tag sewn into the top or bottom seam. Read these labels carefully to make sure that you're getting what you want before you take it home. Some farm shops have a policy of not allowing returns for medicated feed, and so getting it wrong can turn out to be an expensive mistake.

Understanding the feeding process

Place chick feed in fairly shallow containers or dishes at first. Dishes need to be long and narrow or have a slotted cover to prevent chicks from walking in the feed or scratching it out all over the place. As chicks grow, so should their feed containers.

Have several feed containers or one large enough so that all the chicks can eat at the same time. You may need to add feeders as well as changing to bigger ones as chicks grow. Many chick feeders and water containers are coloured red, which is thought to attract chicks, but they find out how to eat and drink from containers of any colour. Figure 14-2 shows some common feed and water containers. (See Chapter 7 for more ideas.)

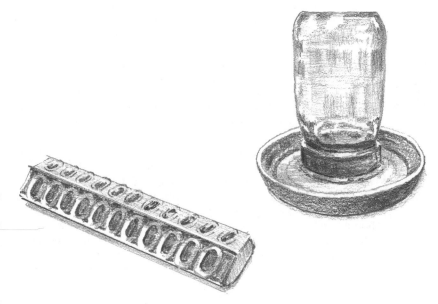

Figure 14-2:
Examples of typical chick feed and water containers.

Meat-type chicks need to have feed in front of them at all times. They need a constant food supply to maintain their rapid growth. For other chicks, a dish that's empty for a few hours is okay, but if your chicks act ravenously and swarm the feeder when you add feed, you're not giving them enough.

Keep feed dishes clean. If you're using a self-feeding, hopper-type feeder, check it frequently to see that it contains feed and is working correctly.

Leading a chick to water . . .

Always have water available for your chicks. A special water dish for baby chicks with a shallow, narrow opening is best. You can buy this type of inexpensive plastic drinkers. Even if you have only a few chicks, buying a small drinker where you can refresh the water frequently is better than leaving stale water in a warm brooder environment. Chicks don't like to drink really warm water, and so place the dish or drinker as far from the heat source as you can.

Baby chicks drown in deep water containers; don't use anything more than 5 centimetres (2 inches) deep.

If you use open containers, such as saucers, lids and so on, add marbles or small stones so that very little open surface for tiny chicks to drown in or get their down wet. Keeping the watering station clean is important, and so try to prevent your chicks from walking through water or perching on top of water containers. Doing so can be difficult, however, and you need to change them often: the more chicks, the more mess. The area around the dish generally becomes wet and messy, and water dishes often get shavings or other bedding kicked into them. Scrub out any algae or scum that develops in containers.

Placing the water container on a large, shallow tray or lid to contain spills helps keep the litter dry and bedding out of the water container. Some people also elevate the water container on a small block of wood, but even when you do so, you may still need to change the bedding under the water area more frequently than the rest of the brooder.

As they get older, baby chicks often try to hop on top of water containers. Doing so may help them escape, and usually results in fouled water and water containers that you hate to handle. If the water container has a flat top, try gluing a plastic funnel on top to prevent them from sitting there.

Some hatcheries suggest adding a teaspoon of sugar per litre (or quart) of water to make a little bit of a sports drink, or some special mix they've devised, for the first day. Doing so doesn't hurt but generally isn't necessary. If the chicks seem very weak, adding sugar may help them perk up.

Rearing Chicks in Your Brooder

When you've set up a brooder with bedding, feed and water, and you're checking the temperature frequently to see how you need to adjust it for your conditions, you're ready to check out some other considerations for getting your new chicks off to a good start. This point is where you have to play at being mother hen. You can't ever be as good as her, of course, but you can get close.

Knowing what to do in the first hour

The brooder is nice and warm, and you have a box of chirping chicks in your hands. You may have hatched them yourself, in which case pop them into the brooder and leave them to it. If the chicks haven't found the water within one hour, treat them like chicks delivered from an external source.

If an external supplier has delivered your chicks, they may have travelled far and be ready for food and water, and so the first thing to do is to fill your feed and water dishes if you've not already done so. (Always use room-temperature water if you're filling the water dishes just before placing chicks in the brooder so that you don't lower the brooder's temperature.)

Remove the chicks from their shipping container one at a time. Hold each chick next to the water container and gently dip its beak in the water. Just dip – don't hold the beak in the water – and then place them under the heat source. Chicks are often dehydrated from shipping and need water more than they need feed. This taste of water encourages them to drink. Spending a bit of extra time dipping beaks can get your chicks off to a good start. Chicks copy each other and so having at least one chick happily drinking on its own is a good sign.

If you have a chick in the container that looks very weak and doesn't stand on its own, the best thing you can do for it is to try and get just a drop or two of warm water into its beak and then place it under the heat source. If any chicks die, or arrive dead, you need to dispose of them.

Negotiating the first few days

On the first day, place some large, shallow containers (large enough that the chicks can walk on them) on the floor of the brooder with some feed scattered on them. Keep your main feed dishes in the brooder, too. Some people use the lid of the container that the chicks came in (if they come from external suppliers); other people use the clean tops of egg cartons, deli trays or paper plates. Doing so helps the chicks find out quickly what feed is, because pecking at the ground for food is natural for them.

Getting the chicks settled in

The baby chicks are going to be a little confused and noisy at first, but if the conditions are right, they soon settle down. Check on them frequently for the first couple of hours, but don't disturb them if at all possible. For example, the first day or two isn't the time to let the children play with the chicks. They're under stress and need time to adjust to their new home.

One exception to disturbing the chicks is encouraging them to eat. In nature, the chicks watch their mum as she pecks for food and then imitate her, and she throws tasty morsels in their direction. You can imitate this behaviour by sprinkling a bit of feed in front of the chicks to attract them by movement, and by tapping a pencil or small stick gently in the area of the feed. Chicks copy each other, and so after a few start eating and drinking, you can relax. All the other healthy chicks soon follow their lead. As soon as you get the chicks eating well, remove the extra feed containers.

Spotting health troubles

Sadly, a dead chick or two in the first few days is pretty normal and can be attributed to the stress of hatching, genetic weaknesses or, if your chicks were delivered, shipping stresses. After a week, though, if chicks are still dying, something's wrong, and you need to check these conditions in your brooder:

- ✔ **Observe the chicks to see whether they're eating and drinking.** Check that the water isn't too warm; it shouldn't feel warm to the touch. Check the chicks can reach the food, and try putting some flat, shallow dishes in again with feed scattered on them.

- ✔ **Observe the chicks to see whether they appear cold or too hot.** (See 'Getting the temperature just right' earlier in the chapter.)

- ✔ **Examine the propane or gas heat, if appropriate.** Ensure that you have enough fresh air ventilating in the brooder to prevent carbon monoxide build-up.

If you make these checks and feel the conditions are good, you may have set poor eggs. If your chicks were delivered from a commercial hatchery, contact the seller and let the person know that you're having a problem. Don't be accusatory; just say what's happening. He may be aware of the problem and replace the chicks, or he may offer you advice on what to do.

If you do get replacement chicks from the hatchery, don't mix them with the old chicks, if you have any left. If they were diseased, they can infect the new batch. If all the chicks from the first batch are gone, clean and disinfect the brooder before placing the new chicks in it. If some of the old chicks are still using the old brooder, you need another brooder for the new arrivals.

Inoculating for diseases

If you buy your chicks from an external supplier, make sure that they've been vaccinated for Marek's disease. *Marek's disease* is caused by a herpes virus, is carried on the dust and dander created by the chicks that gets onto clothes and other surfaces, is highly contagious and kills unvaccinated chicks. Hatcheries across the country almost always carry out this vaccination at one day of age. Sometimes they charge you extra for the vaccination, but the expense is worth it. If you're unsure whether the hatchery vaccinates chicks for Marek's, enquire before you buy. Mixing vaccinated and unvaccinated chicks is never a good idea, and so if you hatch your own chicks as well as buying in, stagger the times you have chicks around.

Home breeders who hatch their own chicks can vaccinate for Marek's but doing so is difficult and paying a vet to do it – if you can find one who's willing to do so – is expensive. If your chicks are from a very rare or expensive breed, the expense may be worthwhile, but most home flock owners hatching small numbers – by which we mean under 50 – won't have a problem.

To vaccinate the chicks on your own, you need to order the vaccines from a poultry supply company well before you hatch the chicks because they need to be vaccinated in the first few days after hatching and the vaccines have a short shelf life. (See Chapters 10 and 11 for more on poultry diseases and vaccinating at home.)

Trimming beaks

Some hatcheries offer beak-trimming services to keep the chicks from pecking at each other, because after blood is drawn on a chick, the other chicks may peck it to death. This type of behaviour is more prevalent in certain breeds such as game fowl and other aggressive breeds. Crowding and other stressful conditions can also cause pecking.

If you keep stress and crowding to a minimum, as a home flock owner you shouldn't experience deadly pecking – in the brooder or in the pen in later life. Chicks do experience stress from the beak-trimming procedure (wouldn't you?), and so we advise you to avoid beak-trimming unless you've had problems in the past or are having problems with your current brood.

If pecking does become a problem with your chicks, you can trim beaks using an electric de-beaker (if you have a lot of chicks to do) or simply cut off the tip of the upper beak using dog nail clippers. But we suggest that, before taking these measures, you focus on providing comfortable, stress-free conditions with plenty of healthy stimulation for chicks and isolate any that become injured.

Preventing disease

It may be a cliché, but the adage is true: an ounce of prevention is worth a pound of cure. Vaccinations are one form of prevention, but even if your chicks have been vaccinated, immunity to certain diseases takes up to ten days to kick in. In the meantime, carry out the following precautions to help keep your chicks healthy:

- ✔ **Keep baby chicks away from adult birds and older chicks.** Of course, if you have a hen that's hatched chicks, she'll be near them.

- ✔ **Don't mix together batches of chicks bought from different places.** If one batch has a problem, it spreads to all the chicks.

- ✔ **Wash your hands between caring for different batches of chicks and between caring for older chickens and chicks.** Doing so helps prevent the spread of disease among chickens just as it does among humans. Some people like to disinfect boots before walking between different groups of birds, too.

- ✔ **Don't interchange feed and water containers.** This precaution is especially important if you've had disease problems with your other birds.

- ✔ **Keep your brooder and chick-growing areas clean.** Remove wet or mouldy bedding promptly. Keep your feed and water containers clean, too.

- ✔ **Observe your birds daily for signs of illness.** Remove and isolate any that appear ill.

- ✔ **Keep vermin under control.** Rats can carry disease between batches of chicks.

People who show birds or who buy, sell and trade birds are more at risk of carrying diseases to chicks on their hands, clothing, shoes and equipment because they're in contact with all kinds of birds.

Watching the Stages of Growth

When they have the correct nutrition and warmth, baby chicks grow quite quickly. Different breeds may grow at different rates, with the modern meat birds growing the fastest.

After being covered in down at hatching, baby chicks develop their first *pinfeathers* in the wing tips by the end of the first week. Feathers develop in the wing and tail areas first, and then on the back and neck. You can see some down still peeking out between feathers for at least a month.

As your chicks grow taller, they start to look like their adult selves. Don't count on the way they look now, though, to tell you what they're going to look like as adults. Some feather colours and patterns keep changing as they grow. And yes, they grow out of the ugly stage!

Don't put the young chicks with adult birds. The older birds bully and may even kill the chicks. Give them their own secure pen until they're mature – about 6 months' old.

One month: In-between-agers

At about a month, when the birds look fully feathered (though they do still have some 'bald' gaps), they're ready for you to move them out of the brooder to larger and cooler quarters. However, some bantam and slow-growing breeds may need a little more than a month in the brooder if they don't seem well feathered and the temperatures are cold.

If temperatures fall below 10 degrees Celsius (50 degrees Fahrenheit), chicks still need a brooder lamp in their larger quarters. If the outside temperature climbs above this level in the day, you can turn off the lamp; but turn it on again when the temperature falls in the evening.

At this stage, chicks need a minimum of 0.2 square metres or 2 square feet of floor space each. They can go outside if the weather is dry and warm, but make sure that they're well protected from predators, and don't give them total freedom just yet.

Smaller breeds of chicks and bantams still need meal or chick crumb but larger birds begin to be able to cope with grower pellets (see the next section for more).

Change the size or number of feeders and water containers as your chickens grow. In warm weather, 25 chicks that are a month old or older may drink more than 3.7 litres (0.8 gallons) of water a day.

Six weeks to maturity: Teenagers

At six weeks, all types of chicks need to be eating grower feed of about 18 per cent protein. Meat chicks can stay on this feed until the time comes to dispatch them for the table at 12–16 weeks, as can any spare cockerels that you bred or reared yourself for the pot. (See Chapter 16 about which meat birds to choose for the best results.) The faster-growing strains of meat chicks may be ready for dispatching much earlier, but they're more prone to sudden death, blisters on the breast and other health problems.

At 18 weeks, or when laying begins, switch pullets that are going to be layers from grower feed to layer feed, with 16 per cent protein and additional calcium and other nutrients. If the chickens are from large, slow-maturing breeds such as Cochins or aren't meant for heavy egg production, you can delay switching to layer feed until you see them laying eggs.

Most chicks start to *roost* (perch off the floor at night) as soon as they get feathered, so when you move them out of their first brooder to other quarters, provide them with perches that are off the floor but low enough for them to hop on to. For most chicks, that's 30–60 centimetres (a foot or two) off the ground. Meat birds don't require perches. (Check out Chapter 7 for more about roosting and perches.)

Young cockerels and pullets

When chickens reach full size and come into sexual maturity, which varies from 18–25 weeks or more, depending on the breed, they're ready to move to their adult home – if they aren't already there. If they're crowing or laying eggs, they're mature.

Don't suddenly throw young birds into housing with older birds – who still bully, peck and even kill the newcomers, especially when you put young cockerels with older males. Introduce the old and young birds to each other gradually. If you allow your birds some free-range time, that's a good point to introduce new flock members. (See Chapter 10 for more on introducing new birds.) ⸱

After a cockerel is crowing well (and maybe even before), he's capable of mating with hens and fertilising eggs. If you don't want this to happen, separate cockerels from hens by the age of 20 weeks. (For more about mating chickens, see Chapter 12.)

Young pullets start 'playing house' as they near the 18- to 22-week mark, especially in spring or summer. Winter weather and long nights may delay maturity, and some breeds are slow to mature. Pullets explore nest boxes, sit in them and croon and arrange the nesting material before they begin laying.

Installing nest boxes for young hens – if they don't have any – by 18 weeks is a wise move. The practice of playing house at this early age tends to encourage good egg-laying habits later. The first eggs the hens lay are very tiny and may be oddly shaped, but in a week or two, the eggs are normal sized. (For information on training young layers, see Chapter 15.)

Getting Along: Chicks and Children

Children love fluffy baby chicks, and young children often think of them as toys. But, of course, chicks aren't toys, and so never make the decision to buy chicks for children on a whim. If your children are begging for the baby chicks they see in the pet shop or poultry auction, don't buy them that day (unless that was your secret intention).

If the children are old enough to help you plan, involve them in looking up information on caring for chicks and making a brooder together. You should have already made some decisions – several of the chapters in this book can help you – about why you want to keep chickens and how you're going to do it. Buy the chicks only then.

Older children who want the chicks for school projects need to be very involved in planning and setting up a brooder and choosing what breeds of chickens to raise. But don't fool yourself: we can tell you in advance that no matter how motivated and responsible children seem to be at times, you have to supervise the care of, or actually care for, the chicks.

If children want to show the birds at poultry shows, the pet shop or auction house is probably not the best place to get the chicks, unless you can pre-order the breeds you want. Seek out knowledgeable people who can explain exactly how much time, effort and money goes into this hobby. If you or your child is still keen they may be able to select premium chicks – those that appear to be robust and have no obvious show faults, but be prepared for the expense of rare breeds.

If, however, you've already succumbed to temptation, your children have surprised you or a relative thought chicks would make a good birthday present, read the whole of this chapter. And good luck!

Facing up to human health issues

Think of all those cute photos you see of children kissing chicks or rubbing them on their faces. Those pictures always make us cringe. Several diseases can be spread from chicks to children with such close contact.

Bird flu, or *avian flu,* is often the first disease people think of when they think about diseases they can get from chicks. However, unless you live in a country where a certain strain of the virus (H5NI) has been confirmed, in domestic poultry, the chances of you or your children getting avian flu are extremely slim. (We discuss avian flu in more detail in Chapter 11.)

Your children are much more likely to get *Salmonellosis* than avian flu. With Salmonellosis, the chicks don't look or behave as if they're ill, and yet they can transmit the bacteria in their faeces, Salmonellosis causes gastrointestinal distress to chickens and humans, and can become quite serious or deadly in some people, particularly the young, aged or immune-compromised.

Everyone needs to wash their hands thoroughly in hot soapy water after handling chicks or chickens, their food or water dishes, soiled bedding, perches – or anything to do with chickens. People in constant contact with poultry can build up some immunity but visitors are very vulnerable. Children are more likely to eat or touch their mouths or eyes without washing their hands and, therefore, are more likely to become ill from handling chicks. When toddlers are around chicken coops, prevent them from putting anything in their mouth that may have been contaminated. After handling chicks or adult chickens, children especially need to wash their hands thoroughly before doing anything else.

Instructing how to handle chicks properly

Tiny chicks are very fragile, and even the loving squeeze of a small child can prove deadly. If someone holds the chest and rib cage of chicks so tightly that the chest can't move, the chick suffocates even if the mouth and nostrils are clear. Children don't mean to squeeze, but they're often afraid of the chick getting out of their hands. So be sure to tell them: no squeezing.

Other problems result when chicks fall from the hands of children and are injured, or when someone picks them up by their legs or necks. Adults need to pick up chicks by scooping them up from underneath and carefully transferring them to a small child who, preferably, is sitting down. Show the child how to cup both hands to hold the chick and not to hold it with one hand.

If you have children around, ask them to be quiet around the chicks – no banging on the sides of the brooder or doing other things that can disturb the chicks or make them run around. Chicks respond well to soft whispering or whistling. The more stress the chicks experience, the less likely you are to raise them successfully, and male chicks who have bad experiences with children can become aggressive when they grow up.

Don't even allow children who are old enough to pick up and handle chicks gently to play with them too often. Taking the chick out of the brooder is stressful to it, and if chicks are too frequently handled, they may not grow as well, and they may become more susceptible to disease.

Part V
Considering Special Management Issues

'Finished trimming the nails and claws, dear?'

In this part . . .

In this part, we cover more specialised care for chickens that you may want to keep for specific uses, such as laying hens and meat birds. In these chapters you can find information on producing eggs for your table and how to go about dispatching chickens.

Chapter 15 covers the layers — how to manage hens that you keep for egg production. Chapter 16 looks at managing chickens you keep for meat and Chapter 17 explains the dispatch and dressing process, as well as suggesting some great ways to store the meat.

Chapter 15

Looking After Layers and Collecting Eggs

. . . .

In This Chapter

▶ Understanding your hens and their life cycles

▶ Managing hens for egg production

▶ Collecting and inspecting eggs

▶ Handling egg production problems

. . . .

Getting your hens to lay well and keeping them laying takes a little more work than some people expect, but the appeal of producing their own food makes many people want to try. In this chapter, we discuss how to manage laying hens – whether you have two hens or many. We also address the problems you may encounter with your layers.

Our focus is on producing eggs to eat rather than incubating them to produce more chicks. However, if you're trying to get a certain breed of chicken to reproduce, you may also want to read this chapter. Getting hens to produce eggs for hatching and getting them to produce eggs for eating have many similarities. Here we discuss hen management and problems you may encounter with egg-laying; the only thing that breeders of eggs for hatching need to do differently is add a cockerel!

Knowing What to Expect from Your Hens

How many hens you need to keep to satisfy your egg requirements depends on your breed of chicken, how long you intend to keep the hens and your appetite for eggs!

If you have production egg-layer breeds, such as Isa Browns, Warrens, Lohmann Browns and so on, and you manage them well and keep them in good health, they probably produce five to six eggs a week in their first year. Other breeds, such as Orpingtons or Light Sussex, probably produce four to five eggs per week. Some fancy breeds, such as Polish, Modern Games and Houdans, produce fewer eggs in a more seasonable pattern – higher in spring and summer and lower or none at all in autumn and winter. If you have mixed-breed birds, your egg production is going to vary widely.

Because an egg takes more than 24 hours to work its way from being released from the ovary to being laid, a hen can't lay an egg 365 days a year. Some high-producing hens have been known to lay 300 eggs the first year, but your home-bred hens probably lay between 200 and 250 eggs in their first year of laying if they're a breed developed for egg production. In the long days of late spring and early summer, egg production may be slightly higher in all breeds. And in the deep darker days of winter, egg production may cease for a bit, or at least decrease.

Young pullets lay smaller than normal eggs, and very irregularly, for the first few weeks. As they settle into a routine of laying, their own individual egg size, colour and production rate develop. Some hens are better layers than others, even in egg-laying strains.

As hens age, they produce fewer eggs, and so you need to decide whether you want to keep them or replace them with better layers. Most home chicken owners are kind-hearted and become attached to their ladies, and so even after they stop laying well, their owners keep them around. If you expect your egg collection numbers to remain the same, you need to increase the size of your flock as your hens age.

So, if you keep four hens from a good egg-laying breed you can expect to collect about 18 eggs a week, at least in the birds' first year. From time to time you may even collect up to two dozen eggs a week. Remember, though, that some eggs you collect may be cracked or otherwise unusable.

The layer's life cycle

When a hen hatches she has all the eggs she's ever going to produce in her ovary as immature cells. As she nears sexual maturity, hormones tell those eggs to begin developing. In laying breeds, laying begins about 18 to 24 weeks after hatching, whereas other types of chickens may take slightly longer. A hen lays these eggs whether a cockerel is present or not.

Eggs-amining the 'egg'

The word 'egg' can refer to the *female reproductive cell,* a tiny bit of genetic material barely visible to the naked eye. In this chapter, 'egg' refers to the large stored food supply around a bit of female genetic material. Because eggs are deposited and detached from the mother while an embryo develops, the embryo isn't able to obtain food from the mother's body through veins in the uterus. Their food supply must be enclosed with them as they leave the mother's body.

The egg that people enjoy for breakfast was meant to be food for a developing chick. Luckily for humans, a hen continues to deposit eggs regardless of whether they've been fertilised to begin an embryo or not.

Children (and even some adult friends) may have lots of questions about how the eggs are laid, such as 'Where do they come out?'. They may get confused if they see baby chicks hatching from some eggs in an incubator and then see you cracking eggs in a pan for breakfast. Take time to explain that not all eggs turn into chicks and how the whole egg-production thing works.

After hens begin laying eggs, laying and dual-purpose breeds continue to lay fairly regularly for at least a year – usually the first two years – unless they're affected by severe stress from very bad weather, illness, poor nutrition or other factors. Some ornamental breeds, however, may lay eggs for only a short time, even in the first year. They may have a break from laying during the first moult, and then laying begins again. After the third moult, egg production drops to low levels or, in some birds, ceases. Each hen is an individual, however, and her egg-laying ability varies as time goes on. Many older hens still lay an occasional egg in their golden years.

Internal factors that influence laying

Many factors influence the laying ability of a hen. Internal factors are beyond your ability to control, with the exception of the moult period, which you can manage. These internal influences include:

- ✔ **Age.** Hens begin laying anywhere from 18 weeks to a year old. Old hens may stop laying altogether or lay just a few eggs. All types of hens lay more eggs between 20 weeks and 2 years of age than they do later.

- ✔ **Genetics.** Some breeds have been selected over the years to lay more eggs than others. Laying breeds of chickens are also selected to moult quickly with the minimum of disruption, and then to resume laying quickly.

All breeds of chickens lay eggs that taste the same, but if you're serious about producing eggs, choose your hens from breeds that lay well. Ex-battery hens are of this type, but be aware that they've already had their most productive year before they're re-homed and so never lay at maximum production for their breed type.

Even if you don't need a top-producing strain, you do want hens that lay reasonably well all year, and so dual-purpose breeds may be right for you. Some breeds, including all bantam (miniature) breeds, lay very small eggs, and some may only lay in the spring and summer. For more about breeds, check out Chapter 3.

✔ **The moulting period.** Although external influences trigger it, moulting is a biological process, and so we include it here. Moulting is triggered by light and generally happens once a year, usually in the autumn. The hen loses her feathers gradually and then replaces them over a period of a few weeks. Growing new feathers is energy intensive and, as a result, hens stop laying while moulting. Some breeds don't resume laying after a moult until the days start getting longer again. To some extent you can manage moulting by using extra lighting, which we discuss in the later section 'Providing supplemental lighting to keep hens laying'. We discuss moulting in more detail in Chapter 10.

✔ **Sex.** This point may sound obvious, but if you want eggs, you need hens. You don't need any cockerels to get eggs. Sexing some breeds of chickens can be difficult (see Chapters 2 and 12).

External factors that influence laying

You can control, at least to a certain extent, some of the environmental factors that influence egg-laying, such as seasonal daylight variations, stress and nutrition. Domestic hens are less sensitive to these factors than their wild relatives, but they still have some seasonal variation in laying unless you manage all these factors effectively.

The most influential external factors are:

✔ **Temperature.** Extreme hot or cold weather can delay the start of laying or cause laying to stop. Managing the temperature can optimise laying. We discuss heating and cooling the coop in Chapter 6.

✔ **Exposure to light.** Day length is a major environmental factor affecting laying. Chicken hormones that affect sexual activity, fertility and egg-laying are strongest when the days draw out – through spring and early summer. Shortening days, such as in late summer, tend to trigger broody behaviour in some hens: they want to sit on eggs and raise a family before it's too late, and when a hen starts to sit on eggs, she no longer lays eggs. (Some hens have this desire to raise a family more strongly than others and, thus, try to sit on eggs at any time of the year.)

The seasonal effect is most pronounced in young pullets just beginning to lay and in older birds, but the season affects all breeds and ages. Basically, increasing day length stimulates laying, and decreasing day length depresses laying. We discuss managing lighting in the section 'Using lighting to encourage laying to start' a bit later in this chapter and also in Chapter 6.

✔ **Stressful environment.** A stressful environment causes pullets to delay the start of egg-laying and may stop laying hens from laying. Stress from escaping predator attacks, being harassed or handled too much, fighting in the flock, crowded conditions, moving to a new shelter and so on, can all slow or stop laying. Just like people, some breeds and individual chickens handle stress better than others. We discuss managing stress a bit in the later section 'Helping your pullets avoid stress' and in Chapter 10.

Managing Your Hens' Laying Years

Whether you use the eggs for table or for hatching, as a flock owner you can do several things to optimise egg production. All breeds of egg-producing chickens benefit from these management strategies. Some of the steps you take to manage layers are similar to what you do to manage chickens for meat or show, but you also find some differences.

Getting young hens ready to lay

To get good egg production you need good birds. And to get healthy young pullets to the verge of 'henhood' you need to pay attention to their nutrition and overall health and wellbeing. Healthy, happy birds begin laying sooner than stressed and poorly maintained birds.

Growing pullets need a diet that contains about 16 per cent protein. They don't need the calcium and minerals at the levels contained in layer feed, and so don't begin feeding laying rations until you start getting eggs or at 22 weeks, whichever comes sooner. (See Chapter 8 for more on nutrition.)

After the pullets that are designed to be layers leave the brooder, separate them from other chickens, such as meat or show-type birds, so you can manage them specifically for egg-laying.

Helping your pullets avoid stress

Chickens establish a pecking order, which entails a lot of fighting and stress as that order works itself out. As soon as an order is established, though, every hen then knows her place, which helps reduce stress.

If your pullets have been together since they were hatched, a pecking order has been in place well before they begin laying. If, however, you're assembling a group of pullets that weren't all raised together, try to get them together by week 15 so they have time to establish a pecking order before laying begins. If you throw young pullets into an established group of older hens, the pullets are going to be bullied and under a lot of stress for a time, which delays laying.

If possible, move pullets directly from the brooder to the housing they'll occupy as adults. Doing so gives them a long time to adjust to their surroundings and reduces stress.

We don't recommend allowing pullets total free-range until they establish a good laying pattern. (We talk about why establishing good laying patterns is important in 'Using lighting to encourage laying to start' later in this chapter.) Instead, give pullets some access to the outdoors in an enclosed run or area of pasture within flexi-netting.

Providing encouragement

Young pullets like to play house as their hormones begin to prepare them for laying. If you make the proper nest boxes available to them – dark, comfortable and secluded nest boxes are attractive – they try the boxes out by sitting in them, arranging nesting material and practising crooning lullabies. Have the nest boxes in place by week 18.

Although a cockerel isn't necessary for keeping laying hens, if you have one (particularly a mature and experienced one), he can be a great coach for young pullets. Cockerels often find a good nest box, sit in it and call the girls over. Then they scratch and turn around in it, showing the hens just how much fun it is. When a hen enters, the cockerel stands outside talking to her. Playing house with a boy is much more fun than playing with just girls!

Your pullets need to discover how to perch too, and so enable them to practise with low perches from around six weeks' old. At first they only play on them and continue to sleep at floor level but as they become more adult the instinct to perch at night kicks in. Laying hens that haven't discovered to perch become dirty and messy around their vents, which affects egg quality.

Using lighting to encourage laying to start

If your pullets are 'coming of age' when the days are still long, exposure to natural daylight probably triggers laying between weeks 18 and 24. Ideally pullets should be hatched early in spring so that they reach that age before the days get short. Getting this timing quite right isn't always possible, but if you have the ability to light your hen house artificially, you can use it to encourage pullets hatching later in the spring to start laying the same year. This approach applies much more to the less prolific layers than their high-production sisters that are bred to produce eggs, come what may, in their first laying year.

Light the hen house for pullets just as you do for adult hens – 16 hours of light and 8 hours of darkness. (You can read more about how to manage adult hens in the section 'Providing supplemental lighting to keep hens laying' later in the chapter.) Begin supplementing natural daylight with artificial light when the pullets are taken from the brooder, or as soon as you get them. Doing so allows laying to start as early as the pullets are physically ready.

Encouraging Egg Production After it Begins

After your hens begin laying, you want them to continue while remaining healthy and happy. Hens that don't lay well still eat as much as ones that do, and so you want to encourage each hen to produce to her full potential. In this section, we discuss management techniques you can use to achieve that goal.

Providing supplemental lighting to keep hens laying

The length of the day and the intensity of light are what stimulate the hormones of hens and prompt egg-laying. In the winter, hens getting only natural light may stop laying, but by adding artificial light inside their house you can keep your adult hens laying more reliably. This lighting can mean anything from one household energy-saving bulb for a small flock to an array of fluorescent tubes in a massive commercial laying flock.

Sixteen hours of bright light followed by eight hours of dim light or darkness is the ideal lighting ratio to keep hens laying. The artificial lights should come on only when it's dark outside and your layers are indoors – they don't stay outside in the dark. The lighting needs to be reliable – not just when you remember to turn on the lights in the morning. Many people buy a timer to regulate the hen house lighting so that they don't need to get up early or go out late to turn lights on and off; you can buy complete timing systems that are all set up for the job.

We prefer to have a dim light in the shelter all night so that if hens fall from the perch, they can find their way back; plus, they can better defend themselves from predators. This lighting is a luxury though and many home flock owners manage well without extra lighting. (We explain the difference between dim and bright light in Chapter 6.)

If you opt for supplemental lighting, make it fit your schedule. For example, if you attend to hen house chores early before work, have the lights come on to fit in with your early morning routine – abrupt light in the morning doesn't bother hens – and use the natural dusk as normal at the end of the day to make up 16 hours of light in total. If you tend to your chickens late in the morning and late in the evening, use the natural waking up time and extend the day with artificial light in the evening – totalling 16 hours – and you have light to do your chores by. You have to use a dimmer switch to organise the lights to allow the hens a 'natural' dusk in the evening, though – the fading light prepares them for roosting. Remember, in late spring and early summer you don't need to leave the lights on as much, unless your birds aren't exposed to natural light.

Keeping up a routine to minimise stress

Just as most animals are at their best in calm, comfortable surroundings, your hens lay better in those kinds of situations, too. So, make sure that your hens have all the things they need to be comfortable – good feed, clean water, dry, clean surroundings, nest boxes, a safe roosting area and maybe a nice dustbath to bathe in. Give them enough room to move around comfortably, scratching and pecking, flapping their wings and conversing with friends, but also to avoid their enemies. Yes, in large egg factories, the hens are packed in small cages, wing to wing, and they do lay. However, as a home flock owner, you don't need or want to subject your birds to those kinds of conditions.

Laying hens can be pets, but educate children not to chase them, and how to handle them gently. Kids should stay out of the shelter, especially in the morning when hens are laying. When they do enter the shelter to collect eggs or help you feed, ask them to be quiet and calm. Don't allow other pets to chase hens or to annoy them from outside their pen.

Another thing to avoid is disturbing hens too much in the morning before laying. Morning isn't the time to bring in visitors, catch birds or clean the house. You can do your normal feeding and watering, but you may want to avoid providing extra treats that cause everyone to come running.

Try not to add and subtract hens from your flock too often. Not only do you risk bringing in disease, but also the pecking order needs to be re-established each time, which can be a stressful process. If you suddenly put together a group of older pullets or hens, they need to do some squabbling and fussing to establish a pecking order and may not lay well for a few weeks. (We talk about behaviour and the pecking order in Chapter 10.)

Retiring old birds when the laying days are done

As your flock ages, you have a decision to make. In earlier days, old hens were sent to the soup pot, but few people are hungry enough to eat these tough old birds today. Instead, many people keep these old hens and feed them, adding young hens as new egg layers. Doing so may be a problem for you if you're limited in the number of birds you can keep. As an alternative, you may be able to sell or give old birds away for pets, or you may have to make the decision to cull them.

In many small flocks, losses to predators and accidents make the decision easier. If you're not sentimental or you need to keep egg production high, getting rid of all your old birds and replacing them with new birds sometime after the second moult is a good plan to follow. Commercial egg farms often get rid of all hens after the first moult, which is why so many ex-battery hens are around, but for home flocks, the second year of laying is usually good enough to keep them.

Mixing young hens with older ones can cause some disturbance in the hen house. We discuss some ways to handle introducing new flock members in Chapter 10.

Collecting and Cleaning Eggs

Harvesting your crop of hen fruit is one of the joys of chicken-keeping. Fresh eggs do taste better than shop-bought ones and they have a different texture – just ask a friend to compare supermarket eggs with your fresh eggs.

This section looks at collecting eggs and what to do when you get them back to the kitchen.

Putting your eggs in one basket

In the earlier section 'Providing encouragement', we talk about training hens to lay in nest boxes, which makes collecting eggs easy. Get yourself one nice basket, pan or bowl with a soft lining that can handle your largest egg collection and faithfully collect your eggs each morning. If you have free-range hens and you can't confine them until after they lay in the morning, get used to checking several locations each day where you come to know they lay, bearing in mind that your hens are always going to be on the lookout for new hidey holes to lay in.

Most hens lay their eggs within a few hours after sunrise, which of course varies with the season. If your chicken house is lit for egg production, most eggs are laid before 10 a.m. Try to collect eggs soon after your hens have finished laying. Doing so keeps the eggs from being broken, which stops hens from discovering that they're good to eat and gives other scavengers less time to raid nests. If picking up the eggs soon after they're laid is impossible for you, at least collect them up once a day, preferably as soon as you can. If you have a lot of hens, you may want to schedule morning and evening collection times.

Bear the weather in mind, too, when collecting eggs:

- ✔ **In very cold weather** eggs left in the nest too long can freeze and crack. These cracks can be hard to see when the egg is brought to room temperature. Picking up eggs soon after they're laid avoids this problem. If you're not able to do so, check all eggs you collect in very cold weather and discard any cracked ones; the cracks allow disease organisms into the egg. Even if frozen eggs don't crack, the quality goes down because of the cold. Yolks may thicken and become tough, and the whites watery. Buying or making a *candler* (a light that illuminates the inside of the egg) helps you to see these flaws (Chapter 13 has more information on candling).

- ✔ **In warm weather** eggs can sit without spoiling for many hours. Even in very hot weather, eggs collected later in the day should be fine. However, those free-range eggs that sat in the sun for several days can be another matter. They're probably spoiled.

- ✔ **In mild weather** eggs remain fresh for many days, even in a nest. Shops store eggs at the ambient temperature for up to four weeks.

If you have free-range hens, you may find eggs popping up in unexpected places (see 'Bringing order to hens that lay all over the place' later in this chapter). The problem with this situation is that in many cases you don't know when the egg was laid. If you're going to produce farm fresh eggs, you need to know that they really are fresh.

If you crack open a spoiled egg, you never want to open another – *ever*. When you're unsure about how long an egg has been in a location, discard it – carefully. Rotten eggs full of sulphur gas can explode like a mini stink bomb, leaving you gagging your way to a faraway location.

If you find eggs and you're not sure how old they are, fill a large bowl with water and gently put the eggs in it. Eggs that sink on their side are probably fresh. Those that stand upright or float are probably old and should be discarded. As an egg ages it loses moisture and the air space gets bigger, which causes the egg to float.

Cleaning your cache

Chicken eggshells are porous, but the shells have a natural coating or 'bloom' that helps prevent bacterial contamination. Hot water and soap remove this natural coating. Water that's cooler than 20 degrees Celsius (68 degrees Fahrenheit) below the internal temperature of the egg causes the insides to shrink and may draw surface contaminants into the egg. For these reasons, some people recommend that you store eggs without washing them, leaving the 'bloom' intact. You can always wash them just before you use them.

Even if you work hard to keep fresh, clean nesting material in your nests, some eggs are going to get dirty. Usually, eggs get contaminated by chicken poop, but sometimes a broken egg, muddy hen feet or other things soil them. To clean dirty eggs:

1. **Rinse all sides in running, mildly warm water, instead of soaking them.**

   Avoid using soap – you don't need it, and strong soapy scents may cause the eggs to have an off flavour – and don't scrub too hard. Sometimes eggs have rough spots or 'pimples', and if you scrub these off, you damage the shell and allow the inside to be contaminated.

   Sometimes eggs have flecks of blood or pigment on the surface – washing removes these marks, which is fine.

2. **Use a paper towel to wipe and dry eggs (never abrasive pads) and discard it afterwards for food safety.**

   Wet eggs can stick to cartons or other containers and then, when you pick them up, a piece of the shell may be pulled off.

If eggs are really heavily soiled, discarding them may be better. If you have cracked eggs or eggs with soft shells, very rough shells or other oddities, you can cook and feed them to pets. We talk more about storing eggs in 'Storing and Handling Eggs', later in this chapter.

Assessing Egg Quality

If, up to now, you've always bought your eggs from the supermarket, you may think that you've a pretty good idea what eggs are supposed to look like. Supermarket eggs are similarly sized and all pretty 'egg-shaped'. In most areas of the country, supermarket eggs are a uniform colour. Eggs marked 'Class A' are clean, and not washed. When you collect eggs from your own hens, however, you're going to notice many differences among the eggs in size, colour, shape and even in the surface texture. These differences occur in commercial eggs too, but imperfect eggs are cracked and sold in huge batches of liquid eggs, and so you rarely see them.

Just because an egg looks different, though, doesn't mean it isn't good to eat. Most egg oddities mean nothing – they're superficial differences – but if you're going to sell some of your excess eggs, you probably want to save those odd eggs for home consumption.

Eggs can also have internal quality issues. Most of these are equally harmless, but can be shocking to novice fresh-egg users. Other internal and external egg-quality differences indicate that you have a management problem, disease or other problem to correct.

In this section, we discuss egg quality, both internal and external.

Identifying the parts of an egg

Take a look at the diagram of an egg in Figure 15-1. It helps to know what each part is as we talk about it. If you have an egg to spare at home, look at it with a scientific eye – dissect it and find the parts. A shop-bought egg works just as well. When you understand the normal parts of an egg, you're better able to determine what's abnormal.

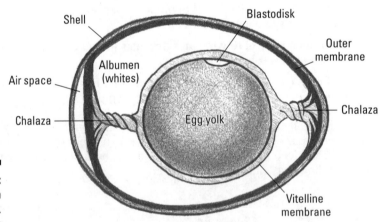

Figure 15-1:
Parts of an egg.

Looking at the outside

When you think about it, the egg is quite a miracle of nature – inside that shell is a food supply meant to sustain an embryo through three weeks of development. That firm shell protects the developing embryo, and yet allows gases to pass through minute pores. So when assessing the quality of an egg, begin by looking at the outside – the size, the shell, the shape and the colour.

Egg size

Egg size varies tremendously among chicken breeds, from the tiny eggs of bantam breeds to the jumbo-sized eggs of some production-laying strains. All eggs are equally good to eat, regardless of size, and even in a flock of the same breed, hens lay slightly different-sized eggs.

When pullets first begin laying, they lay small eggs, sometimes the size of a large marble. These eggs are practice ones, and as the hen's body matures, the eggs become normal-sized for the breed and for that hen. As hens age, the eggs become fewer but larger. When a hen hasn't been laying for a time, the first few eggs she produces when she resumes laying may also be smaller than normal.

To give you some idea of how your home-produced eggs compare to the ones you may be used to buying from the shops, here's how those eggs are graded:

- ✔ 53 to 62 grams: 'Medium'
- ✔ 63 to 72 grams: 'Large'
- ✔ Over 73 grams: 'Extra large'

Some shops sell eggs with an 'eggs of mixed sizes' notice on the box and a minimum pack weight. This labelling helps smaller producers who have fewer hens and a less consistent supply of sizes.

Rough shells

Occasionally, chicken eggs have rough patches, 'pimples', ridges or other strange textures in their shells. These problems are usually temporary, simply quirks in the egg-formation process. The eggs may look ugly, but are fine to eat. If these eggs occur throughout the flock or very frequently, however, you may have some problem with management, nutrition or disease.

Diseases such as *infectious bronchitis* (a respiratory complaint) can cause eggshells to have rough, odd textures or odd shapes. The unusual eggs occur frequently and throughout the flock. If you have other symptoms of illness or suspect disease, have a veterinarian test the flock.

Soft or thin shells

Sometimes eggs have a rubbery, soft shell or a paper-thin shell that cracks easily. These commonly occur when pullets first begin to lay and are nothing to worry about. Soft or thin shells also occur when hens experience long periods of hot, humid weather, which increase the hen's metabolism rate, making calcium in the blood less available for shell formation. Hens nearing the end of their laying days or going into moult (see Chapter 10 for more on moulting) also lay soft-shell eggs more frequently.

If soft-shell eggs occur at other times, it may indicate a nutritional problem, generally a mineral imbalance (Chapter 8 has more on feeding your flock). Sometimes it can also signify a disease problem (in which case, see Chapter 11), especially if other symptoms appear. Making sure that your hen has access to plenty of soluble grit at all times is one way to minimise this problem.

Cracked shells

Small, fine cracks may occur as a hen lays an egg. Cracks can also occur from hens jostling and fighting over a nest, an egg freezing and the egg collector dropping them or slipping them into a pocket and forgetting about them.

Cracked shells enable bacteria to enter the eggs and so you need to use your judgement on whether you use them and for what purpose. If you're unsure, you can always cook them and feed them to pets.

If the eggs crack just before you use them, of course they're okay! Check your eggs carefully before you use them, though. With frozen eggs, you may need to bring them to room temperature to look for cracks, or cracks may be very apparent, with bulging, frozen egg white. You can cook these cracked eggs – just don't eat them raw, in mayonnaise or as soft-boiled eggs. Use them for dishes where you thoroughly cook the egg.

Be sure to crack eggs individually into a cup first to look at and smell them, to judge whether they're useable. Trust your eyes and nose, and use your common sense if you suspect a problem. Checking eggs individually avoids the situation where one spoilt egg contaminates a whole dish of perfectly good eggs.

Odd-coloured or odd-sized eggs

The colour of eggshells ranges from chalk white to deep chocolate brown. Breeds such as Araucana, Ameraucana and some mixed-breed hens with genes from those breeds, lay greenish or blue-looking eggs. Whereas some brown eggs have a rosy tint, no eggs are red, orange or pink, unless they're left by the Easter bunny. Even within the same breed, each individual can have a different shade of egg. As hens age, eggshell colour tends to lighten in brown- or coloured-egg layers. If you notice an odd-coloured egg in an established laying flock from time to time, it's probably just an odd quirk in the egg-laying process.

Chicken eggs aren't normally spotted or flecked. Sometimes eggs have small dots of blood or pigment on them. Often you can wash these spots off the shell.

The important thing to remember is that with one exception, eggshell colour doesn't affect its taste or nutrient content. That exception is eggs you find in a newly discovered nest that you don't know the age of – eggs that look darkened or greyish are probably spoiled. Eggshell colour also has nothing to do with the internal yolk colour.

If you especially like the look of a particular egg colour you can select a certain breed to produce the types of egg you want at home (Chapter 3 helps you to find the right breed or hybrid for the job).

As well as sometimes being odd-coloured, eggs can sometimes be laid with odd shapes: long and narrow like a torpedo, very round or lop-sided. You don't want to use these eggs for hatching, because they aren't the right shape for a developing embryo, but they're fine to eat if fresh. These odd shapes are just another quirk, and most are just a one-off, but they can be a sign that a hen has a defect in her oviduct that causes her to lay odd-shaped eggs. If the hen is happy and healthy, she can easily live with this and you needn't worry. If, however, she looks ill, take a look at Chapter 11 for more information.

Looking at the inside

Hens that are fed a commercial diet produce eggs that have pale yellow yolks. Hens that free-range or are fed a lot of greens or foods such as carrots and yellow squash, which have lots of carotene pigments, produce yolks with a golden yellow to almost orange colour. Some commercial egg producers feed their chickens marigold petals or other sources of red and yellow pigments to darken the yolk, and they sell these eggs for premium prices. Twenty years' ago producers commonly put artificial yellow dyes in the chicken feed to achieve this result – don't worry, it doesn't happen now. If you give your own hens access to natural food, their eggs have that nice golden colour. The colour of a yolk doesn't tell you whether an egg is fertilised or not.

Around the yolk of each egg is a *membrane,* which keeps the yolk's shape. You may also notice the white spot on the yolk, which is the *blastodisk* or *blastoderm.* This blastodisk contains the hen's genetic information. On each side of the egg yolk you may notice a white, cord-like strand called *chalazae,* which keep the egg yolk suspended in the centre of the egg.

The two layers – one thick and one thin – of gelatinous 'whites' (or *albumin*) of an egg surround the yolk and appear almost clear in fresh eggs, but as an egg ages they look whiter, due to evaporation of moisture from the shell pores. The outer layer of the egg contains two membranes, which become more apparent when the egg is cooked. One is close to the white; the other lies just under the shell and is thick and tough.

Commercially, a 'Class A' egg is in a clean, unwashed shell with a small air sac denoting its freshness, when sold. A top-quality egg has a yolk that's large, compact and distinct, the whites contain more thick, rather than thin, albumin and the egg holds its shape well when cracked and slid onto a smooth surface.

Determining a fertile egg

Many people want to look inside eggs to see if an embryo has started growing. You can look inside an egg by shining a bright light on it in a darkened room, although doing so isn't necessary if you're using the eggs at home. This practice is called *candling,* and we talk more about it in Chapter 13. Unless an embryo has been growing for a few days, however, you can't tell a fertile egg by candling it. If the egg was fertile but was then refrigerated, any early embryo was killed and isn't noticeable; it can't harm you.

Remember two things here. First, fertilised eggs don't taste different from unfertilised eggs and they have no extra health benefits. Second, just because a cockerel was with a hen doesn't mean that every egg she produces is fertilised.

We've heard stories of people cracking an egg and seeing what they thought was an embryo. If you're collecting your eggs promptly and storing them correctly, seeing anything resembling an embryo developing is impossible. If it is a chick embryo, the egg wasn't collected soon enough after it was laid, or you have a very warm refrigerator! What people suspect to be an embryo is in fact usually a foreign object in the egg, the chalazae or a blood spot.

Blood spots

Blood spots can occur in any egg, fertile or not. They don't mean the egg was fertilised. What they actually mean is that a blood vessel broke somewhere along the hen's reproductive tract before the shell was applied to the egg. This happens most often in young pullets or older hens but doesn't seem to affect the hen's health in any way.

Blood spots may look nasty, but they don't hurt you. You can eat these eggs or use them in baking if they don't offend you too much.

Other things in eggs

Occasionally, other things turn up in eggs. Little pieces of debris, usually from the lining of the reproductive tract, sometimes appear. Items a chicken eats can't pass from the digestive tract into an egg, and so eating these eggs is safe.

Only very rarely do you find a worm in an egg. If you do, it means that your hens have worms (a worm or worms penetrated the reproductive tract by moving into the *cloaca,* and then up the oviduct) and need to be treated. You don't eat the egg, of course, although you can safely eat other eggs from the flock if they're worm-free until treatment begins. You need to treat the entire flock and discard eggs until the directions on the worm medication say you're safe to stop and you can resume eating them.

Different yolks for different folks

Sometimes, you crack open an egg, to be confronted by two egg yolks. This happens when two yolks pass from the ovary at the same time and are enclosed in a shell. Double-yolked eggs are generally larger than normal, as you'd expect, and are perfectly fine for eating but not so good for hatching, although in rare cases two chicks do successfully hatch from an egg.

Occasionally, a yolk colour looks odd. This colour may be from something that a hen ate, but it would have to consume a lot to change the yolk colour. Lots of greens and foods with orange or red pigments make the yolk appear deeper yellow to orange. Grey, mottled or brownish yolks probably indicate an egg that isn't fresh, and so you're best off discarding it. Cooked egg yolks occasionally look green or dark around the outside, but that's a normal oxidation process and is harmless. Cooling hard-boiled eggs immediately by putting them under cold running water reduces this effect.

Storing and Handling Eggs

When you've collected all that glorious, fresh hen fruit, you need to know how to store and handle it to maintain that farm-fresh flavour. Even just a few good-laying hens can provide you with more eggs than you need at times and so we discuss ways to deal with excess eggs, too.

How to store eggs

Storing eggs may sound simple, but a few tricks and some dos and don'ts can help you to keep the eggs at their best for longer. Storing eggs with the large end – where the air sac is – up is the place to start. Doing so keeps your eggs 'relaxed'.

Egg cartons provide a good way to store eggs. Designed to keep eggs fresh and safe, cartons enable you to stack eggs to save room. Cartons come in two distinct types – trays that hold 30 eggs, or boxes for 6 or 12 eggs, plus a few that come in in-between numbers. As a home flock owner you can recycle cartons from shop-bought eggs or purchase new cartons from poultry supply shops to store them. If you're going to sell eggs, you need to buy new cartons – the authorities frown on the use of old ones. You can also get other containers for storing and protecting eggs; you can even buy an 'egg skelter' to keep your eggs in date order.

Try to avoid storing your eggs near strong-smelling or coloured foods. Because an eggshell is porous, eggs can pick up flavours if you store them with foods such as onions, cabbage and fish. And as anyone who has ever dyed a hard-boiled egg knows, the shell can also absorb colours if you place it on coloured paper or other surfaces that bleed colour.

If you're producing nice fresh eggs at home, letting them get old in storage is a waste. So that you use the oldest eggs first (and unless you want to put some by for a special purpose), mark the date you collect the eggs on the shell with a pencil (pencil writes well on eggshell without contaminating it) and use your eggs in rotation. Eggs remain edible for up to six weeks if you store them properly away from strong smells and direct sunlight.

Fresh eggs can be very difficult to peel after cooking, and so you may want to use your oldest ones when you hard-boil eggs for salads or devilled eggs. After two weeks or so, eggs peel like shop-bought eggs because they've lost moisture and the inner membrane has shrunk away from the shell. This shows just how old shop-bought eggs are before you get hold of them.

The easiest way for home flock owners to store eggs for a long time, when you have more than you need at any one time, is to freeze them. Don't ever freeze eggs in the shell – the shells crack and bacteria can contaminate the inside of the egg. Instead, crack the eggs into a bowl with a pinch of salt or sugar (discard shells), and lightly beat them before freezing. You can also combine eggs in twos, threes or other amounts you find useful, and freeze them in bags or containers. Frozen in this way, eggs remain good for about six months, and you can use them in cooking or for making scrambled eggs after thawing.

Eggs to discard

Always discard badly cracked eggs or soft or thin-shelled eggs that look or smell bad. Really dirty eggs have a heavy bacterial load and probably should be discarded, too. With stored eggs, discard any that are more than six weeks' old, or six months' old if you've kept them in the freezer.

Eggshells are porous, and so dispose of any eggs that have been in contact with a poisonous substance. If you're feeding hens any medications, read and follow the label directions carefully as to when you can safely eat eggs from the hens, and increase the withdrawal dates if you have any doubts. Don't feed these eggs to pets.

If you have many ill chickens or lots of sudden deaths in your flock, discard eggs until you consult with a veterinarian or other expert. Don't feed these eggs to pets, either.

What to do with excess eggs

If you have lots of excess eggs, one thing you can do is to feed them to your pets – dogs and cats, that is, not your pet chickens! – but remember to cook them first. You may also be able to sell some. Don't expect to get rich this way, though – you need a lot of hens and a lot of work caring for them to earn a living from eggs – but your excess eggs can help to support your chicken hobby.

If you want to sell eggs, you need to observe the law. You're allowed to sell uninspected eggs from your home or where the eggs are produced – at the farm gate, for example. You're also allowed to sell 'door-to-door', which includes friends and work colleagues who are the end users, but you're not allowed to sell uninspected eggs to a shop, café or bed-and-breakfast establishment, or anywhere where your eggs will be sold on to a third party, cooked or uncooked. You're not even allowed to give such eggs away to hostels or as raffle prizes.

If you want to make the leap from keeping a few egg layers to supply you, your friends and family to increasing your output to a commercial scale, which brings lots of rules and regulations into your life, think carefully before going ahead. Getting eggs inspected to sell at market or other outlets is a complicated process and isn't a project for a home flock owner with just a few dozen eggs. If you want to know more about the rules, check out the Department for Environment, Food and Rural Affairs (DEFRA) website (`www.defra.gov.uk`) or phone the help line (08459 33 55 77).

To sell 'free-range', 'organic' or even 'barn' eggs, you first have to be inspected by the Egg Marketing Inspectorate (EMI, a department in DEFRA). All these terms have a legal definition, and to be able to use them you need proof from the EMI that you meet certain standards. You're then required to buy an egg-stamping kit and keep food safety records.

As soon as you have more than 50 poultry you're required to register your flock with DEFRA, whether you sell eggs or not.

If you don't want the bother of selling eggs, simply giving them away to friends and neighbours can go a long way to 'paying' in advance for their help in looking after your flock when you want to take a holiday. Chickens are an all-year-round project and sometimes you need friends' help.

Dealing with Production Problems and Bad Habits

In this section, we discuss why hens may not produce eggs at all, or at least less often than you want. We also address the problems of hens that lay eggs outside of the nest boxes, hens that prefer to sit on eggs rather than lay them and hens that consider eggs a tasty treat.

When we talk about bad habits here, we mean the ones in connection with laying eggs, not pulling up all your germinating peas or tasting your strawberries as they ripen. In most small home flocks, you have a few hens that display unladylike behaviour, but occasionally a real rebel arises.

Addressing the failure to lay

Home flock owners who have the space and means to keep a large flock of hens may not really care whether all the birds produce eggs at an optimum rate. If you have just a few hens, however, you're more likely to want their egg-production rate to remain reasonably good. So when this falls, you need to determine why your hens aren't laying and work out which hens are to blame.

Assessing the most common reasons

A hen may not lay at all for several reasons. Just like any type of animal, a hen may suffer from a genetic problem or physical injury that prevents laying, but in most cases, lack of laying indicates a management problem.

Common causes of laying failures include:

- ✔ **You have the wrong sex.** Don't laugh. You have to have hens to get eggs. We've been consulted many times by people who aren't getting any eggs only to find that they have all cockerels. Chickens can be difficult for new owners to sex and some places sell many cheap birds of uncertain sex.

  If you need help determining which of your chicks are pullets, turn to Chapters 2 and 12 for ways to sex chickens, or ask an experienced chicken owner to help you. Any bird that's crowing, definitely isn't going to give you any eggs.

- ✔ **Your hen is too young.** Chickens come in many breeds, and they all mature at different rates. Although hens from egg production breeds can begin laying at 18 weeks, many breeds take longer to mature. Be patient with young pullets.

- ✔ **Your bird is too old.** Hens don't show obvious signs of aging, and so you need an experienced eye to detect older ones. Many people buy laying hens that are simply too old to lay. After three years of age, many hens stop laying or lay only a few eggs.

- ✔ **Your hens are suffering from poor nutrition.** If a hen has to struggle to meet her basic metabolism requirements, she probably doesn't lay. If the feed fails to have the correct balance of vitamins and minerals, her hormones probably don't function correctly, and hormones control the laying cycle. Check out Chapter 8 for more information about chicken nutrition.

- ✔ **Your hens are plagued by poor health.** Sick hens don't begin laying or stop laying altogether. New chicken owners can have a hard time spotting when their chickens are ill because birds try to hide the symptoms until they're near death. Some diseases produce few symptoms anyway, other than a lack of vigour and a lack of laying. Check out Chapter 10 for tips on how to keep your flock healthy.

✔ **Your birds are stressed out.** After you eliminate the source of stress or the chickens adjust to the situation, normal egg-laying generally resumes. See the earlier section 'Keeping up a routine to minimise stress' for details on how to reduce the stress in your chickens' lives.

If you have hens that aren't laying and you're sure that none of the aforementioned reasons is to blame, you may want to have an experienced chicken owner, or a vet who's familiar with poultry, take a look at your flock.

Working out which hen has the problem

If you have a flock of six hens and you never collect more than three eggs a day, one of them probably isn't laying. How do you know which bird that is? And if you're offered laying hens to buy, how do you know if they're still laying?

Checking appearance

If you look carefully at a laying hen next to a non-laying hen, you can usually tell which one is laying. The comb, ear lobes and wattle of a laying hen appear shiny, plump and brightly coloured. The fleshy parts of a non-laying hen look small, dry, pale and shrivelled. The *vent* (anal area) looks large and moist in laying hens, whereas it looks dry and small in non-laying birds. If you carefully feel the pelvic bones, you can see that the bones of the laying hens are spaced farther apart (about two fingers' width between them) than those of non-layers.

If a non-laying hen is healthy, she may look fat and sleek next to her more productive sisters because she's not expending energy on making eggs. If she's ill or old, she may look fluffed up and drowsy and be slower moving.

Observing behaviour

You can also watch the behaviour of hens in your flock – if you can reliably tell them apart. Hens that aren't laying don't spend much time around the nest boxes – unless they've become broody, in which case they spend all their time in a nest. Hens cackle loudly when they lay, and so if you see a hen cackling near a nest box in the morning, she's probably laying.

Penning up your hens

If you still can't tell which hens aren't laying, you can take turns penning the hens up with a nice nest box until about noon. Remember even a good laying hen doesn't lay an egg every day, and so give the hen several chances before deciding she isn't laying.

Checking for hidden eggs

If you have a large hen house or your hens have free-range, you may be missing eggs because the hens aren't laying in the nest box and are hiding their eggs instead. If your egg count drops, take a look around the place for hidden nests. You may want to confine the hens until midday to see whether egg count improves.

Bringing order to hens that lay all over the place

The most annoying problem that crops up in small flocks is when hens lay outside the nest box. In any large group of hens, you're going to get the occasional egg laid in the wrong place. Some hens never seem to learn where to lay eggs – they lay them off a perch or on the floor – but most hens love to snuggle down in a purpose made nest box so if all your hens, or many of them, seem to be missing the box, something is wrong.

Living with – and without – the hormones

In the normal course of events, up to an hour before an egg is to be laid, a rise in hormones causes the hen to suddenly have motherly feelings. She wants to seek out a dark place and do a little nest-making. Soon after she lays the egg, in most cases, the hormones drop, and the hen loses interest in the nest box.

The breed of chicken you keep has much to do with this behaviour. Traditional breeds of the type more likely to go broody (see Chapter 13) love to seek out cosy dark nesting spots and so are likely to lay exactly where you want them to lay. The problem of hens laying all over the place is more associated with high production egg-laying breeds. After all, they've been selected to have fewer hormonal urges. If, however, you keep your nest boxes comfortable and clean, in the darkest place in the hen house, and you have a sufficient number of them to prevent a queue of busting egg layers from forming, the nest box should be inviting enough to lure your hen – even one with a minimal interest in motherhood – to the nest box to lay.

A hen whose maternal hormones are telling her to seek out a place to lay can delay egg-laying for a time if a suitable nest isn't available because they're all in use, or she's waiting for that special box. But after a certain amount of time, if the nest doesn't become free, the hormones drop anyway and she no longer cares where the egg is laid – it comes out anyway. She may lay it from the perch or as she's walking around in the outdoor run. If hens lay eggs this way occasionally, it's not really a problem, but if a hen is always laying willy-nilly in this way, she's probably particularly low in that maternal hormone and isn't often inclined to wait very long for a free nest box.

Keeping your hens shut in until most of the laying is done in the morning (most egg-laying is done by 10 a.m.) is one way to solve this problem. You can usually allow the hens to use an enclosed outside run, but don't turn them out to free-range until they've laid. If your hens use the outside run too often for egg-laying, confine them to the shelter or check to see if the problem is with your nest boxes.

Dealing with 'problem' nest boxes

Another possible cause of willy-nilly laying can be that your nest boxes aren't quite 'right'. Check to see whether the nest boxes are dark enough. If too much light is coming in, hang a cover over the front to create shade (a feed sack does the job nicely) and tear it into strips so the hen can push her way through into the nest box. If she feels that she's found a really secretive place she uses it again and again.

Perhaps another hen higher up the pecking order (we explain the pecking or ranking order of chickens in Chapter 2) has commandeered the nest box, or a hen has been sleeping in one and made it too dirty to be inviting. You may need to provide another box so that lower-ranking chickens have a chance to use one, or clean it out more frequently.

Another thing to check is access to the nest box. If the box is too high, or the approach is obscured, by perches for example, the hen may not be able to get a 'good run up' to land safely. A preliminary step or alighting rail in front of the nest box entrance solve this problem. (See Chapter 6 for more chicken house design tips.)

Fixing bad habits

Laying eggs outside of the nest can also be partially due to habit. Confining your hens in their shelter until afternoon and ensuring that you provide plenty of comfortable nest boxes usually fix this bad habit. When bringing in new hens, keeping them in their shelter until later in the day until they become used to laying in your nests helps to prevent the habit from forming in the first place.

Because habit plays a part in where hens lay eggs, we recommend that you don't give young pullets total free-range until they establish good egg-laying habits. Free-ranging young pullets may decide that under the porch is the proper place to lay eggs, and you may not even realise that they've started laying at all!

Understanding broody behaviour

The one time that habit and good nest boxes may not keep a hen laying in the right place is when she goes broody and decides that she wants to raise a family. If the hormones don't drop after the egg is laid, the hen may want to stay on the nest – that is, she's 'going broody'. This state can occur whether a cockerel is present or not, and the eggs she's sitting on don't have to be fertile eggs.

If you allow a hen to free-range, when she goes broody she sneaks off and finds a secret spot to lay in. If a cockerel's in the flock, he may aid and abet her in this decision. When she first lays eggs in her secret hidey-hole, she doesn't sit on them all day and may come back to the shelter at night to roost for a few days while she fills her nest with eggs one by one. When she has her clutch – around 12 eggs – she sits tight on the secret nest day and night. Hens can be very sneaky, making sure that no one is watching them coming and going, thus making these nests hard to find. Not only are you unable to find and collect the eggs, but also the hen is vulnerable to predators when she stays out all night. (We discuss natural incubation and what to do to help a hen hatch eggs in Chapter 13.)

If you confine hens and they can't choose another place, they may just commandeer a nest box and remain on it all day. They may allow other hens to lay in it, too, but they're likely to attack your hand fiercely as you try to collect eggs, all the time squawking in a harsh, warning tone.

Getting a broody hen to go back to laying

If you want your hens to produce eggs, a broody hen isn't a good thing.

A broody hen only lays the amount of eggs that she can cover properly with her body. This collection of eggs is called a *clutch*. Twelve or so eggs is the average size of clutch, and so at the laying rate of one egg per day 12 days are necessary for her to prepare. Then, when she's ready, she 'sits tight', squatting low over her eggs with her wings held out slightly to increase the warm area under her body. When this sitting phase starts, she stops laying to concentrate on brooding the eggs. If all goes to plan – from the hen's point of view – she's now off egg-laying duty for three to four months while she sits on her eggs and spends the following three months looking after the chicks. After that, depending on her breed, she may begin to lay or may even wait until the following spring before she lays her next egg – not good news from your point of view if you're keeping your hens to produce eggs.

The type of breed you keep is key in this respect. In laying breeds, the urge to brood eggs almost never happens. In dual-purpose breeds, it's somewhat more likely, and in some breeds, it's quite frequent. We recommend choosing breeds that don't go broody if you're keeping chickens for egg-laying.

The best way to stop a hen being broody is simply to keep collecting her eggs, which are no different from 'normal' eggs – never leaving them for more than a few hours if you can. When you remove the eggs, the hen keeps laying for a while to try and replace the eggs you've taken. Then one of two things happens:

- ✔ She abandons the nest, stops feeling broody and goes back to normal egg production after a short break.
- ✔ She continues to sit on the nest until the time comes when her eggs would've hatched – in about three weeks. Let her sit it out until this time, after which she should give up. Be sure to check each day to check that other hens haven't laid in the box.

When an egg hunt isn't fun

Rob says: 'My brother bought some adult hens that were supposed to be layers at an agricultural show. He brought them home and let them range freely in his yard. After a month without eggs, he was beginning to think that he'd made some bad choices. Then, while cleaning the yard, he stumbled upon a treasure trove – about 20 eggs hidden near some old tyres. Pen up those new hens to avoid having egg hunts all year-round.'

If the hen starts another broody cycle right away, you have three options: replace her with a laying breed hen; let her raise a family if you have a cockerel or access to fertile eggs; or just humour her until she gives up.

Handling hens that break and eat eggs

Some hens break and eat their own eggs or those of other hens, though fortunately this problem pops up less frequently than the other causes of low egg production. After one hen begins this behaviour, however, it quickly spreads through the flock because hens quickly copy rewarding behaviour. Make sure that you know the hens are the ones breaking and eating eggs before you accuse them, though. Many animals love chicken eggs. Dogs, crows, magpies, stoats and even rats eat eggs, and when these predators are feeding young or food is scarce, some of them raid the henhouse whenever eggs are available.

If you feed eggshells to your hens, be sure to crush the shells into small pieces so they don't resemble eggs. Promptly clean up any eggs that get dropped or have been smashed in the nest box, and don't feed raw, discarded eggs to the hens. If you don't take these precautions, you may well end up with an egg-breaking and eating problem on your hands.

Hens that you allow to roam in a large enclosure or to free-range are less likely to hang around in the shelter and eat eggs than hens that are bored and confined to a small area. To alleviate boredom and reduce the chance of egg eating with confined hens, add interest to your chicken run by hanging a cabbage or other vegetation up for your chickens to peck at. Feeding a balanced laying ration and providing crushed oyster shells helps hens get the protein, vitamins and minerals they crave, and makes them less likely to source these essentials by breaking eggs and eating them.

Commercial laying flocks have 'rollaway' nest boxes, which aren't very cosy for the hen but hide the egg from her straight after she lays it. If you're keeping an ex-battery hen, be aware that these rollaway nest boxes may have deprived her of the chance to see her own egg in the first year of laying and she may want to investigate it. If she manages to break the egg, it may start an egg-eating habit. Rollaway nest boxes may be the only solution for such hens.

Some people have successfully broken hens' habits of eating eggs by putting plastic or ceramic eggs in the nest. When the hens can't break them, they give up, but this trick doesn't work in all cases.

Chapter 16

Managing Meat Birds

. . . .

. . . .

*N*ot everyone is interested in rearing animals for meat but if you are, raising meat chickens is a good way to start. Chickens are small and easy to handle, and anyone with a little space can successfully rear all the chickens a family can eat. Unlike looking after beef animals or pigs, you can be eating home-reared chicken a mere 12 weeks after you get the chicks, or even sooner if you want to. Plus, chickens don't require large, upfront outlays of cash and time.

If you decide to keep meat birds, you have to manage them a bit differently from chickens you keep for other reasons. In this chapter, we give you the information you need to look after meat chickens well.

Rearing Chickens for Meat

If you're considering rearing your own meat chickens, don't expect to save a lot of money even if you regularly pay a premium price for organic, free-range chickens at a shop or farmer's market. Most home owners rearing chickens for home use end up paying at least as much per kilo/pound as they pay to buy chicken from their local supermarket, and usually more. But people don't raise meat chickens to save money. They do so to control what they feed their chickens and how to treat them; to take responsibility for the way they produce some of their food; and to feel that sense of pride that comes from knowing how to do it yourself.

Looking at the Three Main Approaches to Rearing Meat Birds

You can approach raising meat birds in three different ways:

- ✔ **Getting the meat fast and cheap.** This approach means using Ross Cobb broiler chicks, confining them inside and using commercial feed. If you push them to gain weight as fast as possible, your meat tastes like supermarket chicken. If you're going to make the effort to raise your own chicken meat, we hope that you don't use this method.

- ✔ **Raising chickens humanely but conventionally.** This approach may mean using the same hybrid breeds that are used for commercial free-range chicken. You give the chicks commercial feed and cosy housing, but with plenty of access to the outdoors and more time to grow before you dispatch them. If you feed them organic food they're as good as (or better than) the best supermarket organic chicken.

- ✔ **Raising organic, free-range or pastured birds.** This approach involves using chicks from traditional, actively foraging breeds and only feeding them organically grown feeds and organic pasture plants. These birds are usually the males you've bred yourself or old-fashioned breeds reared for meat production. The meat is less fatty than supermarket chicken and the carcass has a leaner look. This original, healthy chicken meat is the kind of chicken your great grandparents ate.

You can blend these three outlooks, too. For example, you can confine birds and feed them an organic feed. Also, you can use Ross Cobb chicks on pasture, but they take longer to grow and you may lose more of them than if you use a more active breed, unless you manage them very carefully.

After rearing many batches of meat chickens in various ways, we've concluded that the single most important factor when raising your own meat is that you have control over the environment in which the chickens grow, what you feed and how you dispatch them. Rear your chickens in the way that feels right to you.

Selecting the Right Chickens for You

When you're choosing chickens for meat birds, selecting a good breed to match with the way you want to raise your meat is important. You can, of course, eat any chicken of any breed, any sex and any age, but for good quality meat that's economical to rear, certain breeds stand out. Sex can be important in some cases too. We discuss those issues in this section.

The controversial commercial chicken

Almost all the chickens available from your local supermarket are Ross Cobb hybrids that were raised and packed wing to wing in huge buildings and fed 24 hours a day. They were *not,* however, given growth hormones as many people think. The fast growth and heavy meat production is the result of genetics, not growth hormones. The use of growth hormones in chickens isn't permitted, and would be too expensive anyway. The chickens may, however, have been given antibiotics in their feed or water. Antibiotic drug residues have been found in some commercially available chicken meat, and for this reason drug withdrawal periods are recommended before dispatching. Some do still slip through the net, though.

Many people object to the way commercial meat birds are produced because of the way they're crowded during growth and the assembly-line manner in which they're dispatched and *dressed* (removing the feet, head and neck, and emptying the chicken of guts). The conditions in many of these processing plants are horrible, both for the chickens and the humans working in them.

Some people also object to the taste and texture of commercially raised chicken meat, which is softer and fatter than naturally raised chicken and has a very bland taste.

Choosing a breed

Most of the chicken consumed in the United Kingdom originates from Cornish–White Rock hybrids. The Cornish–White Rock hybrid is basically a cross between those two breeds, but lots of precise selective breeding work has gone into producing various strains such as Ross Cobb, Hubbards and Sasso, all of which are similar hybrids but with slightly different characteristics. These chickens have been bred to grow and produce meat quickly, use feed efficiently and have lots of breast meat. The fastest-growing supermarket birds are ready in five weeks, and the slower-growing strains that are more suitable for outdoor rearing are ready in around 12 weeks – with all stages in-between.

Commercial and home chicken producers use these breeds. However, the very quick-growing strains of Cornish–White Rock hybrid aren't active birds and aren't suited to a free-range lifestyle, where they're prone to physical problems. Fortunately, and thanks to the free-range movement, you can now source commercial Cornish–White Rock birds that are robust enough to free-range well (though you need to make a good feed available to them at all times). When ordering chicks from a commercial hatchery or a company that sells them, request the hybrid strain that's 'suitable for free-range'.

In addition to Cornish–White Rock hybrids, you can use any of the fairly fast-growing, dual-purpose or heavy breeds for meat (and you get fairly good egg numbers, too): Plymouth Rocks of any colour; Cornish (sometimes called

Indian Game) of any colour; Orpingtons; Faverolles; Wyandottes; Sussex; Maran; Welsummer and Ixworths are good examples. These breeds are all suitable for free-ranging but aren't always easy to get hold of. The very heavy breeds – such as Brahmas and Jersey Giants – also make good meat birds but take a long time to mature.

Producing your own meat-type crosses isn't too difficult to do. For example, you can have a flock of Buff Orpington hens laying eggs for you and keep a Plymouth Rock cockerel with them. Unless you have a willing hen to sit on the eggs, however, you need to incubate artificially the eggs the hens produce to get your meat chicks (see Chapter 13 for more on incubation). For best results, cross two breeds that have good meat qualities.

Technically, although you can eat any breed of chicken, you should avoid certain breeds, simply because they don't give you much of a meal for your time and money. Breeds to avoid are any of the egg production breeds – such as Leghorns; Warrens; Isa Brown and coloured-egg layers – which have a light frame. Also avoid the small ornamental breeds or any bantams.

At certain times of year unwanted cockerels of indeterminate breeding appear on the market at cheap prices, or even free. Although you can raise these birds for meat, you can have no certainty of getting a good meal from them. They may be the offspring of mixed or production-egg layers, and in the long run these cockerels aren't cheap and may cause hassle for you along the way.

Selecting a sex

If you're going to raise hybrid meat chicks, ordering 'as hatched' (which means no sexing, just sending chicks as they're grabbed) is fine because both sexes make good meat birds with little difference in maturation rate. With other breeds, mainly excess cockerels are reared for meat, but eating the females of any breed is fine. Remember, though, that females are ready to lay eggs from about 20 weeks onwards, which makes *dressing* (gutting and cleaning, sometimes also called *eviscerating*) them a little more tricky. If you know that you're using females for meat, dispatching them before egg production gets under way makes your job easier.

Choosing the Best Time of Year

You can raise meat birds at any time. Hybrid meat broiler chicks are usually available all year-round but some other breeds can be hard to find in autumn and winter. You can incubate chicks from your own flock any time you have fertile eggs, but fertility goes down in most breeds in the cold, dark months.

Meat bird hybrids, if you choose the strain suitable for free-range (recommended for home flocks), take about 10 to 12 weeks to mature from the day you get the chicks to dispatching day. Other breeds need up to 20 weeks to mature. All breeds come out of a brooder at 4 to 5 weeks. (See Chapter 14 for information on raising chicks in a brooder.)

When choosing a time of year to raise chickens for meat, you need to take two factors into consideration: the weather and your housing method.

Meat birds, especially the fast-growing broilers, are very susceptible to heat stress. To avoid problems with birds suffering from heat stress and heart problems when summers are very hot and humid, order slower-growing strains only. If you're going to raise free-range or pastured meat birds, aim to have them out of the brooder ready for their pens when the weather is normally dry and mild, and ensure that your pasture is growing well. Avoid muddy times of the year. Chickens enjoy life less in muddy conditions.

You can raise meat birds in a dry, enclosed area in the winter as long as ventilation is good. Cold doesn't hurt them unless it's really severe – -10 degrees Celsius (14 degrees Fahrenheit) – for long periods, but they have to eat more feed to make each kilo/pound of meat compared to birds raised in milder weather, and your costs for heating also increase. Another problem is that they can't go outdoors if snow is on the ground, making them more problematic to manage. Chickens bred to forage aren't happy in confinement and may develop bad habits such as feather pecking. (In Chapter 10 we give tips on how to keep your flock happy and healthy.)

As well as the time of year, consider your own time when planning your meat-bird project. If you're very busy at certain times of the year, you may want to schedule the project for when you have more time.

The traditional time of year for slaughtering meat animals is in the autumn – a good time for your first dispatching session. Temperatures are low, cold weather reduces the smell, fewer bugs are present and you don't get feathers stuck to your bare, sweaty skin. Plus, if the temperatures are comfortable, you're less inclined to hurry through things.

Deciding on Quantity or Quality

As a beginner rearing meat chickens, start with 10 to 25 meat chicks. That's a good-size batch for the typical family. After you rear and dispatch a group or two successfully, evaluate how well your family likes the meat, how you like the rearing process and how the dispatching went, before purchasing a larger batch (say, 100 meat chicks). These numbers may seem large, but raising only three or four meat chickens is just not efficient in terms of time or money.

For your first small, trial batch of meat birds, you may want to share a hatchery order with a friend or order from a company that combines orders, because large hatcheries don't always want to deal with very small orders. If you also want layers or fancy chickens, you can combine chicks from those breeds with meat chicks, in one order.

After your 'trial run', how many meat chickens you then want to raise in a year depends on how many your family wants to eat. If your family eats two chickens a week, 100 chickens are just about right for the year. Although you can raise 100 chickens in one large batch, you're better off dividing them into several batches raised over the year. Here's why:

- ✔ **You don't put all your eggs in one basket (so to speak).** Large batches provide some economy in time and even in larger purchases of feed, bedding and chicks, but if something goes wrong (such as a predator getting into the pen), you can lose a year's supply of meat in one go.

- ✔ **Meat chickens have a living 'shelf life' – you need to dispatch them before they get too old.** For most meat chickens that means before 20 weeks (more like 12 weeks for hybrid meat birds), if you like tender meat. You don't want to keep a bunch of them in a pen and just go out each Sunday and dispatch one. People did that before they had freezers, but then they expected to eat tougher, stronger meat, too. If you do want to keep larger numbers for some reason, don't use hybrid meat birds. The longer hybrids live, the more likely they are to die suddenly from a heart attack – and not from seeing you pick up the axe.

- ✔ **Dispatching, plucking and dressing takes time and is hard work.** Smaller batches are more manageable and produce smaller quantities of waste to deal with.

- ✔ **For a large batch of meat birds, you need a large housing unit, a large cash outlay and a large freezer.** The housing unit stands empty for a long period of time each year. If you send the birds out to be dispatched, you need a larger supply of cash to get it done than if you have a small group, and suitable crates to carry them in. Plus, if you dispatch all of them at once, you need a lot of freezer space. And, as meat gets older, it loses some quality – even when you package it correctly for the freezer.

  Smaller groups of meat birds need smaller quarters and put less of a strain on your freezer – and your wallet.

If you raise meat birds in batches, you can have several lots around at the same time in various stages of growth or you can skip several months between batches, giving yourself a break and avoiding cold wet times of the year.

Caring for Meat Chickens

Caring for meat birds is a bit different from caring for other types of chickens. Their nutritional requirements are different, and so you need to separate your meat birds from the rest – at least after they leave the brooder. Plus, because the meat birds aren't around forever, it doesn't make sense to get them situated in a pecking order, only to have that all change when they leave. In a mixed situation, the big, slow Ross Cobb broiler-type chicks often get picked on by more active chicks, and that stress may be enough to kill them. Even if you're raising a traditional breed with the intent of eating the males and keeping the females for eggs, separate them when you can tell the sex. Males without hens fight less and spend more time eating.

Getting the housing right

Meat birds need about 0.2 square metres or 2 square feet of floor space per bird. More space is nice for active breeds. Meat birds don't need perches or nest boxes and, because meat birds are here on a temporary basis, the housing can be temporary, too, but it must protect them from predators and weather. See Part II for housing ideas.

Some keepers use wire floors for meat birds for cleanliness, but they cause breast and leg sores in broiler-type birds and so we don't recommend them. Use a deep litter to cushion the birds and keep it clean and dry. Because meat birds are prone to leg problems anyway, avoid types of slippery flooring such as tile, paper, metal and so forth.

If you're putting meat birds in movable pens on pasture, you must take great care not to crush or run over birds when moving the pens. Meat birds often move very slowly and aren't always quick to learn.

Focusing on nutrition

From the time they start eating, meat chicks need a high-quality, growers ration feed. The protein percentage should be around 18 per cent. Broiler feed is also high in energy (fat), and using it reduces the amount of feed required per kilo/pound of grain compared to other types of poultry feed. To feed your meat birds, you can purchase commercial meat bird, game bird or broiler feed (it goes by different names in different places). *Starter* or *chick crumb ration* refers to the feed with the higher protein content that's been 'kibbled' down to a suitable size for chicks to peck at, and *grower* or *finisher ration* refers to the feed with the slightly lower protein content usually in a pellet form for the birds to have when they're bigger.

If broiler-type birds don't get enough protein, their legs and wings can become deformed and they may become unable to walk; you then need to dispose of those birds because they don't eat well. This is another very good reason for avoiding the very fast-growing strains and opting for the free-range types that are bred for activity as well as a good growth rate.

If you're brooding layers and meat bird chicks at the same time they all need chick crumb feed. Don't provide small chicks with layers' rations – this feed is too high in calcium.

Some meat-bird starter feed contains antibiotics. However, if you maintain a good management system (see Chapter 10) and your birds are happy, healthy and not stressed, you shouldn't need to bother with it. Fortunately, feed containing antibiotics isn't so common nowadays, but buying medicated feed unnecessarily can be an expensive mistake because sellers generally have a no-return policy for it. Grower and finisher rations shouldn't contain antibiotics because these substances can be carried into the meat.

In some areas you can buy vegetarian and organic feeds. If these feeds aren't available where you live, you may be able to have feed specially made for you, although certified organic feed has to be milled under special conditions and from organic grain. Generally, to have your own feed blend mixed for you, you need to order a fairly large amount, often a minimum of 2 tonnes. This amount can be hard to transport, store and use before it goes stale and loses nutrition, and so it isn't a good option for small-scale chicken-keepers. (For more about chicken feed, read Chapter 8.)

Feeding whole grains isn't a good idea for meat birds and feeding kitchen scraps isn't allowed by law. We prefer to feed our meat birds pellets, crumbs or mash, because it blends all the ingredients they need and prevents birds from picking and choosing what they like and wasting the rest. Garden trimmings can add interest to a meat bird's diet and are perfectly legal as long as the food hasn't been into a kitchen area.

Meat birds need to have food in front of them at all times. When days are long, this practice may not make a big difference, but when you're raising meat chickens on pasture in early spring and late autumn, it can make a big difference in the time needed to get your birds to the size you want.

Choosing pastured or free-range poultry

We don't recommend total free-range conditions for meat birds. Managed pasture is better if you want birds raised on grass and allows you to rear meat chickens in a decent-sized garden situation. With a well-managed pasture (not just a grassy spot in the yard) where the birds are moved onto fresh grass regularly (Chapter 5 gives tips on how organise housing to make this easy), the pasture can furnish a large part of your meat birds' diet. If you want a good rate of growth in free-range situations you need to provide high-protein broiler feed at the same time, but you should reduce the amount greatly.

Restricted access to grass is the most usual way to produce free-range chickens. If you introduce your chicks to regular snippets of grass in the brooder, they develop the gut flora to make best use of your pasture as well as enjoying the taste of grass when they leave the brooder. Remember to give them grit too. Chicks need to eat well, and good management strategies can make all the difference.

For help in determining what type of pasture grasses to grow in your area, consult with a good seed firm. You need the proper machinery to plant this pasture and maintain it. In principle you're producing a crop on the pasture and your animals are harvesting it.

You must move your chickens to clean pastures frequently. How often depends on the weather, the rate of vegetation growth and the number of birds and how much room you have to use. Move them before they eat all the grass to the roots or the pen gets too dirty and can't regenerate quickly. When moving pastured chickens, take care not to harm them or stress them too much. Make water available to them at all times and ensure that the birds have a shady place to go to when the sun is too hot.

Getting pasture-raised birds to a good eating size may take a little longer than confined ones. Also, because pasture-fed, free-range birds move around more, you see more dark muscle meat and less fatty breast meat in their carcasses, and the meat has more substance than from a less active life style. The skin of the birds may be yellower due to beta-carotene pigments in the grass.

Free-range meat chickens that just have the run of your land with no particular care taken regarding the type of vegetation they eat vary greatly in how fast they grow and how tender they are when eaten. Apart from a few lucky farmyard chickens this mode of chicken-keeping is pretty well outdated.

Managing stress

Stress affects how fast your meat birds grow; stress may even kill them. The hybrid broiler birds, for example, may keel over dead with heart problems if their stress levels get too high. Stress can come from temperature, crowding, disease, predators or too much noise and confusion in the immediate area. All birds eat less when stressed and, as a consequence, grow more slowly. Even the stress of dispatching should be kept to a minimum: the meat tastes better if the animal hasn't been subjected to a lot of stress just before death.

The fast-growing, broiler-type birds are subject to what's known as *sudden death* (also called 'flip over'). Keepers usually find the chickens on their backs, feet in the air. The cause of death is usually a heart attack due to a combination of a genetic weakness and strain caused by the rapid weight gain. Outside stress, such as heat, or bullying from other birds may also be to blame, or a bird may simply die.

Don't eat any meat birds that you find dead. The chance always exists that the bird died of disease, and a carcass that sits around – even for a short time – may be contaminated by flies and other insects.

As a home chicken-rearer, try to avoid using medication to treat the stress-related problems of meat birds. Instead, concentrate on improving your management techniques to keep stress to a minimum. Don't use antibiotics as a preventative against diseases – use them only to treat diseases when a veterinarian advises it. (One of the reasons for raising your own meat is to avoid meat contaminated with antibiotics.)

If you do need to medicate your meat birds for any reason – whether by injection, putting things in the water or feed or any other method – make sure that you read and follow directions on the product label regarding the length of withdrawal period before dispatching the birds for meat. Follow those directions exactly. Certified organic producers double the withdrawal times, just to be sure that no medication passes into the meat.

Here are some tips for raising stress-free meat birds naturally:

- **Place your meat-bird pens in a place that's out of the way of noisy, busy conditions.** Chickens may seem to adjust to busy, noisy locations, but their bodies still feel stress, and that translates to less weight gain and more problems with illness and sudden death.

- **Provide your meat birds with a clean space, and keep the litter dry.** Large, heavy meat birds are prone to developing breast sores or blisters from the amount of time they spend sitting down. Even birds in clean conditions develop them, but wet litter makes them more common. Keepers often find the blisters after dispatching. Usually, you can cut this area out and the bird is safe to eat, but if blisters appear infected, or raw areas get bigger than a 10-pence coin on live birds, you may want to destroy them. We don't recommend eating these birds. If many birds are affected, you have a problem with your management system.

- **Make sure that your birds' environment has a comfortable temperature.** Heat is a big killer of heavy meat birds, so make sure that you ventilate your meat-bird area well and with plenty of shade.

- **Develop a feeding routine and stick to it.** Routines reduce stress, because the birds know what to expect. This approach includes introducing them to the types of bacteria they meet on the range by feeding them grass and grit from early on.

- **Don't let pets and children scare the birds.** They may literally scare them to death.

- **Keep predators out of the pens.** Predators, obviously, are another source of instant death.

Chapter 17

Taking Meat Birds through D-Day

- -

In This Chapter

▶ Planning for dispatch day

▶ Outsourcing the job and doing it yourself

▶ Choosing your method of slaughter

▶ Plucking and dressing a chicken

▶ Packaging and freezing your produce

- -

Rearing meat chickens isn't easy, especially at first, but it isn't so difficult that you can't master it. For most people, the hardest part is the actual killing, or even just the thought of it. The good news is that this part of the process needn't stop you from rearing your own meat chickens – you can find people to do that job for you (for a fee).

This chapter explains the process of killing (also called *dispatching* and *slaughtering*) the chickens you've reared, whether you do decide to do it yourself or you hire someone to do it for you. We also look at the steps you need to take after dispatching to complete the process and to ensure that you can enjoy the meat itself.

Planning for D-Day

Slaughter time is often the hardest part of raising any type of meat. These days, people who eat meat rarely participate in the killing of the animal that produced it, and so it can be a difficult step to take. Don't feel bad if you feel very conflicted about killing a living creature that's been in your care in order to provide food for yourself and your family. Everyone must deal with this issue in their own way. The first time is the hardest; after that it does get easier.

You can make the dispatching process a bit easier for everyone involved by taking a few simple steps:

- ✔ **Raise chickens that are bred specifically for meat.** Going into the project with the firm resolve that you're raising chickens for meat can make the dispatching stage easier. This way, all the birds are destined to die from the beginning – you don't have to pick and choose who lives and who dies. Your purpose for raising the animals is for meat and you're committed to this goal from the time you buy the chicks.

- ✔ **Keep in mind that broiler-type chickens are an end product or 'terminal breed' – they aren't meant to grow to adulthood.** If you do spare their lives, they're unlikely to live a long and happy life anyway.

- ✔ **Approach the day of the dispatching in a calm, orderly manner, with the proper planning and equipment.** The occasion is a solemn and purposeful one, not a fun-filled day.

- ✔ **Keep very young children or people who are emotionally attached to the chickens away on dispatching day, especially the first few times.** After you have some experience, you're sure of what you're doing and you can dispatch calmly and neatly, introducing older, interested children to the process so they, too, have a sense of what happens when meat's produced is fine. Taking responsibility in this way is good.

- ✔ **If you have children or other sensitive people in the household, don't tease them about the fate of the birds or when they're about to consume them.** Make clear from the beginning what the chickens' purpose is, but never, ever make someone watch the dispatching or dressing of chickens if that person doesn't want to.

Some people can never kill anything unless it's a life-or-death situation. If you fall into this category, send your birds out to be dispatched. Be understanding if anyone, including yourself, doesn't want to eat the chickens straight away. After a short freezer stay, the eating becomes easier. Sometimes a bit of emotional distance is a positive thing.

Families with children or sensitive people may find that using a professional killing service is easier from the start, having them bring the chickens home in a plastic bag ready to freeze. It may be easier for you, too – and that's fine.

Assessing when your birds are ready

You can eat a chicken at any stage of its life, but most people want at least a little meat on their chickens. If you're raising broiler-type birds on a good commercial feed, you can expect at least some of them to be ready to dispatch at ten weeks and others soon after. Pastured, free-range and more traditional breeds take longer to reach a decent size. Size isn't everything though – if you want a tender carcass, and to avoid the nuisance of having too many cockerels around, opting for a smaller-sized carcass from a younger bird is fine. It tastes just as good.

How large you like your chickens to be for eating is a personal choice. A meat-type bird usually dresses out about 1 kilo (2.2 pounds) lighter than when it was alive, and so a 3-kilo (6.6-pound) live bird yields a 2-kilo (4.4-pound) carcass. Other breeds usually dress out at somewhat less. From your batch of birds you can get several sizes by spreading out when you kill, or you can dispatch all at the same time and use them according to size.

Get to know your stock by feeling them at all stages of growth and weighing a few from time to time. Males usually grow a bit faster; larger females are smaller boned but often a plumper shape. You can dispatch the hybrids with a rounded plump breast first, even if small – they make a nicely shaped carcass – and leave the bonier birds to catch up over time.

Don't wait too long to start your dispatching regime. You need to dispatch broiler hybrids by 16 to 18 weeks at the very latest, or their huge size starts causing health problems. Other types of chickens just get tougher as they get older. When cockerels start crowing or hens begin laying, they've finished growing for the most part and aren't going to get any bigger – just tougher.

Deciding between making home kills or visiting a poultry abattoir

When your birds are ready to be dispatched, the biggest decision is whether to do it yourself. You're not alone if this part of the chicken-raising process is something you dread. And don't feel that you've failed in your job if you decide to hire someone to do it for you – hiring a professional poultry slaughterer is perfectly acceptable. We take that approach ourselves.

In this section, we explain some of the reasons you may want to use a poultry abattoir. Then we walk you through what you need to know if you decide to do it yourself.

Don't try to do the dispatching yourself if you don't have everything you need to do it properly.

Looking at why you may want to hire the job out

Many good reasons exist for hiring someone to dispatch and dress your meat birds for you. One of the main reasons is time: a skilled person can do the complete job – dispatching, plucking and cleaning – in a very short time, whereas you may take hours to do your first few meat birds.

You can pluck a bird by hand, but a machine does it faster, and most professional slaughterers have plucking machines. You can buy them too, but if you're only doing a few birds a year, a plucking machine is a big expense and one you're unlikely to want to incur.

As you may expect, dispatching is messy and smelly. It creates a lot of disgusting waste that has to be disposed of – something that some people have a hard time doing on a small piece of property, especially in an urban setting. And if you live in close proximity to your neighbours, they may object to you killing animals.

For all these reasons, we find the small amount of money worthwhile that it costs to use a professional to dispatch and dress our meat birds for us. No mess, no fuss – and we still have good, wholesome meat that we reared ourselves.

Understanding what's involved in doing it yourself

Many people feel that, as responsible meat eaters, they ought to experience the whole process of producing meat at least once or twice, from the chick to the chicken dinner. And if you believe that you can do the dispatching part, you probably can, even though you may be a little clumsy and slow at first. If you've already dispatched game or other livestock, you probably have a good idea of what you're getting into; if not, read on.

To do your own dispatching and dressing, you need the following:

- ✔ **The proper location:** To dispatch your chickens at home, you need to have the right location – and that isn't your kitchen! You may finish the cutting and packing inside, but you need to dispatch and clean your birds outside your home. (For more on picking the proper location, see 'Choosing the location', later in this chapter.)

- ✔ **Plenty of water:** Having abundant, clean, hot and cold water available at your dispatching site is very important for filling the tubs if you choose to scald birds, for washing your hands and implements and for cleaning up after the dispatching is done. If you have water on tap at the site, you're less tempted to reuse water – something you shouldn't do.

- ✔ **Somewhere to dispose of the waste:** The wastewater has blood, manure and bits of feathers and other things in it. If you have a septic tank in good condition and without leaks or overflow problems, and if you're dispatching ten or fewer chickens, you can pour the liquid waste straight into it. If you can't route the waste into a septic tank but you have space in a large garden, you can pour small amounts of wastewater into the ground – trees love it.

Remove any floating solids from the wastewater before disposal and don't direct wastewater onto food crops or into ponds, lakes, streams or old wells. Doing so is illegal and changes the ecosystem. The Environment Agency polices this kind of action.

As well as wastewater, you also have to dispose of solid waste, feathers, heads and feet, guts and so on from dressing the birds. Keep a big, lined dustbin near your cleaning station for solid waste, with a lid to discourage flies and wasps, to hold the waste temporarily.

Speak to your vet, local hunt kennels or local *renderers* (their job is to dispose of dead farm animals) about disposing of the waste from there. All these professions are licensed to remove this kind of waste. A fee applies for using their services.

If you have a solid fuel enclosed heating system, you can incinerate small amounts of waste at a time.

Don't try burying solid waste; foxes, dogs and badgers dig it up, and composting feathers and guts isn't a great idea, either. It can be done, but is very smelly and attracts pest animals. Feathers take ages to break down in a compost heap and they don't make a good garden bonfire – burning them outside makes a lot of very smelly dense smoke that can easily annoy neighbours. Your general refuse collection isn't geared up for this kind of waste.

✔ **Time:** Dispatching your first chickens takes you quite a bit of time. Plan for a lot more time than you think you need so that you don't feel rushed. Your first few chickens may take at least an hour each, and that may be only half the job done, and so plan accordingly and start with small numbers. When you get the hang of it, things speed up.

✔ **Labour:** Dispatching chickens involves standing, bending, lifting (maybe heavy lifting to empty water containers) and repetitive cutting, but the average person should be able to handle the work. Because the average home dispatching involves small numbers of chickens, you probably don't have to worry about carpal tunnel syndrome from handling the knife, but people with limited hand function from arthritis or injury may need some help.

If you have difficulty standing for long periods of time, you may be able to adapt your dressing station to a seated position.

Using a Licensed Poultry Slaughterer

If you're unsure of how to dispatch a chicken or you just don't want to, you can still have your meat and eat it too. Some areas have licensed poultry slaughterers who dispatch poultry for you, meaning that you can still enjoy all the benefits of raising your own meat chickens but without having to handle the dispatching yourself. Don't be embarrassed or ashamed if you don't want to do the killing yourself. You may want to watch someone do it just the first time anyway. Some may even show you how to do it properly.

Finding a suitable poultry abattoir

Millions of chickens are killed in the United Kingdom each year, most of which go through huge poultry processing plants (another name for a poultry abattoir). Such places are simply too large to want to dispatch your chickens for you, and you're be unlikely to get your own chicken back even if they did.

For dispatching just a few home-reared chickens, seek out a small poultry abattoir instead. These places aren't all that plentiful, but you may be able to find out about one by checking with the Environmental Health Officer for your area, who polices these small abattoirs. Other sources of information include smallholding association groups and advertisements on notice boards in farm supply stores. And if you have a good farmers' market with someone selling poultry, ask how they get their poultry dispatched – they may have their own abattoir and be willing to do yours for a fee or tell you about one.

Don't expect using a poultry abattoir to be cheap. A poultry slaughterer doesn't make much money and won't get rich on your few chickens. If you expect a good job done and want to feel sure that you're getting your own chickens back, you have to pay properly for the service. Small poultry abattoirs don't have the enormous, robotic equipment that huge processing plants do, but the smaller the abattoir, the more likely that your birds are humanely treated and that you get your own chickens back.

Knowing what to expect

If you want to hire out the dispatching of your chickens, arrange to take a quick tour of the abattoir beforehand. Abattoirs should be clean, without piles of waste or blood-spattered walls, and you shouldn't see hundreds of flies buzzing around. It shouldn't smell too bad (although even the cleanest butchering facilities have some smell if they've been used recently) and ideally, the facilities are located in an enclosed building with running water.

The best places have:

- Cement floors
- Walls that are easy to clean
- Stainless steel sinks and tables
- Good lighting
- Provisions for heating water
- Plucking machines
- Cold storage units

When touring the abattoir, talk to the processor and ask these questions:

✔ **Do you guarantee that I'll get my own birds back?** If so, what precautions does the facility have in place to make sure that you do get your own birds back?

✔ **How will the birds be packaged when I come to get them?** Some processors ask you to supply bags to take the chickens home in. Others put several chickens in bags they supply and still others bag each chicken separately. Most processors return whole birds to you, unless you've arranged for them to be cut up. Most poultry processors don't freeze the birds before returning them to you unless you ask them specifically to do that.

✔ **If I want the feet, necks or other odd parts, can you accommodate me?** Most processors give you the liver, heart and gizzards (the *giblets*) along with the carcass, but if you want these delicacies, ask to make sure.

✔ **What are your requirements for drop-off and pickup?** In most cases processors want the birds delivered the evening before dispatching or early in the morning. They may want you to collect the processed birds at a certain time when they're ready.

Drop off the birds as late as you're allowed and make sure that you're on time if you're allocated a morning drop-off.

✔ **What are the processing fees?** Fees vary widely across the country and according to what services, such as packaging, the processor provides. If you have a choice of several processors in your area, you may want to compare prices against facilities, services and convenience of scheduling.

Don't expect the overall cost of producing your own meat to be cheaper than supermarket chicken meat – it isn't.

If your religion requires the chickens to be butchered in a certain way, hire a processor of your faith, or at least someone who's very familiar with the religious restrictions.

After inspecting the abattoir, the day comes when you need to get your birds there. For this, you need secure boxes or crates to transport the chickens. Cat or dog transporting crates are fine for chickens, but if you plan to make this a regular occurrence, invest in some proper poultry crates. If the poultry abattoir is nearby they may even lend you crates, as life is much easier for them when birds arrive in a secure crate rather than a taped up cardboard box. If you do have to improvise the carriers, make sure that you give the birds plenty of air holes in their boxes – the birds will be stressed by the journey and lack of air adds to that stress. Chickens make a mess when being transported and so use a protective sheet in your car if necessary.

If you're planning to have fresh chicken for a special date, such as Christmas, make sure that your slaughterer knows this well in advance. These occasions can be hectic times in the slaughter business and the firm may not want the hassle of a few more chickens just then. Don't assume that just because a company dispatches your birds in autumn, they want to do the same at Christmas. Check first.

Ask your processor what the usual procedure is at the abattoir. In some systems the birds are ready for collection in one or two days. Other systems take longer to complete the process and it may be a week before you can pick your birds up. Either way, check in advance and plan ahead – your processor doesn't want to be spending time trying to chase you on the phone. They're busy professionals and so work with them as much as possible.

When you collect them, your birds should look clean, with few or no pin-feathers remaining, and nothing more than an occasional small tear in the skin. On rare occasions your processor may warn you about something that was noticed when cleaning the birds, such as tumours, abscesses or signs of disease. The carcasses of those birds should be separated from the others, and you may be advised to discard those birds.

Some processors may give you other advice, too. For example, they may advise you to butcher sooner or later or give you other management tips. Listen to them – they're generally trying to help you.

Most chicken processors don't cut up the birds, although some do so for an additional fee. You can cut the birds up or freeze them whole. We discuss packaging and storing meat in the later section 'Packaging Home-Reared Poultry'.

Preparing to Do the Deed Yourself

If you're willing to do the job of dispatching your meat birds yourself or you can't find a suitable small-scale abattoir or slaughtering facility in your area, you need to make sure that you have all the proper equipment and that you're prepared in advance. Doing a little planning makes the job much more efficient. When you've mastered the techniques the job becomes easier. Take heart from the fact that human ancestors had to do it when they wanted to eat poultry meat.

Choosing the location

You have to consider the law when dispatching chickens. If you're killing for home use you must employ one of the two legal methods (neck dislocation or decapitation, as we discuss in the later section 'Humane and efficient home dispatching') and you must do it on the site where the chicken was reared. So, if you keep the chickens in your back garden, you must carry out the dispatching at home; likewise, if you have a field down the road where you keep your birds, you have to dispatch them there. You're not allowed to drive chickens to a strange location or to a friend's house – if you're getting help from someone, they must come to you.

Finding out how to dispatch humanely

As well as reading this chapter, you can attend various courses run by licensed poultry slaughterers that take you through every step first time around in the flesh, so to speak, and familiarise you with different techniques. The practical experience that such courses provide can give you the confidence and knowledge to dispatch and dress your own meat humanely and efficiently, so we heartily recommend them. Even if you don't intend to dispatch chickens yourself, you never know when you may need the skill – perhaps when you have a sick or injured chicken that needs to be put out of its

misery. You need to know exactly where you stand legally when producing your own meat in case you have to explain what you're doing to anyone.

A good course provides birds that are already dead for you to practise the 'neck dislocation' technique on before taking you through the real thing on a live bird. A comprehensive course should impart knowledge and confidence. Look at www.providencefarm.co.uk on its Dispatch and Dressing Course page or search for 'poultry dispatch and dressing' courses on the Internet.

The law does permit you to transport your chickens to a licensed poultry abattoir, however, because a trained, licensed poultry slaughterer does the work. The slaughterer stuns the chickens electrically to make them unconscious before cutting their throats – the accepted method at poultry abattoirs.

The ideal spot for dispatching is a small building with good lighting, electricity and water and some kind of drainage for wastewater. A nice shed, garage or barn is ideal, but you can also set up an outdoor site that does the job. Look for a level, well-drained spot with drainage sloping away from the site.

If you decide to do your dispatching outdoors, ensure that the location is out of sight of neighbours or passers by, however tolerant they seem. You may want to put up a temporary privacy screen. The dispatching site may be attractive to insects and remain a bit smelly for at least a short time, and so try to avoid using the children's playhouse, the back porch or the garage if you're precious about it. Choose a spot where the wind doesn't whistle through – it isn't comfortable for you and blows the feathers all over the place if you're hand plucking.

Gathering equipment and supplies

How much equipment you need to dispatch, pluck, cool, dress, package and freeze your chickens depends on the method of dispatch, how you choose to pluck and whether you dress your bird 'hot' or 'cold'. (See the section on 'Inspecting and dressing the bird' later in the chapter to help you choose your method.) You can find all this equipment at farm supply shops or hardware stores, or in the cupboards of your own home.

Dispatching chickens the legal way

Strict laws apply when you decide to raise your own chicken meat and do the dispatching yourself. The laws exist to create a level of animal welfare that you'll want to aspire to. This chapter gives you the essentials, but if you want more information look at the Humane Slaughter Association website (www.hsa.org.uk) or get in touch with the Department for Environment, Food and Rural Affairs (www.defra.gov.uk).

Killing equipment

You can dispatch a chicken with your hands or a stout stick (a broomstick handle is perfect), or decapitate with an axe, and you need a sharp knife if you want to bleed your bird after neck dislocation (see 'Looking at your options: Dispatching methods', later in this chapter). You also need a way to keep that axe and knife sharp. If you use the axe (or decapitation) method, you need a solid surface to lay the chicken's head on that's a comfortable distance off the ground for you. Many people find a large tree stump or use a sturdy stool or small table. Whatever you use, you'll be hitting it with the axe, and so it must be robust.

After you've dispatched the bird, you may opt to use killing cones (cone-shaped metal, or sometimes plastic, holders with an open bottom; see Figure 17-1), into which you insert the chicken head down to drain the blood and to stop the chicken from running around with its head cut off or splattering blood everywhere. You nail or hang the cones somewhere off the ground. You can make cones from thin sheet metal if you can't find them to buy. As a guide, traffic cones are the right shape for this even if you do need to modify them a little. A couple of sizes, to hold large and smaller chickens, are useful to have.

Figure 17-1:
Chicken in a
killing cone.

Instead of cones, some people have a rack or a rafter with nails or hooks in it. You tie the dead chicken's legs together with a piece of rope or wire, and hang it from a nail to bleed out. The rack or cones need to be at least 1 metre (3 feet) off the ground. Whatever you use, you need some sort of bucket under each chicken to catch the drained blood. Any kind of bucket is fine, but light-coloured plastic may be stained by the blood.

Plucking equipment

You can pluck birds using one of two methods – dry plucking and wet plucking. You can do both by hand or with the aid of machinery, but most home chicken-rearers pluck by hand. Unless you can find a good second-hand one, plucking machines are usually too complex and expensive for the numbers of chickens home raisers usually dispatch and they need some practice to use skilfully.

You need minimal equipment for dry plucking by hand. You may want to wear gloves but it needn't be a messy job. If you have a blowlamp or a lighter, keep it handy because some birds need a light scorching after dry plucking. You may want to suspend the dead birds by hooking their feet into shackles or tying them up at a comfortable height to work from, and for this you need shackles or suitable rope. If you'd rather sit as you work, anything that's the right height for you will do – small straw bales are a better choice than your best dining room chair! We recommend wearing overalls or old clothes for this job and throwing a little straw on the ground to stop the feathers blowing around. An old towel or rag for hand wiping and bags to collect up feathers (and straw) also come in useful. This method generates the least amount of waste to deal with.

The wet-plucking method involves hot water, and for this you may want to wear rubber gloves. You need a tank big enough to immerse a whole bird in without overflowing the water and a method of heating the water at the plucking site, so you avoid having to carry hot water to the tank. This water becomes bloody and dirty very quickly, and if you're doing more than one bird you need to replenish and possibly reheat it. You need drains nearby for disposing of all the dirty hot water and a container to gather wet feathers into. Handling wet birds can be messy, and so we definitely recommend wearing protective clothing as well as keeping some hand-wiping cloths close by. You may also want a table beside you to place the birds on after you scald them so you can reach them easily while removing the feathers.

If your heating unit is propane, you need propane tanks and propane to run them. If your plucking machine or heating element is electric, you need access to electricity. The latter method involves using a lot of water, and as water and electricity don't mix, make sure that you place any extension cords or plug outlets where they can't get wet.

Cooling and dressing equipment

After dispatching and plucking comes the cooling and dressing (or *evisceration*) process.

Here's a list of all the equipment you need for dressing your chickens:

- ✔ **A work surface.** To do the job efficiently you need a clean table to work on. Make sure that the table is at a comfortable position for you to work. You can use a picnic table, a folding table, a trestle table, boards on sawhorses or any other platform that works for you as long as it's sturdy enough. You may want a cutting board too, in which case clean it thoroughly and disinfect it before putting it back into general kitchen use.

- ✔ **A good knife.** You need at least one good, sharp knife such as a boning or a butcher's knife; a boning knife with a 12-centimetre (5-inch) blade and a butcher's knife with a 20-centimetre (8-inch) blade are common choices. You may want both, and a meat cleaver is also handy at times. Other useful tools that you may need include chicken secateurs or garden secateurs that you need to scald in boiling water before using.

  Sports shops sell good knives for gutting and butchering. Your knife should fit your hands well. A strong, well-made knife is worth the additional cost.

- ✔ **A knife sharpener.** Dull knives cause accidents more often than sharp ones. A butcher's steel is a piece of rounded metal that you hold in one hand and stroke the blade of your knife over to keep it sharp. It doesn't sharpen a dull knife, but it helps keep an edge on a sharp one. Use one after working on each bird.

- ✔ **A refrigerator.** If you plan to dispatch your own birds on a regular basis, investing in a larger fridge kept especially for chickens is a good idea. You can cool your chickens after plucking by hanging them in the fridge. Doing so buys you time because you can 'hang' a dispatched and plucked chicken for up to a week in a good fridge and dress it another time. Some people prefer to do the dressing with a cold bird especially if they're squeamish or need an emotional distance after dispatching chickens they've looked after.

- ✔ **Soap and water.** Be sure to wash your hands frequently in piping hot water at all stages of the process to avoid bacterial cross-contamination of the meat. Take time over hand washing. Many people are less likely to scrub their thumbs and fingertips thoroughly, and so pay special attention to those parts of the hand. Keep an anti-bacterial hand soap, a scent-free washing up liquid and plenty of hot water in your sink or in clean buckets nearby. Wash between the dispatching stages as well as plucking and dressing. Wash if you use the loo and, of course, before you eat or smoke. And wash well when you finish the job.

  Keeping knives and cloths clean and rinsed while you work is equally important.

- ✔ **Plastic gloves.** If you're squeamish, you may prefer to wear gloves. You need several pairs if you use disposable ones.

- ✔ **A spray bottle with a mixture of one part unscented chlorine bleach and two parts water or an anti-bacterial surface spray.** Use this mixture to clean the table if it gets contaminated with faeces from a punctured intestine or the contents of the crop and for disinfecting at the end of the process.

✔ **Paper towels and disposable cloths.** Have paper towels or kitchen roll to hand, especially if you don't like wiping blood on the back of your trousers. Mopping up is best done with disposable cloths.

✔ **An apron and rubber boots.** Essential kit includes a workmanlike apron or a waterproof apron. Wear old clothes or overalls that you don't mind getting stained, even with your apron on. Rubber boots keep your feet dry and keep you from slipping.

Packaging/freezing supplies

Plastic freezer bags are generally used to package chickens. Some people now use vacuum-sealed bags. If you want to do that, you need a machine that does the vacuum sealing. These machines can be bought in kitchen appliance shops or online stores. Buy bags that are sized according to how you want to store the chickens. A whole chicken generally needs at least a 30-x-35-centimetre (12-x-14-inch) bag; if your birds are big you may need a 30-x-45-centimetre (12-x-18-inch) bag. If you cut up the chickens before freezing, you can use smaller bags.

Buy a waterproof, permanent marker to write the date on each bag or on labels (which may be easier to read). Choose labels manufactured for freezer use, otherwise they fall off in the cold. You can seal freezer bags by tying a knot in the end or with a twist of wire or tape.

Dispatching and Dressing at Home

D-day has arrived. You have all the necessary equipment, you've set up your dispatching station and everything's in place and you're ready to begin. This section explains exactly what to do on a dispatching and dressing day.

Humane and efficient home dispatching

When you dispatch at home, you have the chance to end a bird's life in a calm and dignified way, unlike the chaotic mess of a commercial poultry processing plant, and you're taking responsibility for the food you eat – be proud of that.

Choosing the best time of day

We recommend that you dispatch your birds in the early morning. Sleeping chickens are easy to pick up, and so gather and dispatch them while they're still in a drowsy stupor. Decide how many birds you'll be dispatching in each session – go for just one or two for your first time – and keep live birds away from the dispatching site until the last moment. Chickens are empathic and can 'feel' for other birds, and so hide their fate from them until their turn comes. If you can do the dispatching before daylight your chickens never know what happened.

If the before-daylight dispatch isn't practical for you, at least keep the process simple by catching the birds the night before, picking them up in the pen after dark or catching them just before nightfall, and confining them in crates or cages. Put them in a comfortable, preferably dark place and don't give them food overnight. Dispatch them as soon as you can in the morning.

Catching the birds the night before has the advantage of letting them and you calm down. Chasing chickens around is never a good idea; they get very agitated, and you get hot, tired and maybe covered with mud and chicken pooh. You lose time that you could have spent on the job in hand. You're calmer and less likely to rush things or make mistakes if you follow any chasing and catching with a good night's sleep before the dispatch.

Birds held in a confined, darkened area remain calm and easy to handle. Move slowly and talk calmly while removing each bird from the cage, and disturb any others as little as possible. Frightened birds fight back and frantically try to escape. Research has shown that animals killed when in a calm state bleed out and clean more easily; they taste better too. Providing a calm and relaxed environment is the humane way to dispatch your birds.

You need to take the chickens' food away the evening before you dispatch them – chickens with empty crops makes your job of dressing much easier and cleaner. Do, however, leave water for them at all times.

Looking at your options: Dispatching methods

You can dispatch chickens in several ways, but only the two we cover in this section are legal for home use. Different cultures have different ways of dispatching chickens, too. Some religions have strict rules that must be followed for the dispatching and butchering of animals. If you're Jewish, Moslem or Hindu, please speak to a religious leader to find out how you should dispatch and butcher your chickens.

For the most part, the methods of dispatching described here are equal in terms of the amount of pain or distress they cause to the bird. Some methods, however, may not appeal to you. Physically dispatching a chicken isn't difficult, but is often very hard mentally for the first-time slaughter person. You become accustomed to it or you don't. If dispatching is too hard for you, remember that you may be able to find people who can do it for you.

Here are some methods to avoid:

✔ **Using electrocution or a stun gun** before bleeding is the way most animals are dispatched commercially, but is rarely viable in a domestic situation. You're only allowed to do this if you've trained as a licensed slaughterer and can prove it.

✔ **Using gas** of various types contaminates the meat and is a slow death.

✔ **Shooting** a chicken is a senseless waste and is hard to do humanely.

✔ **Drowning** is inhumane.

✔ **Using 'poultry dispatching tools',** unless you know how to use them properly. Used badly, these tools are inhumane and cause a slow death by crushing the neck. Don't be fooled into thinking these tools are a good idea.

No matter which of the legal methods you use, shackling or tying birds in any way before dispatching is something else to avoid – it's illegal.

Choosing decapitation

To dispatch chickens by *decapitation* (in other words, by removing the head), you first need a stout, sturdy stump or table. Put two large nails in it just far enough apart to hold a chicken's neck without the head sliding out. Some people prefer to stretch a wire or heavy cord across the stump and slide the chicken's head under it. You also need a sharp axe. It doesn't have to be a large, heavy axe, especially if you're a small person. A meat cleaver or machete also works – just make sure it's sharp.

Here are the steps to take:

1. **Pick up the chicken by its legs and hold it for a minute until it stops struggling.**

2. **Place the chicken on the stump, with its head between the nails or under your wire.**

3. **Pull back on the legs slightly with one hand, stretching out the neck.**

4. **Pick up the axe in one hand while still holding the bird in position.**

5. **Gripping the axe firmly, strike the neck quickly and decisively just below the head.**

   You need to cut through the major artery, but you don't need to completely cut off the head for the bird to die. Most people, however, remove the head just to be sure.

In the past, people would then release the birds' feet, and the birds jumped up and ran around with their heads cut off – yes, they can do that by nervous 'memory' until they bleed out. Rest assured, the birds are dead and don't feel anything. But this is messy, and is why finding an area where you don't mind blood spattered everywhere is important. The sight of a chicken running around this way is traumatic to many people, too.

Instead of letting the birds run around, we prefer to place them neck down in a killing cone (see 'Killing equipment', earlier in this chapter) to bleed out with something under the bird to catch the blood. Or, when they're dead, you can tie the legs together and hang them over a hook by the legs or suspend them in an empty feed sack. They may twitch or flop their wings a bit, but they're dead: any movement is just residual nervous activity. This method is much cleaner and less traumatic to the observer.

Choosing neck dislocation

A chicken's head is joined to the neck vertebrae. What you're aiming to do is to make a space between the head and the top of the neck, or *dislocating* the neck. Dislocation stretches the blood-carrying vessels to breaking point, meaning that the blood can't reach the brain, which causes death. Done well, neck dislocation is a quick, simple and humane method of dispatching a chicken, and isn't at all messy.

You carry out neck dislocation by hand. Some people call it 'wringing a chicken's neck', which is a bit misleading because no 'wringing' is involved. Here are the actual steps to take:

1. **Using your left hand in a knuckles-down position, pick the chicken up firmly by the legs and let it hang upside down, breast towards you. (If you're left-handed, do the opposite at each stage.)**

   Most people put one finger between the legs while gripping firmly at the end of the thighs nearest the drumsticks.

2. **Hold the bird across your body and with your fingers of your right hand in a 'vee' (for victory) position slide your index and middle fingers down the back of the neck with the bird's lower beak facing towards the ground.**

   The back of your hand is closest to the back of the chicken's neck.

3. **With the two fingers tucked down neatly behind the skull, restrain the legs, lever the knuckles round and at the same time push down with the right hand to dislodge the head.**

   The head should give way and you can close your fingers in the space behind the skull. The neck skin can stretch to leave a good 5-centimetre (2-inch) gap behind the head where the blood pools, this does exactly the same job as bleeding the bird out, giving you white chicken meat.

   The bird continues to flap at this point, but is now insensible – the nervous system is causing the flapping.

   When you first attempt this technique, you're better to be overzealous than not to complete the job. The downside is that the head probably comes off, making it similar to decapitation.

4. **If you want to remove the blood, do so straight away by holding the bird over a bucket and cutting the throat with a sharp knife. This action isn't necessary for any reason other than making your job of dressing the chicken slightly less messy.**

   The blood congeals surprisingly quickly.

A bird that you've raised especially for meat is still young and tender, making it easy to dispatch in this way. You can use this method to dispatch an older, tougher bird, but if you're not strong or are very short (which makes getting enough leverage to be effective difficult), or you have large birds to dispatch, using the 'broomstick method' is a safer bet. When you first use this method it's better to have an assistant or two.

Here are the steps to follow:

1. **Put a thick stick – a broomstick is ideal – on the ground.**

2. **Hold the stick still under your feet while rocking back slightly on your heels.**

   You may find it helpful to lean back against a wall for stability.

3. **Lift one end of the stick and place the chin and lower underside of the chicken's beak on the ground under it.**

4. **Hold on to the chicken's feet with enough tension to keep the head still while making sure that the stick is exactly behind the skull.**

   Your assistant can really help here, but make sure that neither you nor your assistant places fingers under the stick while you're doing this.

5. **Bring both feet down on the stick firmly while pulling upwards on the chicken's feet.**

   This step pushes the skull off the vertebrae and dislocates the neck. If you have plenty of help at hand, ask another assistant to hold the bird's wings against its body to stop it flapping after you've dispatched it.

   You now have a gap behind the head from the stick where the blood pools. You can put the bird in the killing cone or hang it up by the feet and slit the throat to remove the blood that naturally pools into this area.

Removing the feathers – plucking

After dispatching the bird and before you dress it, you have to *pluck* the bird (remove its feathers). You can use the dry pluck or wet pluck method but whichever you choose, do it as soon after you dispatch the bird as possible, while it's still warm.

Dry plucking by hand is the most common method for small numbers of birds. From a hygiene point of view, dry plucking keeps your bird clean and dry. It uses less bulky equipment and no water, and so cleaning up afterwards is much easier than with wet plucking. The process may, however, take slightly more time, until you're skilled, than wet plucking.

Here's how to go about it:

1. **Hang the dead bird in shackles or sit and place it over your knee.**

2. **Pluck the breast first. Take just one or two feathers between your fingers at a time and pull in the direction in which the feathers lie.**

   Fat lines run under the skin down both sides of the breast and across the top of the thighs and wings. These lines are visible as slightly raised distinct patterns. The skin is most tender here and likely to rip when you pluck feathers. You soon work out how rough you can be before the skin rips; each chicken is different. Ripped skin doesn't affect the bird in any way except cosmetically, but to be able to show off a fantastic looking bird at the table is the ideal to aim for.

3. **When you've carefully plucked the breast, move on to the legs and back where you can speed up and pull out bigger clumps of feathers, and work against the lie of the feathers. Work quickly and methodically over the body and well up the neck towards the head.**

4. **Move on to the wings, where the hardest feathers to pluck are. Do them one at a time.**

   Allow the discarded feathers to mingle with some straw on the ground under your feet to keep them in one place, making clearing up easier.

Wet plucking can be quicker than dry plucking, but remember that you have the problem of discarding a large quantity of smelly, dirty water after the job's done and that you've dipped your future dinner in some very dirty water! If this method appeals, follow these steps:

1. **Prepare a large container big enough to immerse a chicken, and heat or add heated water of about 55 to 60 degrees Celsius (130 to 140 degrees Fahrenheit).**

   If the water is too cool, the feathers are hard to remove. If the water is too hot, the skin starts to cook and tear as you pluck.

2. **Holding the bird by its feet, dip it into the water to cover all feathered areas and count to ten slowly.**

3. **Remove the bird and try pulling out the wing feathers by grasping and pulling them in the direction in which they were growing. If they come out easily, begin plucking out the other feathers, working as quickly as you can. If they don't slide out easily, dip the bird again.**

   You may need rubber gloves to handle the hot, wet bird!

4. **Change the water when it looks dirty or becomes cool.**

   Be careful not to drip the water on yourself, spill it or get your fingers in the pot while dipping the bird. This job is quite smelly.

5. **Keep plucking until you've removed all the feathers.**

Some immature feathers, called *pin-feathers,* may be hard to remove. They look like a thick pin in the bird's skin. Most broiler-type pin-feathers are white, although coloured birds may have dark pin-feathers. Grasp them with your fingernails and pull them straight out. If you don't have fingernails, place a flat blade, like that of a butter knife, on one side of the pin-feather and a finger on the other to pull it out.

Some birds appear 'hairy' after plucking, but look closely and you can see differently shaped feathers remaining. A light scorching with a blowtorch or even a lighter removes them. Blowing sharply shifts the burnt residue.

Inspecting and dressing the bird

Before you go ahead with the next stage of the dispatching and dressing process, taking a good look at your bird is a good idea, just to make sure that it's one that you'd like to eat. Sometimes it can be obvious from the outside that something's wrong with the bird, but hopefully your good flock management has brought a wonderful healthy plump juicy chicken to this stage. You're then ready for the last few steps of the process before tucking into a chicken dinner that surpasses all others you've ever eaten!

Checking out the flesh

How much flesh a chicken has and the size of its breast vary by breed and age. If you're used to shop-bought chicken and are dispatching a breed that's not generally raised for meat, the breast may look small and skinny to you. The drumsticks of birds raised on grass may look darker and larger than those of hybrid meat broilers, which are what commercial chicken comes from. Also, the legs of other meat breeds may look longer. All this is normal.

The age at which you dispatch a bird and its feeding programme determine how fleshed out it becomes. Some breeds never achieve a lot of flesh on their bones. Hybrid meat birds should be quite plump and heavy by 12 weeks; if not, something's wrong with your feeding and management. Most heavy breeds of traditional chickens stop growing at 20 to 25 weeks.

Examining the bird for signs of disease

When you're dispatching your birds, examine the carcass for signs of disease as well as signs that your management programme is working well. Modern meat birds aren't alive long before you dispatch them and thus have less chance to pick up a disease or develop other problems, but it doesn't hurt to at least look over your carcasses and their organs carefully.

A healthy carcass has white or yellow skin. Chicken skin is normally loose and thin. The skin and fat of chickens raised free-range or pastured usually have a deeper yellow pigment.

Home chicken owners rarely eat Silky chickens, but in some cultures Silkies are a delicacy. Their skin is a bluish black no matter what colour their feathers were.

A healthy carcass may have a few bruises on the skin. These marks can happen at the dispatching time, but a lot of bruises means that your birds aren't being handled correctly. As bruising a dead bird is impossible, any bruises you see date from when the bird was alive and generally come from rough handling.

Breast blisters on meat birds are fairly common and may indicate the chickens lacked enough clean bedding. They look like a blister on the skin with clear fluid. You can cut out these superficial blemishes and the meat is safe to eat.

Some conditions, however, render a chicken unsafe to eat. Be especially careful to keep an eye out for the following:

- **Abscesses** are pus-filled lumps and aren't a good sign. They can be on the outside or deep inside a body. Abscesses can happen because of an injury or from disease, but either way they make a carcass unsafe to eat. If many birds have abscesses, something's wrong in your management. Check your housing for rough objects and don't overcrowd your birds.

- **Tumours** can be hard or soft and appear anywhere on the body or an organ. Some may contain fluid. Don't mistake forming eggs for tumours. In older hens, eggs may be forming in the oviduct, near the backbone of the bird. They can range in size from tiny dots to an almost fully formed egg. If you look inside the lump and see egg yolk, an egg it is. Some cultures consider these eggs a delicacy and save them to be eaten, but most people toss them out.

- **A very pale or mottled-looking or white-spotted liver** tells you that the chicken was diseased. The liver of a healthy chicken is reddish brown. Most other chicken organs don't have much to tell you.

- **Open sores or wounds** where the bird has obvious weeping or bloody patches on the skin are a bad sign. The sores and wounds are either from a predator attack or signs that the bird probably wasn't very healthy.

- **An outpouring of straw-coloured fluid when you open the tiny hole in the back of the bird to remove organs during the dressing process** means that the body cavity filled with fluid because of a disease problem.

- **A dispatched bird that was left out of the fridge for flies to lay eggs on** is one to avoid. Flies are attracted to dead creatures. You don't know where the flies have been before they land, and they will lay eggs if you leave the bird unattended for even a short time. The eggs turn into maggots within twenty-four hours.

> ✔ **Any bird that was partially cooked from being scalded too long or at too high a temperature when being wet plucked** should be completely cooked immediately or discarded.

Don't eat chickens that died of unknown causes or that were killed by predators. Birds that looked or acted ill before dispatching are also unsafe to eat.

Dressing your chicken

People use various terms for the evisceration process – gutting, cleaning, dressing and drawing – but they all essentially mean the same thing: removing the feet, head and neck and emptying the chicken of all the guts, some of which – the *giblets* – are useful items to save for your stock pot. The giblets consist of the neck, heart, gizzard and liver.

In the commercial world, most chickens are gutted straight away after plucking – called *hot evisceration* because the work is done when the birds are still warm. In a domestic set up, if you have to get all the work out of the way at one time, hot evisceration is for you. However, chilling your bird before gutting has practical, hygienic and emotional advantages that make *cold evisceration* a better option for some people.

When you eviscerate chickens warm, many more live bacteria are around, especially if you've wet plucked the bird and immersed it in tepid mucky water. The guts may still be full of warm pooh as well. (This stresses the importance of taking food away from birds the night before dispatching.) Chilling, however, 'sets' the flesh, firms up the guts and their contents (so that you're less likely to have spillage accidents when you cut the bird open, making the job more hygienic), reduces smells and kills most bacteria – especially if you first dry plucked the bird. On an emotional level sometimes leaving a bird to cool for a day or so and coming back to a piece of 'meat' may suit you better than having to cut up a warm bird straight away that you knew as a living creature.

If you want to chill a bird before dressing:

1. **Tie a plastic bag over its head to catch any last drops of blood.**

2. **Hang the bird upside down in a refrigerator set between 0 and 4 degrees Celsius (32 and 40 degrees Fahrenheit). If you can't hang it, lay it on its back with its head hanging over the shelf.**

3. **Leave the bird for at least 24 hours to chill completely or up to a week if your fridge is a good one with a fan to move the air around.**

The 'hanging' process involved in cold dressing helps to develop the flavour of the meat and tenderises the flesh of older birds. This improvement is another great difference between home-reared and commercial chickens, which never have the opportunity to be hung.

Here are the steps to take to dress your chicken:

1. **Remove the feet. With secateurs chop them cleanly part way down the shaft below the drumstick.**

   An alternative method is to run a sharp knife around the joint, grab hold of the feet and then twist and pull. You may also remove sinews this way. On big tough birds removing sinews with this method makes sense, but it's not necessary for young tender meat hybrids. Discard or keep the feet as you prefer.

2. **Lay your chicken breast down on your table and with the sharp knife slit the skin from the back of the head to between the shoulder blades.**

3. **Peel the skin open and pull out the neck.**

   The neck should be dislocated from the head. Yank the neck bone up to free it from the connective tissues and cut round the flesh of the neck as close to the body as possible. A twist removes it altogether. Put the neck to one side in a clean place, to go into the giblet pile.

4. **Cut off the head if it's still attached, along with any bruised and bloody neck skin, but leave as much clean neck skin as you can.**

5. **Open the neck skin wide and look for two tubes, the windpipe (or *trachea*) and the oesophagus.**

   The windpipe is stiff with cartilage, and so loosen it off and remove it. Pull it away by hand or cut it with a knife where it enters the body. Only air has passed through this tube and so it's clean.

6. **Starting from one edge of the open neck skin, peel away all the small meaty glands and the second tube – the oesophagus – which lies flat against the skin.**

   This step is trickier on a warm bird than a cold bird, and if you've not starved your bird out properly before dispatching it can be quite messy, and so keep some paper towels handy. The oesophagus carries food and as you peel it away it increases in size to reveal the crop, which looks like a deflated balloon.

7. **Follow where the oesophagus enters the body of the bird. Insert your finger and loosen any connective tissue that you can reach, paying particular attention to the backbone area.**

   The more you loosen at this end, the less work remains to do at the other, more 'dangerous' pooh-filled end. You should be able to feel the ribs, but be careful – these bones are quite sharp.

8. **Cut off the dangling oesophagus if you want, or leave it if it's still filled with any food or fluids. The oesophagus is joined to the rest of the innards which you will remove eventually.**

9. **Place the bird on its back on the table, with the vent end facing you. Press the legs down into the body, just like you see trussed chickens in the supermarket, so that you can now work easily on the vent end.**

   The chicken is stiff at this stage from rigour mortis or from chilling.

10. **Pull up on the flesh and skin over the top of the vent. Holding a sharp knife across the bird – not pointing down or poking into it – make a shallow slice across the backside of the bird, just above the vent.**

   Don't make your slice too wide at this point and don't push too deeply – you want to cut through the skin only to make a small hole through to the body cavity.

11. **When you have a small hole, put your fingers in it on both sides and gently pull apart the skin and flesh to find the division between the innards – which on a hybrid meat bird has a coating of fat – and the muscle meat.**

   If pooh comes out at this point because the bird wasn't empty before you dispatched it, or if at any time while you're removing the intestines you break them and pooh comes out, use some disposable kitchen towel to mop up.

12. **Now that you've widened the hole and can see down each side of the vent, carefully cut all the way round the vent with a very sharp knife. The vent is joined to the intestines and you want to remove the vent with them.**

   Avoid breaking or puncturing anything with your knife.

13. **Slide your whole hand, with your fingers out straight (yes, your whole hand can fit inside all but tiny hens), in under the breastbone as far as you can, moving your hand from side to side.**

   You feel the connective tissues breaking, the tissues nearest the backbone being the toughest.

14. **When you've loosened most of the connective tissues, slightly curl your hand over the top of the mass and draw – don't squeeze and pull – the organs back towards the rear.**

   The organs come out in a big clump, especially if you've chilled the bird. Put them to one side for a moment.

15. **Look to check that the bird is empty inside.**

   The lungs may still be attached at the top end of the cavity, tucked in among the ribs. Don't worry if you can't get them all out cleanly – they've only contained air and don't affect the taste of the bird when you cook it. The kidneys are a dark patch embedded in the back closer to the vent end; you can safely leave them in place.

Don't rinse your dressed bird in water; doing so only spreads bacteria. Wipe up any messes as you go.

Sorting out the giblets

Very few commercial chickens are sold with their giblets, and so having them available to you is another one of the unique aspects provided by rearing your own birds. Here's how to sort through them:

1. **Put the chicken carcass to one side in a clean cool place and look at the mass of organs that you've removed.**

2. **Carefully loosen the liver – the dark, shiny organ – at the top of the 'clump' from the bulk and cut it away, taking great care not to break the dark green gall bladder that adjoins it. Put the liver with the neck.**

You're better off losing a bit of the liver than bursting the gall bladder, which holds a very bitter substance that can taint your giblets. Don't worry if a green stain is present on the liver where it has touched the gall bladder – this is normal.

3. **Pop the heart – which is at one end of the mass – out of its sac. Cut cleanly across the fat line and save the best bit.**

You're now left with all the intestines and the gizzard.

4. **To find the gizzard, search out the only hard, rounded organ, cut off the inlet and outlet tubes and remove any fat surrounding it and discard it all.**

This kidney bean-shaped organ, which has done the bulk of the work of grinding up your chicken's food, should look shiny, healthy and strong.

5. **Cut into the flesh at the edge of the gizzard until you see a white line. Peel it open to reveal a yellow leathery sac, which contains stones and the remains of the last meal. Throw away the contents and add the gizzard to the other giblets.**

This completes your giblet pile. Bag the pieces up and keep them with your chicken. Your chicken is now ready to go into the fridge or freezer.

During the process wash your hands frequently and when you've finished, clean up very well and disinfect the work station and all implements.

Packaging Home-Reared Poultry

Whether you dispatched and dressed your birds yourself or had someone do it for you, you need to package the meat properly so that it can be stored safely and without spoiling until you're ready to eat it. The most popular way to store home reared poultry is by freezing it. In this section, 'packaging' refers to birds that are going to be stored by freezing.

After you dress your birds, you need to chill them to bring the body temperature down to between 0 and 4 degrees Celsius (32 and 40 degrees Fahrenheit). If you're starting with a cold bird this process doesn't take long. When the birds are cool, you can begin to package them for storage.

Jointing and filleting

Cutting up chickens conserves space in your freezer and makes preparation for cooking easier. Of course, if you like a yummy, whole roasted chicken, you may not need to do any trimming before storing, but if you just use necks, backs, tails, wings or other parts for soup or stock, cutting them out and storing them separately makes sense.

Some people prefer to separate parts of the chicken with breasts in one package, legs in another and so forth. Other people cut up a chicken and store all the parts from one chicken in one package. Think about the ways in which you like to cook and let this dictate how you package the parts. The number of parts per package is also something you need to consider.

To cut chicken, use sharp butcher or boning knives as follows:

- **To remove the legs:** Cut through the skin in the crease between body and thigh, pull the leg away from the body and pop out the ball from the socket joint; cut through the meat between the body of the bird and the top of the thigh. Leave the leg whole or bend the knee joint at the top of the drumstick and use the dip it leaves to guide your knife to take off the drumstick end of the leg. With a bit of practice you can avoid cutting through any bones – you only need to cut through meat to remove the legs.

- **To remove the wings:** Wiggle each wing about until you work out how it joins the body and cut the skin and *ligaments* (the 'stringy' bits that hold the wing joint together) only. Doing so helps to keep your knives sharp for longer.

- **To remove the breast fillets:** Lay the chicken on its back and run the knife just to one side of the central line of its breast. Follow the bones of the ribcage, taking as much meat as possible. Do the same on both sides or, using chicken secateurs or a cleaver, split the whole chicken into quarters for meat on the bone.

You're left with the bones of the ribcage and back with the tail or 'Parson's nose' attached. You can use this to make good stock.

Selecting suitable packaging products

The most common packaging for putting meat in the freezer is the plastic freezer bag. Not all storage bags are made for freezing, so be sure to choose the bags that are. They're thicker than non-freezer bags and protect the meat better. Get bags just big enough to hold the portions you want to freeze. For a whole chicken, you probably need a 30-x-45-centimetre (12-x-18-inch) bag. The number of parts you plan to store dictates the size of each bag. For giblets, necks and so on, small bags are best. Some people prefer to freeze giblets in plastic containers.

Always use new, food-grade, freezer bags to store your meat. Don't save the bags your newspaper came in, plastic shopping bags or bread bags for this purpose. If you recycle plastic containers for giblets or small parts, wash them with hot, soapy water first, and make sure that their lids still fit tightly.

Many new vacuum-sealing bags are on the market (you can find them online or in kitchen and catering shops that sell small appliances). You put your meat in the bag and remove the air with a vacuum machine that sucks the air out through a specially designed hole, and then heat-seal the bag. These bags are very good for preserving the quality of meat, but they're more expensive to use than ordinary bags.

Some freezer bags have zipper-type seals; others you seal with a twist-tie. Buy a waterproof, permanent marker to put the date on each bag. Some bags have a special area for this info. If you use stick on labels, make sure that they have special freeze-proof glue otherwise all your carefully written labels end up in the bottom of the freezer.

Packing it in

Place your selected parts or whole chicken in the bag. Fill the bag as full as you can without making it hard to seal. The less air space in the bag, the better the meat keeps. If you're using a zipper-type closing bag, try to keep the seal area clean so it seals well.

Lay the bag on the table and partially seal it; use your hands to push as much air out as possible. You want your bags to look flat and moulded to the meat, not puffed with air. Then use a waterproof marker to label the bag with what's in it – for example: three chicken breasts – and the date you packed it. Labelling is very important so that you can rotate your packaged meat and know its use-by date.

Avoiding freezer overload

If you're going to dispatch very many chickens at a time, you probably need more than the freezer space that comes with a refrigerator. Chest and upright freezers work equally well. How many cubic metres/feet you need depends on what you intend to store. Buy the most energy efficient freezer you can afford. In the long run, it saves you money.

Home freezers aren't meant for freezing huge quantities of meat at a time. Don't start with an empty freezer, stuff it full and expect it to freeze your meat correctly. Turn an empty freezer on a day before you intend to use it, and check your owner's guide for the recommended weight to freeze at one time. If you try to freeze more than that, the meat may take a long time to reach the correct temperature, which gives bacteria a chance to grow, affects the quality of the meat and, even worse, may burn out your freezer's motor.

So what do you do if you have 50 kilos (110 pounds, or 25 or so medium-sized birds) of chicken to freeze and the directions say 15 kilos (33 pounds) are the limit? First, you need to chill the chicken to 4 degrees Celsius (40 degrees Fahrenheit) or lower before you attempt to freeze it. Then, the trick is to put a maximum of 15 kilos (33 pounds) – preferably less – in the freezer, spread out so most of the package surface is exposed, and temporarily keep the rest of your chickens in your refrigerator or in cool boxes with ice packs. When the first batch is frozen hard, which should be in 24 hours, place neatly into the smallest space possible in the freezer and add a second batch. Continue like this until all your chicken is frozen (note, however, that you do need to complete this process within four to five days).

Taking advantage of some freezer tips

If you like to take a few pieces of chicken out to cook at a time or if you want to save space in the freezer by packing chicken pieces in large bags, you can freeze pieces separately, and then combine them in a bag after they're frozen.

Here's how:

1. **Spread your pieces on a baking sheet so that they don't touch each other.**

2. **Cover the chicken with some plastic wrap.**

3. **Freeze the pieces.**

When the chicken pieces are frozen solid, you can combine them in one bag and they're easy to separate.

Enjoying the end product – safely

To prevent any bacteria remaining, always make sure that you thaw and cook chicken thoroughly – home-reared chicken is no exception. Be sure to you cook your chickens until the juices run clear. If you want to be more rigorous, buy a meat thermometer and follow the instructions; they aren't expensive.

If you have empty spaces in your freezer, fill plastic milk bottles or large pop bottles with water – making sure that you leave room for the ice to expand inside – and place them in the empty spaces after all the meat is frozen. Full freezers work more efficiently than half-empty ones, and if the power goes off, your ice bottles help keep the freezer cold.

From time to time, rotate the meat in your freezer so that you always use the oldest meat first. After six months, frozen chicken may still be safe to eat, but the quality starts going down. After a year in the freezer, frozen chicken needs to be cooked and fed to pets or discarded.

Part VI
The Part of Tens

'I think the little boy who accidentally
kicked his ball into our back garden
wants to go and fetch it.'

In this part . . .

*I*n this part we have two short and sweet chapters that give you handy info in a top-ten list fashion. We provide information to help you rear healthy, stress-free chickens and we clear up some misconceptions that you're likely to hear from other people.

Chapter 18

More than Ten Tips for Keeping Healthy, Stress-Free Chickens

*I*n addition to the basic needs of food and rest, chickens require as little stress as possible to be able to perform well. You're probably aware of how stress affects people and how it impacts on their health. Well, stress can affect chickens' health, too. Stress in chickens may lead to fighting and injury, improper nutrition and a lowered immune response to disease. Layers may stop laying, and meat birds may die of sudden heart attacks from stress.

Crowded conditions, chickens being moved in and out of the flock frequently, poor ventilation, heat, cold, irregular lighting, poor-quality feed, lack of water, disease, parasites and predators are the most common causes of stress in chickens. In this chapter we divulge the 10 (okay, 11) most important ways in which you can prevent these evils, keep your chickens happy and healthy and, if you're raising them for eggs or meat, ensure a plentiful supply.

Choose the Right Breed for Your Needs

Not all breeds of chickens perform equally well under the conditions you can provide for them and in your environment, and so before buying chickens, decide what you want them for – laying eggs, providing meat, showing or just for enjoying. Then carefully study breed characteristics and choose a breed that seems to fit your needs. If you do this, you're less likely to give up on keeping chickens, and your chickens are healthier and happier too.

In Chapter 3, we discuss which breeds are good for laying, for meat and so forth and give you some information on the behaviour characteristics of the breeds and whether they're better-suited to certain climates. If possible, visit poultry exhibits at your agricultural or county show or another location before you buy, to see the chickens and talk to their owners.

Be aware that people who like a certain breed can be more than a little partial to it, but their experience with the breed may not translate into a good experience for you. Ask questions and visit chicken forums such as www. backyardchickens.com to get a feel for what breed is right for you.

Set Up Suitable Housing

Having the right housing is not only better for your chickens but also better for you, as their keeper, and so plan your chicken housing and get it set up and well organised before you buy the birds. Chicken housing needs to protect your birds from the elements and from predators, and keep them dry and out of draughts. It doesn't have to be elaborate, but it does need to be clean and functional.

Plan the size of the housing to suit the number of birds you intend to have. Housing usually consists of indoor and outdoor space – see that each large-size chicken has at least 0.3 square metres (3.2 square feet) of indoor space and 0.3–0.5 square metres (3.2–5.4 square feet) of outdoor space for optimum health. The more space you can provide, the happier your chickens are. Also decide how you'll access the housing for cleaning, caring for the birds and collecting eggs, how it's going to fit into your garden or backyard and whether you aim to light it. And make sure that the housing accommodates your needs, as well as the chickens'. The more functional the housing is for you, the happier you'll be with chicken-keeping.

In Part II, we discuss chicken housing in lots of detail. Having good furnishings in the housing, such as nest boxes, perches and feeding equipment, also keeps your chickens happy and healthy and makes things more convenient for you.

Supplement Lighting when Needed

A chicken's life cycle revolves around the amount of daylight or artificial light it receives. When the days are long, chickens are prompted to lay eggs and mate; when the days start getting shorter, they moult. *Moulting* is the process by which all a chicken's feathers are replaced, and it's energy-intensive so that when chickens moult, they usually stop laying.

Chickens have to moult sometime, but you can manipulate the 14 hours of daylight and 10 hours of darkness (or very dim light) that your chickens receive by supplementing natural light with artificial light of daylight strength to generate the right conditions to keep them laying or get them to moult when it suits you. Supplementing the light of young pullets helps them to grow faster and mature sooner and can keep them laying through the autumn or winter.

If you keep your chickens for meat, you don't have to worry about moulting, because the birds should be in the freezer long before moulting occurs. If you keep chickens for showing, you do want to keep your birds from moulting before an important show, because they don't look good enough when they're moulting.

 Supplementing the light for pet birds isn't important – you're unlikely to mind when they moult and you may not need to stimulate them into laying. Leaving a very dim nightlight on in chicken housing, however, enables chickens to defend themselves against some predators. It also helps to stop your chickens causing damage when they panic over things that go bump in the night – particularly important if you're trying to keep show birds in tip-top condition.

Control Pests

Pests are creatures such as wild birds, rats, mice and flies that may hang around poultry housing. Not only are they offensive to neighbours (and a huge pain for you), but they can also be dangerous to your chickens. Wild birds such as starlings and sparrows or wild geese and ducks can carry many diseases to your chickens, as can rats and mice. They also eat a lot of feed, which can become a huge money drain if you aren't paying attention.

Controlling pests means keeping the chicken house clean, storing feed so that pests can't access it, putting out poison bait or traps when you notice signs of pests and having secure housing that limits pests' access. You can find out more about controlling pests in Chapter 9.

Protect Against Predators

Predators are a big concern for chicken-keepers – after all, people aren't the only ones who enjoy chicken for dinner. If you want to have healthy chickens you need to protect them, because they have few defences of their own.

Predator protection works best if you can anticipate problems and protect the chickens with sturdy pens or restricted areas to roam. Chickens running free, even in urban areas, are at the mercy of all kinds of animals and other dangers, including cars and mowers. Supervised play is best for chickens.

Here are some tips for keeping your chickens safe:

- ✔ **Use roofs, wire or netting on outside runs to avoid predation from hawks, crows and owls.**

- ✔ **Clear away overgrown areas around chicken housing.**

- ✔ **Place locks on hen house doors if two-legged predators are a problem.**

- ✔ **Leave a low-level nightlight on in the chicken house at night.** Chickens have a better chance of defending themselves if they can see, and they're less likely to panic, too.

- ✔ **Make sure that fencing is strong and sufficiently tall.** Chicken wire isn't very sturdy – even a medium-sized dog can rip it apart. (Dogs are one of the top chicken predators, by the way.) We suggest using heavy welded wire on outside runs. You'd be wise to bury it in the ground 30 centimetres (1 foot) or so deep to keep out burrowers.

For more guidance on protecting your flock against predators and dealing with them if they do attack, see Chapter 9.

Control Parasites

Parasites not only make birds uncomfortable, they can carry disease and lower a chicken's immune system response to disease. Birds carrying a heavy load of internal or external parasites produce fewer eggs, grow more slowly and eat more feed. Keeping your birds well fed and stress- and disease-free helps their bodies to repel parasites and makes them better able to tolerate any they may still contract.

Some parasites, such as worms, are hard to eliminate entirely because their eggs persist in the environment. A regular worming routine can keep worms down to a level the birds can tolerate. Lice live on the birds, but mites may spend most of their time in some part of the housing, and so to control them you need to treat the housing as well as the chickens. Allowing your birds regular access to dustbaths and organising your housing to minimise infestations is another essential part of keeping these parasites at bay. You can buy modern products for eliminating – or at least controlling – parasites in poultry supply and other pet shops and from veterinarians.

We discuss disease and parasite control in Chapter 10.

Vaccinate Your Birds

Preventing problems is always better than trying to cure them. When you purchase baby chicks, you're often offered the opportunity to have them vaccinated for a small additional fee. Saying 'yes' is a wise idea.

You can give vaccines at various life stages of chickens. Many vaccines have an optimum age, but if the chicken doesn't get the vaccine then it can sometimes be administered later, depending on the disease you're tackling. Vaccines can be given by mouth, as a spray mist or by injection, depending on the disease they're meant to prevent. Some vaccines work in one dose; others require several doses.

Many vaccines exist today to prevent chicken diseases. These vaccines are reasonably priced, and most home flock owners can administer them. Ask at your local vet's office or the hatchery where you buy your birds which vaccinations are recommended, and get your chickens vaccinated. If you don't want to do it yourself, have a vet do it or ask an experienced friend to help you. For more details on vaccinating your chicks or chickens, see Chapter 10.

Feed a Well-Balanced Diet

Well-fed chickens lay more eggs, grow faster, produce better meat and have good immune systems to fight off disease. Chickens are like children, though – you have to supervise their diets. Not only do they eat almost anything, whether it's nutritious or not, but also a chunk of land just doesn't provide the nutrition that chickens need. Even if they have a large area to forage on, they need at least part of their diet to come from commercial feed so they get all the nutrients they need. Unlike domestic chickens, though, wild chickens that get all their nutrition from Mother Nature have plenty of space to roam around and hunt to meet their needs.

Today's commercial feeds are well balanced, with the correct ratios of protein, minerals and so on for the type of bird they're labelled for. They come in pellet, mash or crumb form so the chickens can't pick out their favourite pieces and avoid the rest. Organic commercial feeds are now readily available on the market, too.

In Chapter 8, we discuss feeding chickens and what constitutes a healthy diet.

Provide Enough Clean Water

Having clean water available at all times is one of the best ways to keep your chickens healthy and productive. Chickens need water available to lay well, to grow quickly and to perform all life's functions. Making sure that water is available, even in winter, is essential to their health. If you have trouble keeping water available for your chickens, you may need to use an automatic water system. Check it frequently to verify that it's working properly.

Chickens can be a bit fussy about water. They don't like water that's too warm or flavoured strongly, and if they don't drink freely they don't eat as much, which starts affecting their production and health. Ensure that your chickens always have clean, fresh water.

Beware Disease-Transmitting Dangers

Many chicken diseases are carried on clothing, shoes and hands. When you visit other people's chickens or go to a show, change your shoes and clothes and wash your hands before tending your flock. If you borrow any equipment, such as carriers, disinfect it before and after use. Also think twice about inviting visitors who have chickens of their own to visit your flock. If you have rare or very valuable birds, you may want to limit visits. The more visits, the greater the chance of a disease being carried into your flock.

Use Quarantines Whenever Necessary

One of the easiest but least-practised things a home flock owner can do to maintain healthy chickens is to quarantine all new birds and all chickens that come back home from a show or sale for two weeks, well away from the rest of the flock. If you have sick chickens, move them away from the rest of the flock and quarantine them to try and prevent disease spreading. Injured birds should be quarantined too, so the others don't pick on them.

Feed and care for your quarantined chickens after you take care of the rest of your flock. If they show any signs of disease, treat or destroy your quarantined chickens, whichever is the most effective method of preventing the spread of disease. To find out how to quarantine birds, see Chapter 10.

Chapter 19

More than Ten Misconceptions about Chickens and Eggs

In This Chapter
▶ Discerning the truth in what you read and hear about chickens
▶ Allaying fears about bird flu
▶ Dispelling myths about what determines egg taste and quality

*O*h, the things that are said about chickens! The very word 'chicken' brings up the image of a coward, but chickens aren't really cowards. And when you find yourself throwing around references such as 'bird-brained' or 'hen-pecked', remember that although some are based in fact, most are misconceptions. To remedy the situation, we present a compilation of the most common myths and misconceptions about chickens and eggs that you may encounter as a chicken owner or chicken-keeper wannabe.

Bird Flu is a Risk to Reckon With

Some people are afraid to keep chickens because they fear bird influenza, or *avian flu*. With the amount of coverage given to bird flu in the media, fear can quickly turn to panic, but the thing to remember is that most strains of bird flu don't infect humans and you're more likely to get human flu than bird flu. Another thing to be aware of is that bird flu has been around for a long time already – it's nothing new – but although outbreaks among domestic poultry have occurred in the United Kingdom and Europe, bird flu more commonly occurs in Asian countries where poultry and people live in closer proximity to each other. Unless you're coming into contact with migrating birds, your chances of contracting bird flu are minute.

The H5N1 bird flu virus, at least so far, passes only from bird to bird or bird to human, not from human to human and not through the air. You can contract it from handling infected poultry, eating raw eggs or meat, or handling something contaminated by the virus shed in animal secretions.

Bird flu also spreads from flock to flock through being carried on shoes, clothing and even car tyres, and so chicken owners should limit visitors who own poultry themselves from handling their birds or going into chicken quarters. This precaution can be tough for proud chicken owners. If you exhibit show birds or buy new ones, put them through a two-week quarantine period before placing them with the rest of your flock.

Just as humans carry many strains of human flu, so wild birds carry many strains of bird flu, and are the likely source when bird flu infects chicken flocks. Migratory birds can spread it, so the advice is to keep wild birds away from your chickens as far as possible and to use common sense and good hygiene in keeping people safe. The Department for the Environment, Food and Rural Affairs (DEFRA) constantly monitors the health of domestic and wild birds in the United Kingdom for deadly strains of bird flu, and in the event of an outbreak would announce what steps home flock owners should take. So that they know the location of the biggest concentrations of chickens and can be best prepared, DEFRA has asked chicken owners keeping more than 50 domestic poultry in one place to register their flock with them.

Your flock is extremely unlikely to contract a serious strain of bird flu, and you're even less likely to contract a serious illness from them. However, it can happen, so knowing what to do if bird flu does strike can help you to be prepared. If lots of your chickens suddenly die within a short period of time and without many symptoms, see your local vet or contact DEFRA straight away (the helpline number is 08459 33 55 77) for advice. Always wear gloves when handling dead or ill chickens, and keep your hands washed!

You Can't Keep Chickens in the City

Chickens aren't just for country folk anymore. Anyone who has a small garden or yard can find a place for a few chickens, even if you live in a bustling urban neighbourhood. A few houses do have a clause written into their deeds that disallows chicken-keeping, so check your deeds before getting chickens. This restriction usually applies to rows of similar houses.

If you keep your chickens clean, they don't smell any more than your neighbour's three Great Danes and aren't any noisier than their car stereo or leaf blower (but a cockerel may be!), and they enable urban dwellers to have some useful pets that lay breakfast for them, too. Chickens are easier to care for than dogs, and they're quiet at night, unlike the local alley cats.

You can even keep chickens when you only have a patio or area of hard standing, but make sure that you choose a small breed such as a bantam. If you plan to keep them in cages, bring them treats and entertainment in the form of greenery so they can experience their own 'outside' world inside. Two bantam hens in an aviary on a balcony would be a fun addition to a household but you must put extra effort into entertaining them.

Cockerels Crow Only in the Morning

Cockerels do greet the sun exuberantly, but they also crow all day long and sometimes, if they're awakened at night, they crow then too. Cockerels crow like songbirds sing, to mark their territory and make the hens aware of their presence. Healthy cockerels crow at every chance they get, although crowing frequency and sound varies by individual.

You Need a Cockerel to Get Eggs

A hen is born with all the eggs she'll ever have, and nature tricks her into laying them whether a cockerel is around or not. The eggs are equally tasty, nutritious and abundant, regardless of whether or not a cockerel is present.

Hens are flock animals, so don't keep them on their own. They don't seem to miss a cockerel if they've never known one as long as they have hen friends to chum around with. Of course, their eggs can't ever become chicks, but many breeds don't care to be mothers anyway. Watching cockerels escort and care for their hens is fun, however, if you have a suitable place to keep them.

Keeping Chickens Penned is Inhumane

Chickens like to be able to roam around freely, but doing so isn't always safe for them, even in the country. Most livestock is kept confined in some way for its own safety, and chickens are no exception. Remember, your children aren't the only ones who like to eat chicken for dinner.

Chickens can be just as happy in a good-sized pen with nutritious food, some entertainment (in the form of vegetation hung up for them to peck at) and a warm, dry place to sleep and lay eggs in as your dog is when confined to the backyard or your horse kept in the field. You can, of course, allow them supervised roaming from time to time, just like your pets, but confined chickens don't annoy the neighbours or damage your flower beds. Confined chickens pose less of a health risk too, because they're less likely to come into contact with wild birds that carry diseases, such as bird flu (something we discuss in the earlier section 'Bird Flu is a Risk to Reckon With').

Chickens are Vegetarians

Chickens aren't vegetarians. They love meat, and eat mice, frogs and worms if they come across them, as well as their own eggs. Chickens are *omnivores* that eat just about anything, and they need some of the amino acids they get from consuming animal-based proteins. Home-made diets based on only grain may not keep your chickens at optimum health, especially in the winter when they can't dig some maggots out of the litter or catch insects, and pasture-only diets are just not sufficient for growing or laying chickens. To compensate, producers of commercial poultry feed usually add amino acids that are missing from grain-based diets, or they include safe animal sources of protein.

Big, Brown Organic Eggs are Best in Taste and Quality

Home-laid eggs are generally brown, because brown-egg layer breeds are easier for most small flock owners to care for than other breeds. And if you eat your own eggs, or buy them locally, they're generally much fresher than shop-bought eggs and taste better. Shell colour, however, isn't an indicator of taste or quality and nor is size – green and blue eggs taste the same as brown or white eggs, and small eggs taste the same as large eggs. You may look at the eggs, though, and see a difference besides shell colour.

The quality and freshness of their food determines the quality of chickens' eggs. Eggs can taste different if you feed your hens a lot of certain foods, such as flax seed, fish or onions, or if you don't properly store the eggs. Eggs can pick up flavours if you keep them next to foods with strong odours.

Chickens that have access to grass and greens or are fed things such as marigold flowers produce eggs with deeper yellow yolks than chickens fed other diets. (In the past, dodgy dyes were included in chicken feed to make the yolks go yellow, but this practice is no longer allowed.) To count as being 'organic', though, eggs must come from registered, certified organic farms. Home flock owners can replicate and improve on 'organic' standards by allowing their chickens to range freely in a chemical-free environment, feeding chemical-free (probably labelled 'organic') feeds and keeping small numbers of hens in healthy, happy, spacious conditions and then collecting and eating a really fresh (but technically non-organic) egg.

The term *organic* is a legal definition of a set of recognised standards of animal welfare and ethics and a commitment to chemical-free farming practices. It wouldn't be cost effective to buy into organic registration as a home flock owner, and therefore you can't legally call your eggs organic.

Nutritional claims about certain eggs vary in credibility. Chickens can be fed so that their eggs have less cholesterol and more of certain nutrients, but this process is an exacting science that most small flock owners can't practise. Besides, nutritional gurus now believe that the cholesterol humans get from eggs isn't the kind that builds up in the blood, and so the point may be moot

Fertilised and Unfertilised Eggs are Easily Distinguishable

Only a trained eye can tell fertilised and unfertilised eggs apart, unless they're stored improperly and an embryo begins growing. Some people think that blood spots indicate a fertilised egg, but they're simply the result of a vein rupturing as the egg is released from the ovary; this usually happens in very young or very old layers.

Shop-bought eggs are almost always infertile eggs. Commercial breeders don't keep cockerels with hens. Only a shop selling locally produced eggs from a small flock with a cockerel has a chance of including a fertilised egg. And if you keep a cockerel with your own hens, the chances are very good that the eggs you eat are fertilised. If that bothers you, don't keep a cockerel with your hens – it's that simple.

Fertilised eggs don't taste any different from unfertilised ones. And that tiny bit of chicken sperm doesn't give the egg any nutritional boost, either.

Egg-Box Advertising is the Whole Truth

When buying eggs, beware. 'Cage-free' doesn't mean organically raised, or that the hens range freely. 'Cage-free' usually means the hens were housed in large pens with some room to move around. Growers refer to this environment as cage-free, but really it's just a big cage with lots of chickens crowded into it. This set-up is slightly better than being crowded into cages so small that a chicken can't stand up or flap its wings, the way most commercial layers are housed, but the eggs you buy from cheapo stores aren't going to be from hens that roam freely outside, no matter what deceptive words are used on the box.

'Free-range' on the box means that certain rules do apply, but it doesn't mean the sort of free-range that most people imagine when they see those words. In big flocks, no one can guarantee that all the birds go outside anyway.

In Europe, eggs labelled 'organic' must come from hens that have access to the outdoors, and this label indicates a situation more along the lines of what people think of when they hear the term 'free-range'. (See 'Big, Brown Organic Eggs are Best in Taste and Quality' earlier in this chapter for more about the term 'organic'.)

If you can't keep chickens yourself, buying eggs locally from hens kept in small flocks – whether free-ranging or fed organically or not – gives you the best-tasting eggs, short of collecting them each morning from your own hens. And it probably means that the hens are kept in more humane conditions than commercial, caged layers. With this in mind, try to buy eggs from producers who let you see their hens – they don't have anything to hide. Not many large flock owners want you to see the conditions they really keep their hens in, no matter what the picture on their boxes intimates.

Chickens are Good for Your Garden

The sad truth is that chickens ruin your garden! They till the soil all right, but their aim is to gobble up the seeds you've just planted. And yes, they eat the weeds – along with the lettuce! And although they eat tomato worms, they take a bite out of each tomato. Chickens don't belong in your vegetable garden. Maybe in the autumn before you clear it all out, but at no other time.

Chicken manure is good for the garden, but only after you've composted it. Fresh chicken manure deposited in the garden burns plants and brings the risk of salmonella bacteria contaminating your fresh veggies. (Chapter 7 has information about composting and using chicken manure as fertiliser.)

Chickens are Stupid

Most people who've kept chickens for any length of time strongly defend their chickens' intelligence and can tell you many tales of chicken bravery, too. Not to mention that sometimes running from a threat is certainly smarter than facing it, so don't judge the chicken that retreats from danger.

As birds – or animals, for that matter – go, chickens are pretty intelligent. They can learn to count and understand the concept of zero. You can train them to do tricks and to recognise colours, and they can work out how to get out of almost any pen you put them in, sooner or later. They can recognise a routine and anticipate it. Chickens and other birds have been observed planning future actions or anticipating reactions to an action they're going to take. And chickens learn by observing and copying other chickens.

Chickens have a well-organised social system and culture that limits strife among a flock. Anyone who has ever watched a cockerel coaxing his hens over to some choice food knows that they communicate among themselves. And although the word 'chicken' has come to mean 'cowardly', chickens can be very brave when defending their babies or their flock. Hens sometimes sacrifice themselves for their chicks, and cockerels often fight to the death, even though most people would consider that rather stupid behaviour. Cockerels can be formidable when protecting their girls – just ask anyone who's been chased by an angry cockerel!

Index

• C •

Notes

FOR DUMMIES®

Making Everything Easier! ™

UK editions

BUSINESS

978-0-470-97626-5

978-0-470-97211-3

978-0-470-71119-4

REFERENCE

978-0-470-68637-9

978-0-470-97450-6

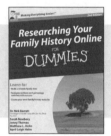

978-0-470-74535-9

HOBBIES

978-0-470-69960-7

978-0-470-68641-6

978-0-470-68178-7

Asperger's Syndrome For Dummies
978-0-470-66087-4

Boosting Self-Esteem For Dummies
978-0-470-74193-1

British Sign Language
For Dummies
978-0-470-69477-0

Coaching with NLP For Dummies
978-0-470-97226-7

Cricket For Dummies
978-0-470-03454-5

Diabetes For Dummies, 3rd Edition
978-0-470-97711-8

English Grammar For Dummies
978-0-470-05752-0

Flirting For Dummies
978-0-470-74259-4

Football For Dummies
978-0-470-68837-3

IBS For Dummies
978-0-470-51737-6

Improving Your Relationship
For Dummies
978-0-470-68472-6

Lean Six Sigma For Dummies
978-0-470-75626-3

Life Coaching For Dummies,
2nd Edition
978-0-470-66554-1

Management For Dummies,
2nd Edition
978-0-470-97769-9

Nutrition For Dummies, 2nd Edition
978-0-470-97276-2

FOR DUMMIES®

A world of resources to help you grow

UK editions

SELF–HELP

Cognitive Behavioural Therapy For Dummies
978-0-470-66541-1

Neuro-linguistic Programming For Dummies
978-0-470-66543-5

Mindfulness For Dummies
978-0-470-66086-7

STUDENTS

Philosophy For Dummies
978-0-470-68820-5

Student Cookbook For Dummies
978-0-470-74711-7

Sociology For Dummies
978-1-119-99134-2

HISTORY

The Tudors For Dummies
978-0-470-68792-5

Medieval History For Dummies
978-0-470-74783-4

British History For Dummies
978-0-470-97819-1

Origami Kit For Dummies
978-0-470-75857-1

Overcoming Depression For Dummies
978-0-470-69430-5

Positive Psychology For Dummies
978-0-470-72136-0

PRINCE2 For Dummies, 2009 Edition
978-0-470-71025-8

Psychometric Tests For Dummies
978-0-470-75366-8

Reading the Financial Pages
For Dummies
978-0-470-71432-4

Rugby Union For Dummies, 3rd Edition
978-1-119-99092-5

Sage 50 Accounts For Dummies
978-0-470-71558-1

Self-Hypnosis For Dummies
978-0-470-66073-7

Starting a Business For Dummies,
2nd Edition
978-0-470-51806-9

Study Skills For Dummies
978-0-470-74047-7

Teaching English as a Foreign Language
For Dummies
978-0-470-74576-2

Time Management For Dummies
978-0-470-77765-7

Training Your Brain For Dummies
978-0-470-97449-0

Work-Life Balance For Dummies
978-0-470-71380-8

Writing a Dissertation For Dummies
978-0-470-74270-9

FOR DUMMIES®

The easy way to get more done and have more fun

LANGUAGES

Spanish
FOR DUMMIES

978-0-470-68815-1
UK Edition

French
FOR DUMMIES

978-1-118-00464-7

German
FOR DUMMIES

978-0-470-90101-4

MUSIC

Ukulele
FOR DUMMIES

978-0-470-97799-6
UK Edition

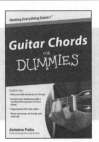

Guitar Chords
FOR DUMMIES

978-0-470-66603-6
Lay-flat, UK Edition

DJing
FOR DUMMIES

978-0-470-66372-1
UK Edition

SCIENCE & MATHS

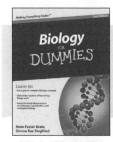

Biology
FOR DUMMIES

978-0-470-59875-7

Algebra I
FOR DUMMIES

978-0-470-55964-2

Genetics
FOR DUMMIES

978-0-470-55174-5

Art For Dummies
978-0-7645-5104-8

Bass Guitar For Dummies, 2nd
Edition
978-0-470-53961-3

Criminology For Dummies
978-0-470-39696-4

Currency Trading For Dummies
978-0-470-12763-6

Drawing For Dummies, 2nd Edition
978-0-470-61842-4

Forensics For Dummies
978-0-7645-5580-0

Guitar For Dummies, 2nd Edition
978-0-7645-9904-0

Index Investing For Dummies
978-0-470-29406-2

Knitting For Dummies, 2nd Edition
978-0-470-28747-7

Music Theory For Dummies
978-0-7645-7838-0

Piano For Dummies, 2nd Edition
978-0-470-49644-2

Physics For Dummies, 2nd Edition
978-0-470-90324-7

Schizophrenia For Dummies
978-0-470-25927-6

Sex For Dummies, 3rd Edition
978-0-470-04523-7

Sherlock Holmes For Dummies
978-0-470-48444-9

Solar Power Your Home
For Dummies, 2nd Edition
978-0-470-59678-4

The Koran For Dummies
978-0-7645-5581-7

30093 (p3)

FOR DUMMIES®

Helping you expand your horizons and achieve your potential

COMPUTER BASICS

978-0-470-57829-2

978-0-470-46542-4

978-0-470-49743-2

DIGITAL PHOTOGRAPHY

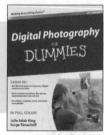

978-0-470-25074-7

978-0-470-76878-5

978-0-470-59591-6

MICROSOFT OFFICE 2010

978-0-470-48998-7

978-0-470-58302-9

978-0-470-48953-6

Access 2010 For Dummies
978-0-470-49747-0

Android Application Development
For Dummies
978-0-470-77018-4

AutoCAD 2011 For Dummies
978-0-470-59539-8

C++ For Dummies, 6th Edition
978-0-470-31726-6

Computers For Seniors For Dummies,
2nd Edition
978-0-470-53483-0

Dreamweaver CS5 For Dummies
978-0-470-61076-3

Green IT For Dummies
978-0-470-38688-0

iPad All-in-One For Dummies
978-0-470-92867-7

Macs For Dummies, 11th Edition
978-0-470-87868-2

Mac OS X Snow Leopard For Dummies
978-0-470-43543-4

Photoshop CS5 For Dummies
978-0-470-61078-7

Photoshop Elements 9 For Dummies
978-0-470-87872-9

Search Engine Optimization
For Dummies, 4th Edition
978-0-470-88104-0

The Internet For Dummies,
12th Edition
978-0-470-56095-2

Visual Studio 2010 All-In-One
For Dummies
978-0-470-53943-9

Web Analytics For Dummies
978-0-470-09824-0

Word 2010 For Dummies
978-0-470-48772-3